Novel Smart Textiles

Novel Smart Textiles

Special Issue Editor
George K. Stylios

MDPI • Basel • Beijing • Wuhan • Barcelona • Belgrade • Manchester • Tokyo • Cluj • Tianjin

Special Issue Editor
George K. Stylios
Heriot-Watt University
UK

Editorial Office
MDPI
St. Alban-Anlage 66
4052 Basel, Switzerland

This is a reprint of articles from the Special Issue published online in the open access journal *Materials* (ISSN 1996-1944) (available at: https://www.mdpi.com/journal/materials/special_issues/novel_smart_textile).

For citation purposes, cite each article independently as indicated on the article page online and as indicated below:

LastName, A.A.; LastName, B.B.; LastName, C.C. Article Title. *Journal Name* **Year**, *Article Number*, Page Range.

ISBN 978-3-03928-570-9 (Pbk)
ISBN 978-3-03928-571-6 (PDF)

Cover image courtesy of George K. Stylios.

© 2020 by the authors. Articles in this book are Open Access and distributed under the Creative Commons Attribution (CC BY) license, which allows users to download, copy and build upon published articles, as long as the author and publisher are properly credited, which ensures maximum dissemination and a wider impact of our publications.

The book as a whole is distributed by MDPI under the terms and conditions of the Creative Commons license CC BY-NC-ND.

Contents

About the Special Issue Editor vii

Preface to "Novel Smart Textiles" ix

George K. Stylios
Novel Smart Textiles
Reprinted from: *Materials* **2020**, *13*, 950, doi:10.3390/ma13040950 1

George K. Stylios and Meixuan Chen
The Concept of Psychotextiles; Interactions between Changing Patterns and the Human Visual Brain, by a Novel Composite SMART Fabric
Reprinted from: *Materials* **2020**, *13*, 725, doi:10.3390/ma13030725 5

Adriana Stöhr, Eva Lindell, Li Guo and Nils-Krister Persson
Thermal Textile Pixels: The Characterisation of Temporal and Spatial Thermal Development
Reprinted from: *Materials* **2019**, *12*, 3747, doi:10.3390/ma12223747 23

Christian Dils, Lukas Werft, Hans Walter, Michael Zwanzig, Malte von Krshiwoblozki and Martin Schneider-Ramelow
Investigation of the Mechanical and Electrical Properties of Elastic Textile/Polymer Composites for Stretchable Electronics at Quasi-Static or Cyclic Mechanical Loads
Reprinted from: *Materials* **2019**, *12*, 3599, doi:10.3390/ma12213599 39

Veronica Malm, Fernando Seoane and Vincent Nierstrasz
Characterisation of Electrical and Stiffness Properties of Conductive Textile Coatings with Metal Flake-Shaped Fillers
Reprinted from: *Materials* **2019**, *12*, 3537, doi:10.3390/ma12213537 57

Orathai Tangsirinaruenart and George Stylios
A Novel Textile Stitch-Based Strain Sensor for Wearable End Users
Reprinted from: *Materials* **2019**, *12*, 1469, doi:10.3390/ma12091469 75

Manuel J. Lis Arias, Luisa Coderch, Meritxell Martí, Cristina Alonso, Oscar García Carmona, Carlos Garcia Carmona and Fabricio Maesta
Vehiculation of Active Principles as a Way to Create Smart and Biofunctional Textiles
Reprinted from: *Materials* **2018**, *11*, 2152, doi:10.3390/ma11112152 93

Lukas Vojtech, Marek Neruda, Tomas Reichl, Karel Dusek and Cristina de la Torre Megías
Surface Area Evaluation of Electrically Conductive Polymer-Based Textiles
Reprinted from: *Materials* **2018**, *11*, 1931, doi:10.3390/ma11101931 115

Xiang An and George K. Stylios
A Hybrid Textile Electrode for Electrocardiogram (ECG) Measurement and Motion Tracking
Reprinted from: *Materials* **2018**, *11*, 1887, doi:/10.3390/ma11101887 135

Guantao Wang, Yong Wang, Yun Luo and Sida Luo
Carbon Nanomaterials Based Smart Fabrics with Selectable Characteristics for In-Line Monitoring of High-Performance Composites
Reprinted from: *Materials* **2018**, *11*, 1677, doi:10.3390/ma11091677 153

Marek Neruda and Lukas Vojtech
Electromagnetic Shielding Effectiveness of Woven Fabrics with High Electrical Conductivity:
Complete Derivation and Verification of Analytical Model
Reprinted from: *Materials* **2018**, *11*, 1657, doi:10.3390/ma11091657 . . . . **165**

László Juhász and Irén Juhász Junger
Spectral Analysis and Parameter Identification of Textile-Based Dye-Sensitized Solar Cells
Reprinted from: *Materials* **2018**, *11*, 1623, doi:10.3390/ma11091623 . . . . **185**

Irén Juhász Junger, Daria Wehlage, Robin Böttjer, Timo Grothe, László Juhász, Carsten Grassmann, Tomasz Blachowicz and Andrea Ehrmann
Dye-Sensitized Solar Cells with Electrospun Nanofiber Mat-Based Counter Electrodes
Reprinted from: *Materials* **2018**, *11*, 1604, doi:10.3390/ma11091604 . . . . **197**

Bahareh Moradi, Raul Fernández-García and Ignacio Gil
E-Textile Embroidered Metamaterial Transmission Line for Signal Propagation Control
Reprinted from: *Materials* **2018**, *11*, 955, doi:10.3390/ma11060955 . . . . **211**

About the Special Issue Editor

George K. Stylios (https://orcid.org/0000-0002-7911-1058) is Professor of Textiles at HWU, where he has been working as a senior research professor since 1999 and as a director of the Research Institute for flexible materials, with a focus on structural mechanics, nanotextiles, smart textiles, wearables, and design/technology. He obtained his Ph.D. from Leeds University working for Marks & Spencer Plc in 1985; became a lecturer at Bradford University (1985–1999); Fellow of the National Institute of Advanced Industrial Science and Technology, Japan in 1992; and a personal chair in Industrial Systems Engineering in 1994 in recognition of his work on the fabric/machine interface. He has led many successful research projects (£20 million) from research councils, the European Union, and directly from industry. As director, he led large government initiatives such as the Regional Innovation Strategy for Yorkshire and Humberside (£14M, 1995–1999) and the Faraday Partnership in Technical Textiles (Technitex £7 million, 2004–2010), working across industry and improving products and processes. For the latter, he received 100% in Research Impact for his work in REF14. He has numerous journal publications and Ph.D. completions with most of his Ph.D. students pursuing successful careers, becoming professors or directors of companies in their own right. GKS has been the founding editor of the International Journal of Science and Technology (1987–present), Guest Editor for many journals, convener and keynote speaker at many international conferences and events, peer reviewer, and advisor to government bodies and industry at home and abroad. Funded Projects:

- EPSRC/ACME Directorate 1989–1992 £250,000;
- DTI 1989–1995 £0.5 million;
- EPSRC/ACME Directorate 1992–1995 £360,000;
- SMART Award 1997 £100,000;
- EU/ERDF 2001–2003; HOMETEX; £1 million;
- EU 2002–2005; Inneurotex; £460,000;
- EPSRC: Engineering the performance and functional properties of technical textiles. 2002–2005 (GR/S09203/01, GR/S09210/01 & GR/S09227/01, £1.04 million);
- TSB/DTI: Novel Micro-Channel Membranes for Controlled Delivery of Biopharmaceuticals, 100327, award £650,000;
- TSB: 2006–2009: Multi-Scale Integrated Modelling for High Performance Flexible Materials, 100256, Total award £1.6 million (£275,000 HWU), 2006–2010;
- EPSRC FUNDING: Integration of CFD and CAE for Design and Performance Assessment of Protective Clothing, DT/E011098/1, £250,000, 2007–2010.
- MiroLink Ltd £0.4 million (2014–2018) for developing wearable products (ECG, Respiration, temperature).
- EU Horizon20; ArchinTex (2015–2018) £380,000.

Preface to "Novel Smart Textiles"

A measure of the importance of smart textiles is its market size, which will exceed USD $5.55 billion by 2025, with the healthcare and well-being sectors being significant driving forces, and garment sensor-based wearable monitoring expected to exceed 50% CAGR in the next five years.

Research on highly specific applications is increasing, exploring the opportunities offered by manipulating textile materials to the nanoscale for creating new smart adaptive/active functionality, and by the development of e-textiles, which offer intelligent flexible integrated systems capable of sensing, actuation, and wirelessly communicating in the form of intelligent high-tech fabrics and wearable garments. The development of these systems presents a complex set of interdisciplinary challenges in material design, hierarchical integration, control strategies, and manufacturing.

This focused journal collection of highly original papers is underpinning these issues by reporting the latest research progress. These research papers increase our knowledge, enable us to see our own work in context, empower us to improve our understanding, increase the rigour of our research, and encourages us to work collectively. I hope that, to some extent, the publication of the concentrated effort of these researchers in this Special Issue can aid to provide a clearer roadmap for further research advancement.

Adding my own observations about the challenges that smart textiles must overcome, washability, user safety, and reliability are three important factors that need to be addressed in our research. Traditional textiles working with electrical components need a change of culture, which is time consuming and difficult. The supply chain is not yet ready to embrace fast changes that are much needed, and designers and engineers must learn to work together. We have a tired textile industry that is producing consumer products at pence per minute and being for damaging the environment. I think that untapped opportunities exist for this industry and that smart textiles must converge with traditional textiles. My own hope for the future of this field is not to aim for incremental changes but to force a transformational change.

George K. Stylios
Special Issue Editor

Editorial

Novel Smart Textiles

George K. Stylios

Research Institute for Flexible Materials, School of Textiles & Design, Heriot-Watt University, Scottish Borders Campus, Galashiels TD1 3HF, UK; G.Stylios@hw.ac.uk

Received: 14 February 2020; Accepted: 19 February 2020; Published: 20 February 2020

Abstract: The sensing/adapting/responding, multifunctionality, low energy, small size and weight, ease of forming, and low-cost attributes of SMART textiles and their multidisciplinary scope offer numerous end uses in medical, sports and fitness, military, fashion, automotive, aerospace, built environment, and energy industries. The research and development for these new and high-value materials crosses scientific boundaries, redefines material science design and engineering, and enhances quality of life and our environment. "Novel SMART Textiles" is a focused special issue that reports the latest research of this field and facilitates dissemination, networking, discussion, and debate.

Keywords: smart textiles; textile sensors; e-textiles; visual brain; thermal textile pixels; stretchable electronics; conductive textiles; wearables; stitch-based sensors; biofunctional textiles; ECG; hybrid electrodes; motion tracking; carbon nanotextiles; composites; EMS textiles; electrospun solar cells; embroidered e-textiles; targeted delivery; psychotextiles; energy harvesting; multifunctional

---

A measure of the importance of smart textiles can be realized by its market size which will exceed USD 5.55 billion by 2025, with the healthcare and well-being sectors being a significant driving force. The garment sensor-based telemedicine part is expected to exceed 50% CAGR in the next five years.

Research for highly specific applications is increasing in exploring the opportunities offered by manipulating textile materials down to the nanoscale for creating new "smart" adaptive/active functionality, and by the development of "E-textiles" offering intelligent flexible integrated systems capable of sensing, actuation and wirelessly communicating in the form of intelligent high-tech fabrics and wearable garments. The development of these systems presents a complex set of interdisciplinary challenges in material design, hierarchical integration, control strategies, and manufacturing.

This focused journal collection of highly original papers is underpinning these issues by reporting the latest research progress. The first paper by George K. Stylios and Meixuan Chen proposes a new type of SMART fabrics called Psychotextiles [1]. After studying the direct relationship between design and brain waves, using EEG, the characteristics and attributes of patterns that influence specific brain emotions are established, which are in turn designed into four pairs of smart pattern-changing fabrics for investigation. A novel thermochromic process was devised to enable the development of novel yarns which when knitted into jacquard patterned fabrics they can switch from one pattern into another. This process was fundamental in realizing these new types of smart textiles named Psychotextiles. This paper shows for the first time how to design specific patterns for affecting specific human emotions and discusses how this research can be extended towards colour and touch. The concept of a thermal textile pixel is addressed in the next paper, which is based on a textile structure that shows spatial and temporal thermal contrast and can be used in the context of thermal communication [2]. Textiles are flexible and easy to form around three-dimensional surfaces such as our bodies. Novel electrically conductive textiles for stretchable electronic systems that can be bent or shaped around complex curvatures are being developed and the optimization and properties of these structures is reported in the paper by Christian Dils et al. [3]. E-sensors is the topic of the next two papers. Conductive formulations containing micron-size metal flakes of silver-coated copper

(Cu) and pure silver (Ag) were used as conductors on woven fabrics, and fabric flexural stiffness and sheet resistance (R_{sh}), were investigated for durability, performance and reliability, as reported by Veronica Malm et al. [4]. In the next paper by Orathai Tangsirinaruenart and George K Stylios [5], novel textile-based strain sensors have been developed and their performance was evaluated. These sensors are likely to change the way we measure stresses and strains by using entirely the textile itself, and hence finding end uses in garments as wearables for physiological wellbeing monitoring such as body movement, heart monitoring, respiration, and limb articulation measurement. The authors show how electrical resistance and mechanical properties of seven different textile sensors were optimized and measured, and report on their composition. The issues of uncontrolled active molecules in spraying of fabric substrates is addressed in the next manuscript in which defined polymers protect active components enabling controlled drug delivery and regulation of dosage, with promising results for home use and in clothes, and hence creating SMART biofunctional textiles [6]. The problem of surface area that enables better performance of electrically conductive polymer-based textiles has been addressed in the next paper by Lukas Vojtech et al. [7]. They used an electrochemical method to measure the resistance between two electrodes for comparing fabric surface areas. The combination of ECG measurement and motion tracking is achieved by developing a new hybrid soft textile electrode. Systematic measurements have shown that this hybrid textile electrode is capable of recording ECG and motion signals synchronously, which may prove a life changing approach to continuous health monitoring, as reported by Xiang An and George K Stylios [8]. The failure of the performance of composites can have catastrophic consequences and composite manufacturing is compensating for not precise detection by overengineering which is costly. This has been addressed in the paper by G Wang et al. [9], who proposed a carbon nanomaterial SMART fibre sensor capable of in-line monitoring in the manufacture of high-performance composites.

With the rapid expansion of the Internet of Things (IoTs) already affecting our work, our homes, and our communications with others, Electromagnetic Shielding (EMS) is important to guard against emissions of Electromagnetic Frequencies (EMF). Beyond electrical reliability, the protection of health and prevention of hacking are high in this agenda and what better for combating this problem than textile fabrics with their high flexibility and formability, as reported in the paper by Mark Neruda and Lukas Vojtech [10]. In the same area of interest, (IoTs), harvesting of energy for all those devices has been a quest that will continue for years to come and the two papers by L Juhasz and I.J Junger [11] and by J Junger et al. [12] shed some new light into textile-based dye-sensitized solar cells. The first paper deals with the understanding of the physical processes in the cell and its optimization, and the second paper provides electrospun polyacrylonitrile (PAN) nanofibre mats coated by a conductive polymer as collar cells on their own right. Finally, the paper by B Moradi et al. [13] proposes a more practical solution of connecting e-textiles using the embroidery process and they report how signal propagation control maybe achieved, enabling customized electromagnetic properties such as filtering for wearable electronics.

Studying of these research papers, increases our knowledge, enables us to see our own work in context, it empowers us to improve our understanding, it increases the rigour of our research, and encourages us to work collectively. I hope that to some extent the publication of the concentrated effort of these researches in this special issue, can aid us towards a clearer roadmap for our further research advancement.

Adding my own observations about the challenges that smart textiles must overcome; I can say that washability along with user safety and reliability are three important factors which need addressing in our research. Traditional textiles working with electrical components need a change of culture, which is time consuming and difficult. The supply chain is not yet ready to embrace fast changes that are much needed, and designers and engineers must learn to work together. On the other hand, we have a tired textile industry that is producing consumer products at pence per minute blamed for damaging the environment. I believe that there are untapped opportunities for this industry and that smart textiles must converse with traditional textiles. My own hope for the future of this field is not to poise for incremental changes but to force its way for a transformational change.

Acknowledgments: I cannot end this editorial by not thanking all authors for their hard work, discipline and rigour, and to encourage them to continue pushing the boundaries of this fascinating field. The exposure and promotion of our work in this special issue has been facilitated by the editorial and production office of MPDI.

Conflicts of Interest: The author declares no conflict of interest.

References

1. Stylios, G.K.; Chen, M. The Concept of Psychotextiles; Interactions between Changing Patterns and the Human Visual Brain, by a Novel Composite SMART Fabric. *Materials* **2020**, *13*, 725. [CrossRef] [PubMed]
2. Stöhr, A.; Lindell, E.; Guo, L.; Persson, N.-K. Thermal Textile Pixels: The Characterisation of Temporal and Spatial Thermal Development. *Materials* **2019**, *12*, 3747. [CrossRef] [PubMed]
3. Dils, C.; Werft, L.; Walter, H.; Zwanzig, M.; Krshiwoblozki, M.; Schneider-Ramelow, M. Investigation of the Mechanical and Electrical Properties of Elastic Textile/Polymer Composites for Stretchable Electronics at Quasi-Static or Cyclic Mechanical Loads. *Materials* **2019**, *12*, 3599. [CrossRef] [PubMed]
4. Malm, V.; Seoane, F.; Nierstrasz, V. Characterisation of Electrical and Stiffness Properties of Conductive Textile Coatings with Metal Flake-Shaped Fillers. *Materials* **2019**, *12*, 3537. [CrossRef] [PubMed]
5. Tangsirinaruenart, O.; Stylios, G. A Novel Textile Stitch-Based Strain Sensor for Wearable End Users. *Materials* **2019**, *12*, 1469. [CrossRef] [PubMed]
6. Lis Arias, M.J.; Coderch, L.; Martí, M.; Alonso, C.; García Carmona, O.; García Carmona, C.; Maesta, F. Vehiculation of Active Principles as a Way to Create Smart and Biofunctional Textiles. *Materials* **2018**, *11*, 2152. [CrossRef] [PubMed]
7. Vojtech, L.; Neruda, M.; Reichl, T.; Dusek, K.; De la Torre Megías, C. Surface Area Evaluation of Electrically Conductive Polymer-Based Textiles. *Materials* **2018**, *11*, 1931. [CrossRef] [PubMed]
8. An, X.; Stylios, G.K. A Hybrid Textile Electrode for Electrocardiogram (ECG) Measurement and Motion Tracking. *Materials* **2018**, *11*, 1887. [CrossRef] [PubMed]
9. Wang, G.; Wang, Y.; Luo, Y.; Luo, S. Carbon Nanomaterials Based Smart Fabrics with Selectable Characteristics for In-Line Monitoring of High-Performance Composites. *Materials* **2018**, *11*, 1677. [CrossRef] [PubMed]
10. Neruda, M.; Vojtech, L. Electromagnetic Shielding Effectiveness of Woven Fabrics with High Electrical Conductivity: Complete Derivation and Verification of Analytical Model. *Materials* **2018**, *11*, 1657. [CrossRef] [PubMed]
11. Juhász, L.; Juhász Junger, I. Spectral Analysis and Parameter Identification of Textile-Based Dye-Sensitized Solar Cells. *Materials* **2018**, *11*, 1623. [CrossRef] [PubMed]
12. Juhász Junger, I.; Wehlage, D.; Böttjer, R.; Grothe, T.; Juhász, L.; Grassmann, C.; Blachowicz, T.; Ehrmann, A. Dye-Sensitized Solar Cells with Electrospun Nanofiber Mat-Based Counter Electrodes. *Materials* **2018**, *11*, 1604. [CrossRef] [PubMed]
13. Moradi, B.; Fernández-García, R.; Gil, I. E-Textile Embroidered Metamaterial Transmission Line for Signal Propagation Control. *Materials* **2018**, *11*, 955. [CrossRef] [PubMed]

© 2020 by the author. Licensee MDPI, Basel, Switzerland. This article is an open access article distributed under the terms and conditions of the Creative Commons Attribution (CC BY) license (http://creativecommons.org/licenses/by/4.0/).

Article

The Concept of Psychotextiles; Interactions between Changing Patterns and the Human Visual Brain, by a Novel Composite SMART Fabric

George K. Stylios * and Meixuan Chen

Research Institute for Flexible Materials, Heriot-Watt University, Scottish Borders Campus, TD1 3HF, UK; michellemeixuanchen@gmail.com
* Correspondence: g.stylios@hw.ac.uk

Received: 25 September 2019; Accepted: 19 January 2020; Published: 5 February 2020

Abstract: A new SMART fabric concept is reported in which visual changes of the material are designed to influence different human emotions. This is achieved by developing a novel electrochromic composite yarn, knitted into pattern-changing fabrics, which has high response in temperature change and uniform contrast. The influence of these pattern-changing effects on the response of the human visual brain is investigated further by using event-related potential (ERP). Four SMART pattern-changing fabric pairs were used in this experiment. Each fabric presents two patterns interactively with different, but complementary or opposing, pattern attributes. 20 participants took part in the experiment, in which they were exposed to the patterns, while their visual brain activities were recorded. Comparisons of the three prominent ERP components; P1, N1, and P2 that correspond to the two patterns of each fabric have shown significant differences in the latency and amplitude of these components. These differences show that patterns and pattern-changing cause different visual impacts and that these changes influence our level of attention and processing effort. The study concludes that with the pattern changing ability of these thermochromic hybrid materials we can create designs with attributes that can directly manipulate user emotions, which we like to call 'psychotextiles'. Our study also poses much wider questions of our image processing process in relation to design and art.

Keywords: SMART pattern-changing fabric; pattern effect; visual response; visual brain; event-related potential (ERP); psychotextiles; art and design

1. Introduction

Recent advances of SMART textiles enable researchers to explore new ways of interaction with users. There is increasing interest to construct SMART structures, systems and prototypes with tailor-made functionality and aesthetics, which can interact with user's behaviour. Colour, pattern, and shape-changing effects of SMART fabrics are now capable for interaction with our emotional responses [1–8]. Most of these designs however focus on exploring SMART fabrics capability rather than designing them to actively influence any emotional user response. This study aims to investigate whether it is possible to pre-determine the pattern design of a novel SMART fabric with pattern-changing properties which would directly influence human visual response.

Studies of pattern perception have found that the shape and form of pattern can influence human visual response. Consecutive circles, radical lines, and stripes had less effect than checkboard, with the smaller check pattern having higher response [9–12]. A triangular pattern had an even higher response than a circle and a square [13]. Sharp corners in patterns had a quicker response than rounded corners [14,15]. Sharpness of patterns were also important; sharp patterns produce quicker visual response than blurred ones [12]. The same is true with symmetrical patterns were quicker in response than asymmetrical ones [16]. There is a clear indication therefore that patterns can affect our visual responses, but there is lack of extended studies for providing concrete evidence, and if there is, how can these effects be facilitated in new materials. Hence, in this study, we test the hypothesis that patterns may affect our emotions, we concentrate here at the start of the image process (visual brain) and we facilitate this study by developing a new composite material capable of pattern changing in a predetermined way. The yarn had to change the colour quickly and uniformly and also to be durable to withstand stresses and strains imposed during fabric making. This study is particularly interesting because this novel SMART fabric enables us to devise pairs of design changes of the same fabric, to test the hypothesis of switching on-off and altering a specific emotion. The question to why we have decided on the four fabric pairs reported here is because in our preliminary experiments [17] we found that these designs were most important to test our concept and without the influence of colour in the first instance.

Two pattern characteristics were investigated in the current study: repeating/non-repeating and weak/intense. Repeating patterns contain regularly repeating elements and have symmetrical and continuous features; in contrast, non-repeating patterns contain irregularly repeating elements and have asymmetrical and discontinuous features. Repeating patterns have been previously found to have more pleasant effect than non-repeating patterns in one of our studies [17]. Weak patterns are faint, light, and simple compared to intense patterns that are high in contrast, bold, and complex. According to literature, the symmetrical feature of repeating patterns may trigger an earlier visual response from viewers than non-repeating pattern, and the intense patterns could evoke a higher visual response than weak patterns.

We appreciate that this study cuts through two different disciplines; materials and neuroscience. It is necessary, like in most other fields, to work interdisciplinarily. Our aim is however a new material; a new SMART textile which will, for the first time be capable of influencing specific emotions by precisely designing attributes found from studies such as ours. Hence, we call this new class of SMART fabrics that are designed to influence specific emotions 'psychotextiles'. The results of our visual brain study are reported here we are preparing follow on results from our complimentary study of the frontal part of our brain.

Human visual response can be measured from the activity of the visual brain, which is in the most posterior portion of the brain and is the centre for processing all visual information that is received through our eyes. Measurement of the visual brain activity provides an objective insight into an individual's visual response to an external stimulation. The brain activity can be measured by a non-invasive electroencephalogram (EEG) technique.

Our brain consists of billions of neuron cells which work together to perform various functions. They operate on electrochemical transmission. Synchronised neurons produce electric potential (EP) in the brain and by attaching electrodes on the scalp this activity can be measured by amplification of these signals. We call EEG the waveform trace that is recorded from the EP over a known time period [18]. Therefore, the brain activity can be inspected by analysing the EEG traces. When responding to a stimulus, the electrical activity of the brain changes as soon as a response occurs. Some changes are large enough to be identified in the primary EEG trances, but some are rather small and concealed inside the unrelated spontaneous brain activities. In order to discriminate the electrical activity that is corresponding to the specific stimulus from the noise that is generated by unrelated activities, a signal averaging approach is used to increase the amplitude of the event related signal relative to noise, in which the specific stimulus has to be repeatedly presented or conducted for a number of trials, then the time-locked EEG signals corresponding to the stimulus are extracted from the continuous EEG record, aligned, and the amplitudes on each time point in the signal are then averaged. Connecting the averaged amplitude at each time point obtains a wave form, which is the event-related potential (ERP) that presents the discrete brain activity responding to a specific stimulus [19–21]. Figure 1 shows a typical ERP evoked by a visual stimulation. A sequence of prominent picks can be found in an evoked ERP, which are named by their negative (N) or positive (P) polarity. As seen in Figure 1, the component

P1 is the first prominent positive peak; component N1 is the first prominent negative peak; accordingly, component P2 and P3 is the second and the third positive peaks. The occurrence, the amplitude and latency of the ERP component have been found varied depending on the parameter of the stimulus and on the subject's psychological state.

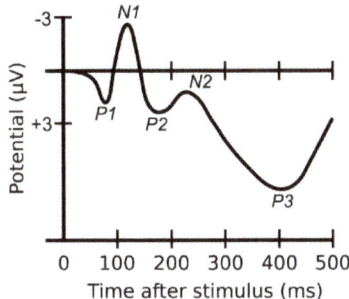

Figure 1. A typical ERP elicited by a visual stimulation.

The ERP measurement is relatively straightforward. It has been use in studies of stimulation effect on viewers' response, such as perception of colour and pattern [22], affective processing [23], and facial expression processing [24]. In our study, viewers' visual brain response to a pattern stimulus is measured by using the ERP technique and analysed on the evoked ERP components.

2. Experimental Methodology

2.1. Materials and Processing Methods

2.1.1. Yarn

The pattern-changing effect of the fabrics is based on a novel composite electrochromic yarn that can change colour by changes of the electric current. As presented in Figure 2, the composite yarn consists of a core and outer sheath. The outer sheath consists of a thermochromic material that can change colour at a given temperature. Colour change is produced by heat being produced in the core of the yarn by electric current passing through it, rendering the changing of the colour on its thermochromic outer sheath. The temperature is dropped by stopping the electric current resulting in heat dissipation in the core of the yarn and causing the thermochromic outer sheath to go back to its original colour. Hence, colour can be changed in the yarn by passing a known electric current to the yarn. After experiments with different diameters, a core diameter of 0.10 mm silver-plated copper wire was chosen, which was pliable enough and thin and it could produce adequate heat.

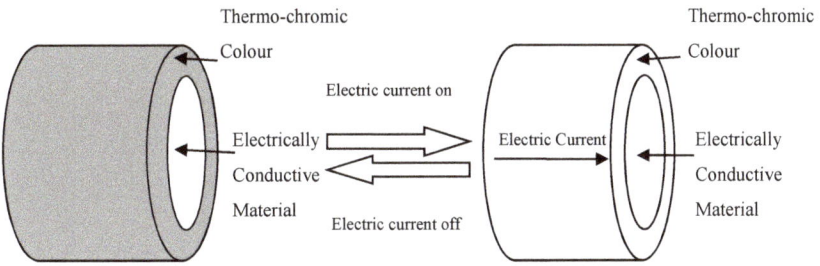

Figure 2. Conceptual model of the SMART colour-changing electrochromic yarn.

To increase its strength and pliability for fabric processing, it was blended with several yarns; 180 tex wool, 30 tex silk, 58 tex spun viscose, 20 tex cotton, using a Gemmill and Dunsmore hollow spindle fancy yarn machine, the conditions of which were optimised after preliminary investigations. Consequently, the yarn produced is a composite hybrid material of textile and copper, and as such non ordinary to use any thermochromic dye. After a number of experiments, it was decided to use a special thermochromic pigment (Chromazone's Water based sprayable system 1510), suitable for metal surfaces, and we found for the first time that it was also successfully applied to textiles, since our composite yarn is composed of a metallic part in its core and a textile part in its sheath. The application of the pigment was done by Chromazon's Sprayable System (LCR Hallcrest Ltd., Connah's Quay, UK). The first step of this process is the preparation of the thermo-chromic colour system solution, which was made up of 55.9% clear lacquer, 24% thermo-chromic pigment, 1.1% adhesion promoter, and 19.0% water, mixed with a mechanical stirrer. The yarn was then wound onto a wooden frame around nails fixed in opposite sides and parallel to each other. The prepared solution was poured into the glass jar of an airbrush (Model 200 Airbrush, Badger Air-brush Company, Illinois, 9128 Belmont Av, Franklin Park, IL, USA) and evenly sprayed over the yarn. Spraying was repeated several times until the whole yarn was covered in black colour. Then, the prayed yarn set in the wooden frame was placed in a pre-heated oven of 100 °C for 5 min, for curing the thermo-chromic pigment. The yarn was left to cool down to room temperature, and then wound onto a yarn cone for knitting. Figure 3a,b shows the composite SMART yarn on the wooden frame before and after colouration.

Figure 3. Composite copper/cotton yarn before; (**a**) and after; (**b**) colouration.

Experiments were then carried out to find the most suitable cotton blend, by uniformly wrapping 4 metres of the yarn around a white cardboard in 20 °C and 65% RH conditioned room, and its colour change characteristics were investigated. The 20 tex cotton yarn was found to be the most suitable which has 9 Ohms electrical resistance and 0.3 A to 0.4 A optimum electric current, and is able to change from black to white in less than 40 s, Figure 4 shows a 35 times magnification image of the yarn and Figure 5a,b its thermochromic performance.

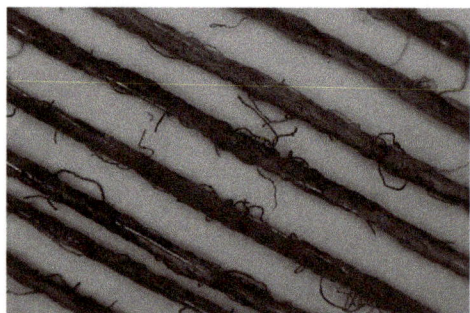

Figure 4. X35-times magnification coloured composite yarn.

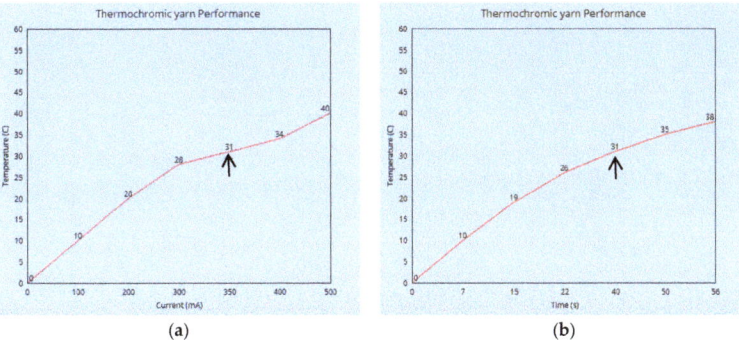

Figure 5. Performance of the composite thermochromic yarn; (**a**) temperature against current and (**b**) temperature against time. The arrows shows the colour change values.

So we have established a hybrid composite yarn of cotton and copper which is now capable of colour changing at an activation temperature of 31 °C by heat being produced from the copper wire and triggers the thermochromic surface to change colour from black to colourless and reverse uniformly and in less than 40 s. Investigations into different knitting and woven processes that followed revealed that the fabric made on an 8-gauge Shima-Seiki SES 122S electronic knitting machine (SHIMA SEIKI Europe Ltd., Castle Donington, UK) had the best design reproduction and response to pattern change, at 0.5 metres per min speed, optimum tension, and with 0.7 mm stitch length.

Having established the fabric several preliminary investigations took place to determine the criteria of designs and their changes that will fulfil our hypothesis, and it was apparent that we should knit four paired fabric patterns. These four pairs were made and used for the brain/material experiments that followed.

2.1.2. Pattern Changing Fabrics

These four pairs of patterns are shown in Figure 6. In order to test the response of our visual brain every pair of the four fabrics has the same design but subtle differences. Fabric 1a,b consists of geometric square and trapezoid small motifs but 1b are bolder (black and white more well defined—sharper) than 1a. Fabric pair 2, consists of diamond motifs both pairs are symmetrical but fabric 2b has larger diamonds making the symmetry more well defined than in the case of fabric 2a. Fabric pair 3 consists of small squares, with fabric 3a being symmetrical with squares of the same repeat, whilst fabric 3b produces random asymmetrical and irregular shapes with darker intensity. Fabric pair 4 consists of large square shapes, with fabric 4a being symmetrical and fabric 4b random asymmetric and irregular squares with darker intensity. Therefore, the predominant effect is geometry, asymmetry, faintness and boldness, intensity and irregularity, all in black and white and without the effect of colour so that there is no colour influence on our tests.

2.2. Material/Brain Interactions

Viewers' visual response to a pattern stimulus is directly measured in the visual brain activity by using the ERP technique, so that the brain/pattern interaction can be established. By analysing the evoked ERP components, the visual effect can be measured and compared between different patterns, from which we can determine the effect of pattern change on viewers' visual response. Firstly, the visual response to each patterned appearance is measured, and the responses of the two paired patterned appearances are compared by analysing the evoked ERP components. Statistical technique: hypothesis test and confidence interval estimation are used to calculate the difference in the amplitude and latency of the ERP components; and hence, to determine the different visual effect trigged by the pattern changes of the fabric.

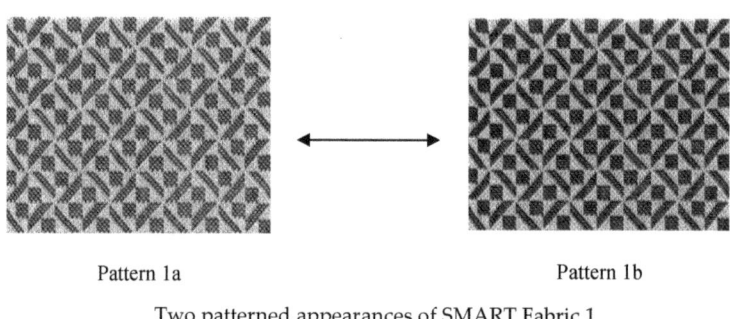

Pattern 1a   Pattern 1b

Two patterned appearances of SMART Fabric 1

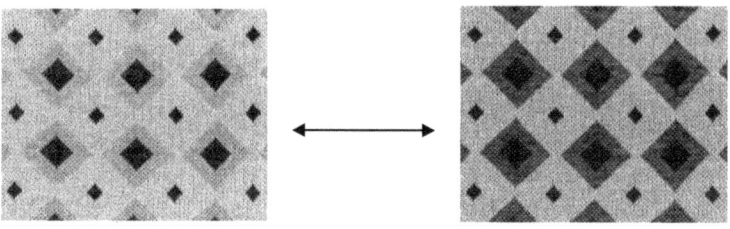

Pattern 2a   Pattern 2b

Two patterned appearances of SMART Fabric 2

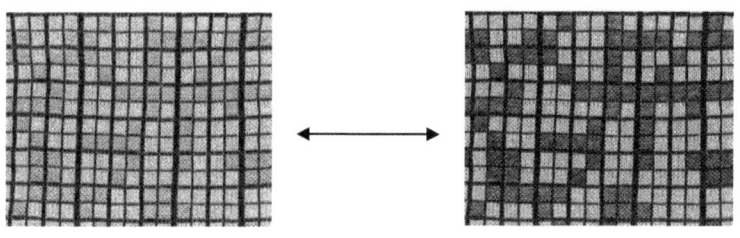

Pattern 3a   Pattern 3b

Two patterned appearances of SMART Fabric 3

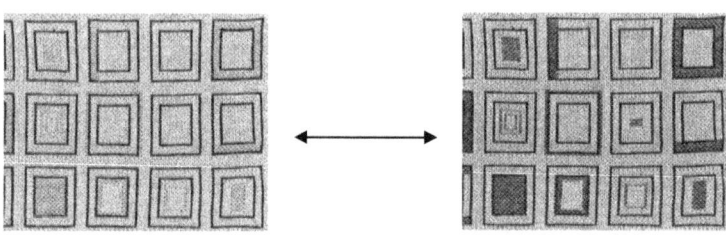

Pattern 4a   Pattern 4b

Two patterned appearances of SMART Fabric 4

Figure 6. Four SMART pattern-changing fabrics investigated in current study.

2.3. EEG Experiment

2.3.1. Participants

Twenty participants; 9 women and 11 men, between 23 and 54 years of age, took part in this experiment. Their mean age was 31.6 years old with 9.0 years standard deviation. All participants had no known mental problems neither any brain operation nor suffering from epilepsy or claustrophobia. All participants had normal-or corrected vision by spectacles and were right-handed. Before the experiment, each participant was explained the experimental procedure without revealing any detail of the experimental target. The ethical procedure was followed by declaring the procedure and obtaining from each participant written consent and approved before starting of the experiment in line with standard procedure. Each participant was alone and there was not possible for any participant to discuss their experience to another.

2.3.2. EEG Experiment Procedure

The two patterns from every of the four fabrics were scanned, and their clear digital image was saved in the PC. During the experiment each fabric pattern was presented to each participant in a 19-inch screen monitor of a grey background. All pattern images were 305 mm in width and 245 mm in height and shown at the same brightness settings. The sitting distance of every participant from the PC monitor was 1400 mm and the visual angle of stimulation of every pattern in the experiment was 10.0 degrees in the vertical and 12.4 degrees in the horizontal coordinate, and the laboratory was equipped with sound-attenuation. Prior to starting the experimental procedure every participant was given a short explanation of the experiment and were reminded the information on the consent form. An Electro-Cap ECI was placed on the head of the participant in order to record their EEG. A pair of electrodes was used to detect eye movement and one is placed 1 cm above and laterally of the corner of the left eye and the other on the left-hand mastoid behind the ear and lower to the skull. Another electrode was placed on the left earlobe of the participant. The ECI was then connected with the EEC system, and the impedance of all electrodes was checked to be less than 20 kΩ prior to starting the experiment, ensuring their proper attachment to scalp for achieving EEG brain wave data acquisition. A PC displayed the fabric patterns having devised a protocol of slides which had the stimulus of each pattern together with specific instructions. In this protocol, the 'presentation' software was used to pre-program in self-written scripts the displaying order and duration of every slide, as shown in Figure 7. At the beginning of the experiment for 8 s a preparatory instruction slide was shown with eye movement commands; 'eyes close', 'eyes open', and 'blink eyes' to relax the eyes of every participant throughout the experiment. A grey screen was then shown for 1.5 s which was the baseline experimental period prior to showing the first pattern for 1 s duration. The repetition of the circles is as shown as in the figure. There is 20 s break between the circles. The pattern stimuli were shuffled in random at the beginning of the presentation and each of them was presented 30 times.

During the experiment, every participant sits on a comfortable armchair opposite the PC monitor with both hands on the arms of the chair, making sure that their eyes were in line with the centre of the monitor. Every participant was informed that a series of slides with fabric patterns will be shown on the monitor and that some of them will give them instructions of closing, opening, or blinking their eyes. Participants were asked to sit comfortably, to be relaxed and to follow the instructions appeared on screen. The operator of the experiment asked each participant to concentrate at the patterns over the period that they appear on the screen without blinking, breathing deep, or moving, and a couple of training familiarity trials were performed by every participant prior to starting the experiment, so that instructions were understood and feeling comfortable. At the end of the experiment the electrodes and the cap were removed from the participants head, the participant was then thanked and asked not to discuss the procedure with anyone else to avoid influencing other participants.

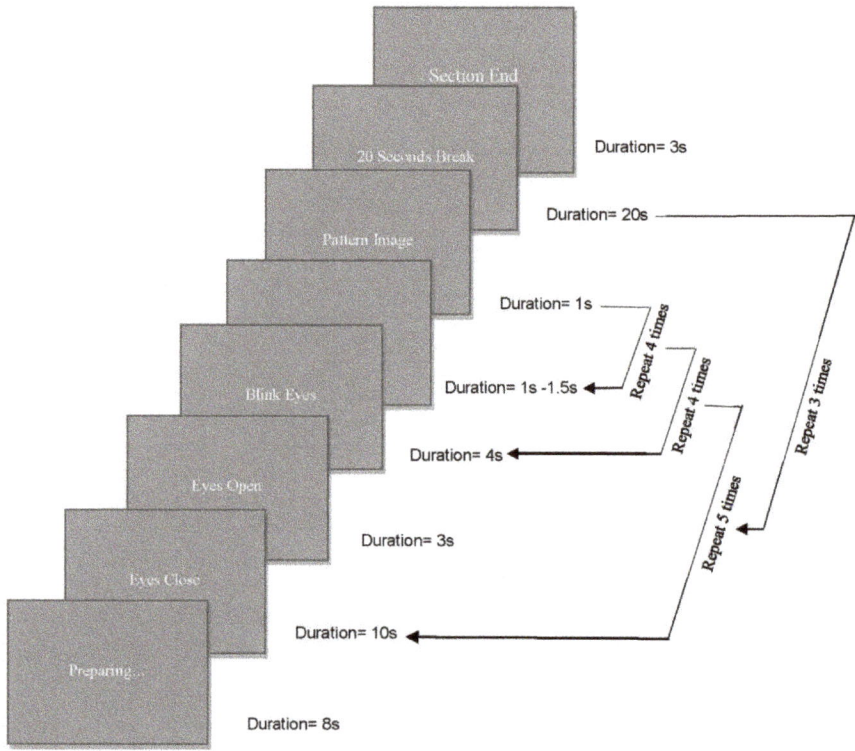

Figure 7. Diagram of the slides with the timing and repeating circles used in the experiment.

2.3.3. Brain Data Acquisition and Processing

The cap consisting of 19 electrodes was worn on the head of each participant so that their EEG signals could be acquired to determine their brain response to each of the 4 paired fabric patterns. The specific locality of the electrodes on the scalp were according the international 10–20 EEG system [25]. The reference electrode is attached on the left earlobe of each participant and the ground electrode is located at the front of channel Fz. The impedance of the electrodes was less than 20 kΩ. Figure 8 shows the whole set up which consists of the EEG system integrated with the presentation slides in the PC through a trigger box, which places an event marker in the EEG signal for every slide displayed on screen. This enables every slide location along the signal to be known marking any changes in brain waves to each slide stimulation. The electric potential generated by the participant's eyes movement, named electrooculography (EOG), was also recorded from active and reference bipolar electrodes.

During signal pre-processing, the recordings of the signals of the EEG data and the log file of the presentation slides were inputted to MATLABs EEGLAB (version 11.0.4.3b) toolbox [26], Then, the data recorded from O1 and O2 electrode channels, which are located in the visual brain (see Figure 5) are selected; and the EEG signals that correspond to the eight patterns were extracted. Each signal contained an epoch starting 100 ms prior to pattern stimulus onset until after 1000 ms. There are 30 EEG epochs corresponding to each fabric stimulus. Any artefact caused by eyes blink, movement, temporal muscle activity, or line noise was extracted and cleaned using the Independent Component Analysis (ICA) routine of MATLABs EEGLAB [27]. The artefact-free EEG epochs were then gathered in a STUDY dataset of the EEGLAB to generate the grand average ERP waves that corresponds to the viewing of each pattern stimulus.

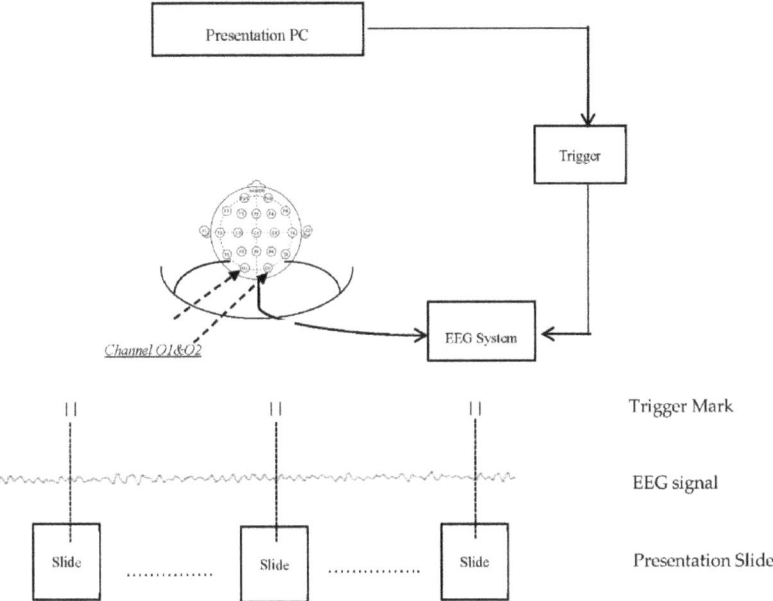

Figure 8. Setup of the EEG signal acquisition in the experiment.

3. Results

The grand average visual ERPs evoked by the viewing of the patterns are analysed in pairs, which are patterns 1a,b, patterns 2a,b, pattern 3a,b, and patterns 4a,b. In each pair, the ERP waves of channels O1 and O2 are presented separately in two graphs as seen in the following section. The waves start at 100 ms before the pattern onset until after 1000 ms. Vertical grey lines in the graphs indicate the significant differences in between the two ERP waves evoked by the paired patterns at 95% confidence level. The following prominent components were observed in all evoked ERP waves: component P1 at around 100 ms, component N1 at around 150 ms and component P2 between 200 to 300 ms. The amplitude (μv) and latency (ms) of each component were measured. There are two methods to calculate the amplitude of the components N1 and P2 in literature [21]. One is based on the local peak value of the component, and the other one measures the distance between the local peaks of the two successive components: the amplitude of component N1 = P1 peak—N1 peak and the amplitude of component P2 = P2 peak—N1 peak. Both methods have their own strengths and shortcomings. The current experiment used both methods so that one can complement the other. Comparison was conducted on ERP components that were evoked by the two paired patterns of each fabric. The difference of the pattern change response of every participant is calculated by the formula

$$\sum \Delta n = Pn_1 - Pn_2 \qquad (1)$$

where $\sum \Delta n$ is the different response of pattern change n.

n = 1, 2, 3, 4 and $Pn_1 - Pn_2$ is the response difference of pattern 1 and 2 of each of the four patterned paired fabrics.

Using the statistical hypothesis test (SHT) and confident interval estimation (CIE), the mean of the difference is calculated from the data acquired from the 20 participants. Then the Ryan-Joiner Test at 5% SL was used to test the normal distribution (ND) of the sample data. The significance of the results (over 80%) is reported and discussed further.

3.1. Significant Differences in the Visual ERPs Evoked by Pattern 1a,b of Fabric 1

Significant differences have been found in the ERP latency and amplitude components N1 and P2, of Pattern 1a,b, shown in Figure 9, which are reported in Table 1. Pattern 1b triggers an earlier component N1 than Pattern 1a in the left of the visual brain, measured at electrode location O1. Pattern 1b triggers a larger amplitude of component N1 than Pattern 1a, which shows that Pattern 1b has higher visual intensity than Pattern 1a, which is in agreement with another study on amplitude and brightness [28]. In component P2, significant difference was found in the latency in the O2 channel location meaning that Pattern 1a triggers an earlier P2 than Pattern 1b in the right side of the visual brain.

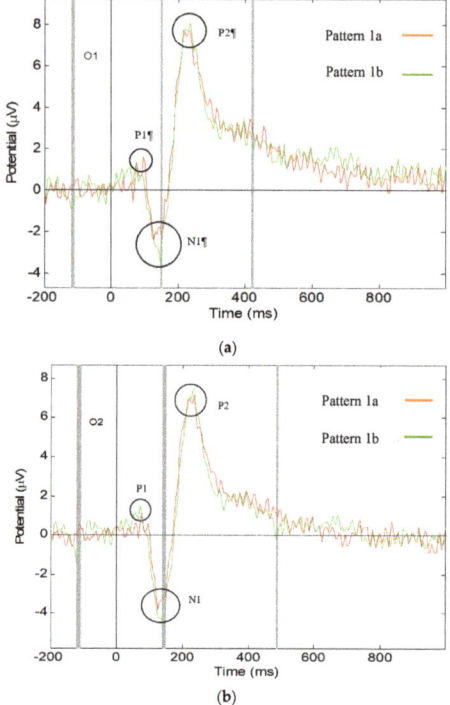

Figure 9. Grand average ERP responding to Pattern 1a,b, the grey lines show the significant difference with p-value < 0.05: (**a**) measured at O1 electrode channel; (**b**) measured at O2 electrode channel.

Table 1. Amplitude and latency differences of N and P components of visual ERP waves evoked by Pattern 1a,b

				O1 Electrode Channel.					
			N	Mean	StDev	SE Mean	90% CI	T	p
N1	Latency (ms)		20	4.5	9.99	2.23	(0.64, 8.36)	2.02	0.058
				O2 Electrode Channel					
			N	Mean	St Dev	SE Mean	90% CI	T	p
N1	Amplitude (μv)		20	1.023	2.527	0.565	(0.046, 2.000)	1.81	0.086
P2	Latency (ms)		N	Mean	St Dev	SE Mean	90% CI	T	p
			20	−9.5	12.97	2.9	(−14.51, −4.49)	−3.28	0.004

St Dev: Standard deviation. SE Mean: Standard error of the mean. CI: Confidence interval. p: p-value.

These results show that the difference in intensity of the fabric patterns causes different visual brain responses, in which clearer and well-defined patterns evoke an earlier and larger ERP component N1 and faint patterns evoke an earlier P2 component.

3.2. Significant Differences in the Visual ERPs Evoked by Pattern 2a,b of Fabric 2

Significant differences have been found in the amplitude and latency of the ERP components N1 and P2, of Pattern 2a,b, shown in Figure 10, which are reported in Table 2. It was found that there are significant differences in the amplitude of component N1 in both O1 and O2 electrode channels meaning mean Pattern 2b evokes a larger component N1 than Pattern 2a on the both sides of the visual brain, in agreement with other work [19]. In component P2, significant difference was found in the amplitude in both O1 and O2 channel locations, showing that Pattern 2b evokes a larger component P2 than Pattern 2a on both sides of the visual brain. A difference was also found in the latency of component P2 in the O2 brain location, which shows that Pattern 2a gives an earlier component P2 in the right of the visual brain.

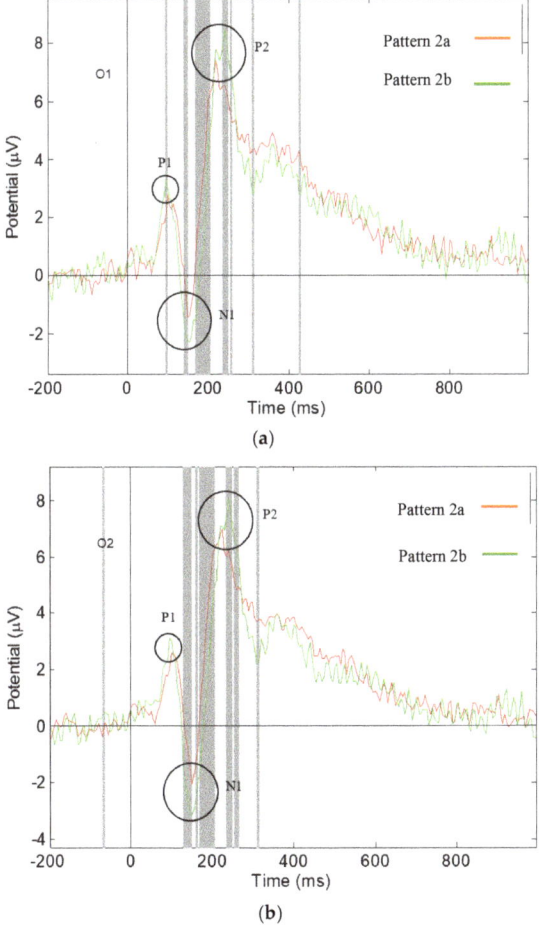

Figure 10. Grand average ERP responding to Pattern 2a,b, the grey lines show the significant difference with *p*-value < 0.05: (**a**) Measured at O1 electrode channel; (**b**) measured at O2 electrode channel.

Table 2. Amplitude and latency differences of N and P components of visual ERP waves evoked by Pattern 2a,b.

			O1 Electrode Channel					
N1	Amplitude (μv)	N	Mean	St Dev	SE Mean	95% CI	T	p
		20	2.143	2.662	0.595	(0.898, 3.389)	3.6	0.002
P2	Amplitude (μv)	N	Mean	St Dev	SE Mean	95% CI	T	p
		20	−3.024	3.271	0.731	(−4.555, −1.493)	−4.13	0.001
			O2 Electrode Channel					
N1	Amplitude (μv)	N	Mean	St Dev	SE Mean	98% CI	T	p
		20	2.201	2.469	0.552	(0.799, 3.603)	3.99	0.001
P2	Amplitude (μv)	N	Mean	St Dev	SE Mean	95% CI	T	p
		20	−2.943	3.048	0.681	(−4.369, −1.516)	−4.32	0
	Latency (ms)	N	Mean	St Dev	SE Mean	95% CI	T	p
		20	−7.75	10.82	2.42	(−12.81, −2.69)	−3.2	0.005

St Dev: Standard deviation. SE Mean: Standard error of the mean. CI: Confidence interval. p: p-value.

The results reveal that pattern changing effects 2a and b of fabric 2 affect the visual response of the participants. The two patterns have symmetrical structures of regular diamond shapes, with Pattern 2a having smaller diamonds to Pattern 2b. The intense effect produced by changing the size of the elements in a pattern causes different responses of the visual brain, in which more intense patterns trigger larger components N1 and P2, indicating a larger brain response.

3.3. Significant Differences in the Visual ERPs Evoked by Pattern 3a,b of Fabric 3

Significant differences have been found in the latency and amplitude of components P1, N1, and P2, of Pattern 3a,b of fabric 3, Figure 11, which are reported in Table 3. In component P1, significant difference was found in the amplitude in the O1 brain location, showing that Pattern 3b evoked a larger component P1 than Pattern 3a in the left of the visual brain. In component N1, significant difference was found in amplitude in both the O1 and O2 channel locations, showing that Pattern 3a evoked a larger component N1 than Pattern 3b in both sides of the visual brain. Significant difference was also found in the latency of component N1 in the O1 channel location, indicating that Pattern 3a evoked an earlier component N1 than Pattern 3b in the left of the visual brain. Our results are in agreement with another study found that the latency of component N1 has been found to be shorter in response to the stimuli with higher brightness [28] and the amplitude of component N1 to be influenced by visual stimuli [19]. The current results reveal different visual parameters between Pattern 3a,b, in which Pattern 3a might have higher visual intensity than Pattern 3b. In component P2, a significant difference was found in the amplitude in the O1 channel location, meaning that Pattern 3b evoked a higher component P2 than Pattern 3a in the left of the visual brain.

The results reveal that our participants were influenced by fabric 3 Pattern 3a,b triggering different responses in the visual brain. Both patterns contain small square shapes of the same size. Pattern 3a has a symmetrical structure of repeating square shapes, whilst Pattern 3b is having asymmetrical randomly arranged squares and rectangular shapes, some filled with intense black colour. Symmetrical patterns with regular repeating elements trigger larger and earlier ERP component N1, indicating a larger and earlier brain response.

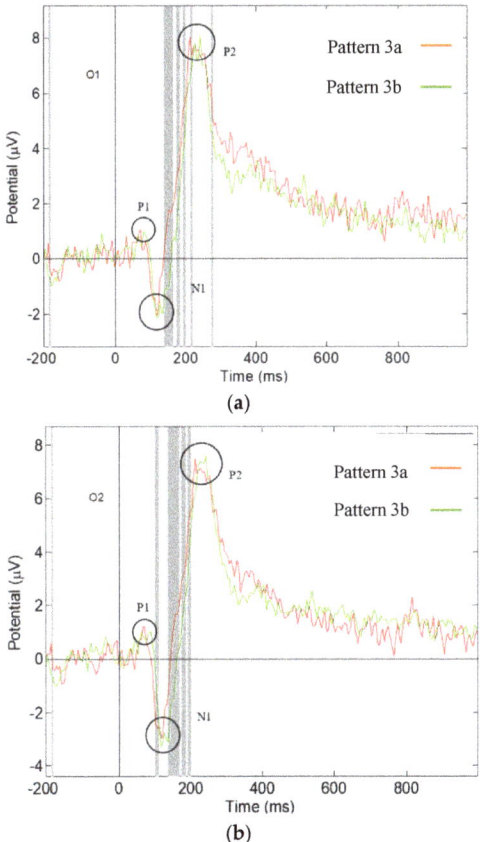

Figure 11. Grand average ERP responding to Pattern 3a,b, the grey lines show the significant difference with *p*-value < 0.05: (**a**) measured at O1 electrode channel; (**b**) measured at O2 electrode channel.

Table 3. Amplitude and latency differences of N and P components of visual ERP waves evoked by Pattern 3a,b.

O1 Electrode Channel								
P1	Amplitude (µv)	N	Mean	St Dev	SE Mean	85% CI	T	p
		20	−0.905	2.641	0.591	(−1.791, −0.019)	−1.53	0.142
N1	Amplitude (µv)	N	Mean	St Dev	SE Mean	80% CI	T	p
		20	−1.333	4.044	0.904	(−2.534, −0.133)	−1.47	0.157
	Latency (ms)	N	Mean	St Dev	SE Mean	89% CI	T	p
		19	−3.68	9.4	2.16	(−7.31, −0.06)	−1.71	0.105
P2	Amplitude (µv)	N	Mean	St Dev	SE Mean	85% CI	T	p
		18	−0.645	1.785	0.421	(−1.279, −0.011)	−1.53	0.143
O2 Electrode Channel								
N1	Amplitude (µv)	N	Mean	St Dev	SE Mean	83% CI	T	p
		20	−1.074	3.319	0.742	(−2.132, −0.015)	−1.45	0.164

St Dev: Standard deviation. SE Mean: Standard error of the mean. CI: Confidence interval. *p*: *p*-value.

3.4. Significant Differences in the Visual ERPs Evoked by Pattern 4a,b of Fabric 4

Significant difference was found in the component P1, N1, and P2 on the O1 channel and in components N1 and P2 on the O2 channel, of Pattern 4a,b, fabric 4, Figure 12, as shown in Table 4. In component P1, significant difference was found in latency in the O1 channel location, indicating that Pattern 4a produced an earlier component P1 than Pattern 4b in the left of the visual brain. In component N1, significant difference was found in amplitude in channel O1 location, showing that Pattern 4a evoked a larger component N1 than Pattern 4b in the left of the visual brain, and significant difference was also found in the latency of component N1 in both the O1 and O2 brain locations, indicating that Pattern 4b evoked an earlier component N1 than Pattern 4a in both sides of the visual brain, in agreement with other work [19,28]. It was also found that component P2 had significant difference in the amplitude of both the O1 and O2 channel locations, and in latency at O1 channel. indicating that Pattern 4b produced a larger component P2 than Pattern 4a in both sides of the visual brain, and latency indicating that Pattern 4b triggers an earlier component P2 than Pattern 4a in the left of the visual brain.

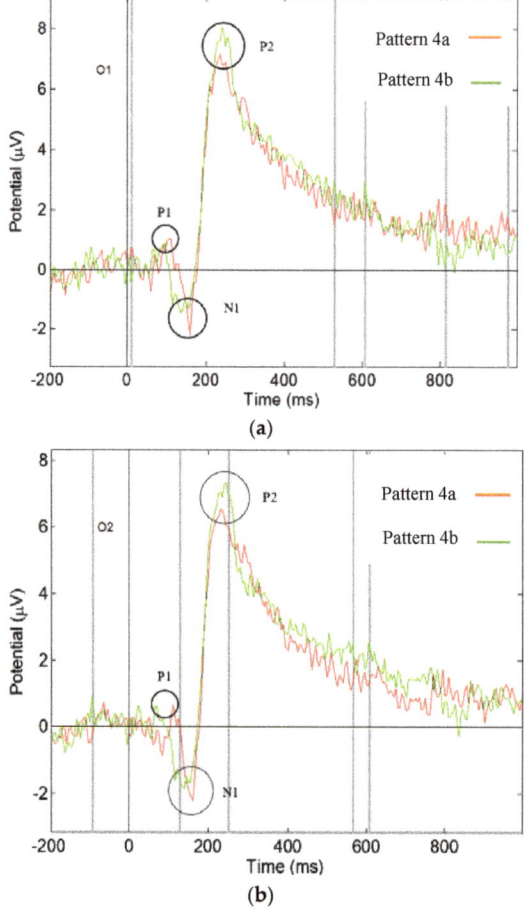

Figure 12. Grand average ERP responding to Pattern 4a,b, the grey lines show the significant difference with p-value < 0.05: (**a**) measured at O1 electrode channel; (**b**) measured at O2 electrode channel.

Table 4. Amplitude and latency differences of N and P components of visual ERP waves evoked by Pattern 4a,b.

O1 Electrode Channel								
P1	Latency (ms)	N	Mean	St Dev	SE Mean	90% CI	T	p
		18	−8.06	16.37	3.86	(−14.77, −1.34)	−2.09	0.052
N1	Amplitude (µv)	N	Mean	St Dev	SE Mean	95% CI	T	p
		17	−11.33	5.91	1.43	(−14.37, −8.29)	−7.91	0
	Latency (ms)	N	Mean	St Dev	SE Mean	95% CI	T	p
		18	10.28	17.1	4.03	(1.77, 18.78)	2.55	0.021
P2	Amplitude (µv)	N	Mean	St Dev	SE Mean	95% CI	T	p
		17	−1.934	2.992	0.726	(−3.473, −0.396)	−2.67	0.017
	Latency (ms)	N	Mean	St Dev	SE Mean	80% CI	T	p
		19	3.95	11.74	2.69	(0.37, 7.53)	1.47	0.16
O2 Electrode Channel								
N1	Latency (ms)	N	Mean	St Dev	SE Mean	88% CI	T	p
		20	5.5	14.95	3.34	(0.06, 10.94)	1.65	0.116
P2	Amplitude (µv)	N	Mean	St Dev	SE Mean	91% CI	T	p
		20	−1.253	3.093	0.692	(−2.488, −0.017)	−1.81	0.086

St Dev: Standard deviation. SE Mean: Standard error of the mean. CI: Confidence interval. *p*: *p*-value.

The results reveal that the visual brain of our participants was influenced by fabric 4 in different ways. The patterns of fabric 4 contain large square shapes, with Pattern 4a having symmetrical structure of regularly repeating square shapes, whilst Pattern 4b having the same square shapes but non-repeating with some filled with black colour and also having smaller squares within larger ones. Consequently, Pattern 4b is more complex and more asymmetrical than Pattern 4a. The results reveal that these differences trigger different responses in the visual brain, in which the relatively simple and symmetrical patterns trigger an earlier component P1 and a larger component N1, indicating an earlier and larger brain response.

In summary the pattern-changing effect of fabrics 1 and 2 produce a larger N1 response in the visual brain because its pattern changes from a light and loose to a darker and higher contrast effect. However, this phenomenon does not apply in the case of Fabric 3, in which, the symmetrical structure of this pattern evokes a higher response in the visual brain. The same result is observed in the pattern-changing effect of Fabric 4. These results are in agreement with the findings of symmetry/asymmetry studies in literature [16], in which the symmetrical patterns have been found to be detected and processed more easily by the human visual system in comparison with the asymmetrical patterns, as well as with a study of pattern intensity in amplitude and latency of the visual brain [19,28]. This could have a direct connection with pleasure and happiness found in complimentary study.

4. Discussion and Conclusions

This study has investigated the influences caused by pattern-changing of four pairs of colourless SMART fabrics on viewer's visual response using the ERP technique. This was possible by developing a novel composite electrochromic yarn of 20 tex cotton wrapped around silver-plate 0.10 mm diameter copper wire in a Gemmill and Dunsmore hollow spindle fancy yarn machine and successfully functionalised using the Chromazon's Sprayable System by a pigment which enables it to activate at 31 °C, in less than 40 s. The yarn was subsequently knitted into four fabric pattern pairs testing geometry, asymmetry, and intensity, of our visual brain, at O1 and O2 electrodes positioned at the back of our brain. The responses of 20 participants were analysed by comparing the amplitude and latency components N1, P1, and P2 in the elicited ERP wave from the visual brain. Our results found significant differences in visual brain responses. These differences were mainly observed in components N1, P1, and P2 caused by the difference of the visual parameters of the fabric stimuli. The properties of N1 and P1 components are regarded as exogenous i.e., of pure sensory processing [29]. This process takes place at about 100 ms from the presentation of the stimulus and is said to be influenced by the physical properties of the stimulus. However, there are some researchers that believe that these components could be endogenous and hence could be enhanced by attention [30]. Although there is a general agreement that the P2 component is associated with memory updating, we will take the view that stimulus is processed in a number of either parallel or hierarchical stages which include pattern encoding, recognition, classification, task evaluation followed by response selection and execution. In summary, there must be two stages; a neutral detection of pattern and a make use of this pattern, recognising that some parallel processing takes place, we deal more with the former in our case. Our findings show that the pattern-changing effect evokes different visual responses in component N1. The pattern-changing effect of fabrics 1 and 2 produce a larger N1 response in the visual brain because its pattern changes from a light and loose to a darker and higher contrast effect. However, this phenomenon does not apply in the case of Fabric 3, in which, the symmetrical structure of this pattern evokes a higher response in the visual brain. The same result is observed in the pattern-changing effect of Fabric 4. These results are in agreement with the findings of symmetry/asymmetry studies in literature [16], in which the symmetrical patterns have been found to be detected and processed more easily by the human visual system in comparison with the asymmetrical patterns, and it is said to be associated with pleasure and happiness. Our point to our findings is challenging our current knowledge on perception. We claim that since N1 and P2 are very early in the visual process and the brain is yet to complete perception, i.e., grouping the images together and separating them from one another. Why have our paired patterns have been separated by stronger N1 for the symmetrical and well defined Patterns 1b and 2b—against their complimentary but nevertheless faint Patterns 1a and 2a? Why do we produce stronger N1 in high intensity and symmetrical Patterns 3a and 4a against their non symmetrical counterparts 3b and 4b? These questions may be supported by the physical study of the eye, which indicate that the primary visual cortex is densely populated with layers of cells which prefer stimuli in the shape of bars or angles of an orientation and direction. With our results one can speculate that although the visual processing mechanisms are not yet completely understood, during the information process which is about shape and colourless (with no movement and no spatial organisation) there is a kind of conditioning of the initialisation of this process that may influence our preferences and emotions, which are detectable in other parts of the brain [17], but are outside the scope of this paper. The results of these experiments have high confidence level, which show that different patterns and pattern-changing effects are highly significant. Whether exogenous or endogenous the question is what is the influence of patterns during the initial processing in the brain and since our results show consistency, does this mean that we can use this knowledge to design materials or to do art? The conclusion of this research is that our brain reacts consistently different to things that is sees. Object details as seen in the case of patterns are important to our brain; the SMART fabrics demonstrate this very well by changing from one pattern to an opposite testing this hypothesis. The consequence is that since we now know what object parameters precisely affect our visual brain and its processing, we are not restricted to engineering this change in objects to textile patterns but can apply this knowledge to the design of any objects. In this context, SMART fabrics can switch emotions as they can change from one pattern to another. These SMART textiles we name 'psychotextiles' because they are purposely designed to change a specific brain emotion. The same concept can be extended to colour, to shape, to touch, and to taste. Finally, this study, however restricted in size poses a philosophical fundamental question: if these consistent visual brain responses are evoked in less than 200 ms, knowing that brain processing is not complete, does this mean that our 'primitive world' is predetermined in our minds, as Aristotelean forms, which get enhanced by memory through perception and experience?

Author Contributions: G.K.S.; conceptualization, methodology and supervision, writing review and editing, M.C. carried out the research investigation and received her PhD. All authors have read and agreed to the published version of the manuscript.

Funding: This research received no external funding.

Conflicts of Interest: The authors declare no conflict of interest.

References

1. Stylios, G. Engineering textile and clothing aesthetics using shape changing materials. *Intell. Text. Cloth.* **2006**, *54*, 528.
2. Baurley, S. Interactive and experiential design in smart textile products and applications. *Pers. Ubiquitous Comput.* **2004**, *8*, 274–281. [CrossRef]
3. Stead, L.J. *'The Emotional Wardrobe': A Fashion Perspective on the Integration of Technology and Clothing*; University of the Arts London: London, UK, 2005.
4. Philips. Philips Design SKIN Probe Receives Prestigious "Best of the Best" in Red Dot Award: Design Concept 2007. Available online: http://www.newscenter.philips.com/main/design/about/design/designnews/pressreleases/skin_reddot2007.wpd#.VLPmvtKsXTo (accessed on 12 January 2015).
5. Sensoree. *GER Mood Sweater*. Available online: http://sensoree.com/artifacts/ger-mood-sweater/ (accessed on 22 December 2019).
6. Yang, D. *The Design of Mood Changing Clothing Based on Fibre Optics and Photovoltaic Technologies*; Heriot-Watt University: Edinburgh, UK, 2012.
7. The Dress that Changes Colour with Your Emotions. 29 October 2014. Available online: http://www.bbc.co.uk/news/technology-29691662 (accessed on 13 January 2015).
8. Pinner, M. Interactive Synapse Dress by Anouk Wipprecht Reveals Wearer's Metal States. 2014. Available online: http://fashioningtech.com/profiles/blogs/interactive-synapse-dress-reveals-wearers-metal-states (accessed on 13 January 2015).
9. White, C.T. Evoked cortical responses and patterned stimuli. *Am. Psychol.* **1969**, *24*, 211. [CrossRef] [PubMed]
10. Spehlman, R. The average electrical responses to diffuse and to patterned light in the human. *Electroencephalogr. Clin. Neurophysiol.* **1965**, *19*, 560–569. [CrossRef]
11. Armington, J.C.; Corwin, T.R.; Marsetta, R. Simultaneously Recorded Retinal and Cortical Responses to Patterned Stimuli. *J. Opt. Soc. Am.* **1971**, *61*, 1514–1521. [CrossRef] [PubMed]
12. Harter, M.R.; White, C. Effects of contour sharpness and check-size on visually evoked cortical potentials. *Vis. Res.* **1968**, *8*, 701–711. [CrossRef]
13. Ito, M.; Sugata, T. Visual evoked potentials to geometric forms. *Jpn. Psychol. Res.* **1995**, *37*, 221–228. [CrossRef]
14. Moskowitz, A.F.; Armington, J.C.; Timberlake, G. Corners, receptive fields, and visually evoked cortical potentials. *Percept. Psychophys.* **1974**, *15*, 325–330. [CrossRef]
15. Ito, M.; Sugata, T.; Kuwabara, H.; Wu, C.; Kojima, K. Effects of angularity of the figures with sharp and round corners on visual evoked potentials. *Jpn. Psychol. Res.* **1999**, *41*, 91–101. [CrossRef]
16. Wagemans, J. Detection of visual symmetries. In *Human Symmetry Perception and Its Computational Analysis*; Psychology Press: London, UK, 1996; pp. 25–48.
17. Stylios, G.K.; Chen, M. *11—Psychotextiles and Their Interaction with the Human Brain*; Elsevier Ltd.: Amsterdam, The Netherlands, 2016; pp. 197–239.
18. Bear, M.F.; Connors, B.W.; Michael, A. *Paradiso Neuroscience: Exploring the Brain*; Lippincott Williams & Wilkins: Philadelphia, PA, USA, 2001; p. 608.
19. Andreassi, J.L. *Psychophysiology: Human Behavior and Physiological Response*, 5th ed.; Lawrence Erlbaum: Hillsdale, NJ, USA, 2007.
20. Cooper, R.; Osselton, J.W.; Shaw, J.C. *EEG Technology*, 3rd ed.; Butterworths & Co.: London, UK, 1980.
21. Luck, S.J. *An Introduction to the Event-Related Potential Technique*; MIT Press: Cambridge, MA, USA, 2005.
22. Rentzeperis, I.; Nikolaev, A.R.; Kiper, D.C.; Van Leeuwen, C. Relationship between neural response and adaptation selectivity to form and color: An ERP study. *Front. Hum. Neurosci.* **2012**, *6*, 89. [CrossRef] [PubMed]

23. Olofsson, J.K.; Nordin, S.; Sequeira, H.; Polich, J. Affective picture processing: An integrative review of ERP findings. *Biol. Psychol.* **2008**, *77*, 247–265. [CrossRef] [PubMed]
24. Luo, W.; Feng, W.; He, W.; Wang, N.Y.; Luo, Y.J. Three stages of facial expression processing: ERP study with rapid serial visual presentation. *Neuroimage* **2010**, *49*, 1857–1867. [CrossRef] [PubMed]
25. Jasper, H. Report of the committee on methods of clinical examination in electroencephalography. *Electroencephalogr. Clin. Neurophysiol.* **1958**, *10*, 370–375.
26. Delorme, A.; Makeig, S. EEGLAB: An open source toolbox for analysis of single-trial EEG dynamics including independent component analysis. *J. Neurosci. Methods* **2004**, *134*, 9–21. [CrossRef]
27. Makeig, S.; Bell, A.J.; Jung, T.P.; Sejnowski, T.J. Independent component analysis of electroencephalographic data. In *Advances in Neural Information Processing Systems 8*; MIT Press: Cambridge, MA, USA, 1996; pp. 145–151.
28. Carrillo-De-La-Peña, M.; Holguín, S.R.; Corral, M.; Cadaveira, F. The effects of stimulus intensity and age on visual-evoked potentials (VEPs) in normal children. *Psychophysiology* **1999**, *36*, 693–698. [CrossRef] [PubMed]
29. Mangun, G.R.; Hillyard, S.A. Allocation of visual attention to spatial locations: Tradeoff functions for event-related brain potentials and detection performance. *Prerception Psychophys.* **1990**, *47*, 532–550. [CrossRef] [PubMed]
30. Paz-Caballero, M.D.; Garcia-Austt, E. ERP components related to stimulus selection precesses. *Electroencephalogr. Clin. Neurophysiol. J.* **1992**, *82*, 369–376. [CrossRef]

© 2020 by the authors. Licensee MDPI, Basel, Switzerland. This article is an open access article distributed under the terms and conditions of the Creative Commons Attribution (CC BY) license (http://creativecommons.org/licenses/by/4.0/).

Article

Thermal Textile Pixels: The Characterisation of Temporal and Spatial Thermal Development

Adriana Stöhr, Eva Lindell, Li Guo and Nils-Krister Persson *

Swedish School of Textiles, Polymeric E-Textiles Research Group, University of Borås, 501 90 Borås, Sweden; smarttextiles@hb.se (A.S.); eva.lindell@hb.se (E.L.); li.guo@hb.se (L.G.)
* Correspondence: nils-krister.persson@hb.se

Received: 28 September 2019; Accepted: 6 November 2019; Published: 14 November 2019

Abstract: This study introduces the concept of a thermal textile pixel, a spatially and temporally defined textile structure that shows spatial and temporal thermal contrast and can be used in the context of thermal communication. A study was performed investigating (a) in-plane and (b) out-of-plane thermal signal behaviour for knitted structures made of three different fibre types; namely, polyamide, wool, and metal containing Shieldex yarn, and two different knitting structures: plain knit and terry knit. The model thermal source was a Peltier element. For (a), a thermography set-up was used to monitor the spatial development of thermal contrast, and for (b), an arrangement with thermocouple measuring temperature development over time. Results show that the use of conductive materials such as Shieldex is unnecessary for the plain knit if only heating is required, whereas such use significantly improves performance for the terry knit structures. The findings demonstrate that the textile pixel is able to spatially and temporally focus thermal signals, thereby making it viable for use as an interface for thermal communication devices. Having well-defined thermal textile pixels opens up potential for the development of matrices for more complex information conveyance.

Keywords: smart textile; thermal textile pixel; thermal communication; non-auditory and nonvisual communication; thermal conductivity; Peltier element

1. Introduction

Thermal communication is perceived through the skin. Textiles are the materials that we often have closest to our skin, which enhances their potential as a medium for thermal communication. Textile materials comprise a vast range of fibres and constructions that possess a wide variety of thermal properties. In order to maximise the functions of smart textiles, it is important to better understand how varying fibre choices and textile constructions affect thermal communication.

This study introduces the concept of a thermal textile pixel and investigates how fibre properties and textile constructions influence signals sent by a thermal source. The thermal pixel represents a counterpart to pixels for visual communication that uses thermal modes (potentially both heating and cooling) for communication. As illustrated in the area denoted (a) in Figure 1, being the overlapping field of thermal engineering and communication.

Numerous studies have examined the thermal properties of textiles in the context of clothing comfort [1–3] and textiles used as thermal insulators [4]. Applications of thermal engineering to smart textiles include devices such as underwear [5], socks and gloves [6] with heating functions. Another example, is the approach used to produce the Embr Wrist band [7], a wearable that consists of a thermal cooling module that is able to cool down or heat up for thermal stimulation. Section (b) in Figure 1 encompasses such devices. Beyond these, a broad range of applications opens up in the area of thermal communication in the intersecting area (c) in Figure 1. One example is Lovelet [8], a wearable device that conveys affection between partners through thermal communication. Research is being made on garments designed to be used as feedback, alerting, or communication systems for daily life use by individuals with deafblindness [9]. Wilson et al. [10] observed that although the research field of thermal feedback for communication remains to be fully explored, the implementation of salient feedback based on thermal input can have immense advantages in noisy and turbulent surroundings compared with vibrating inputs [10]. Lee and Lim investigated heat as a communicatorin the field of haptic technology and robotics [11,12]. They concluded that although there is great potential for using thermal devices to facilitate interpersonal communication, the focus should not be to replace current communication types, but rather to enhance or enrich current means of communication [12].

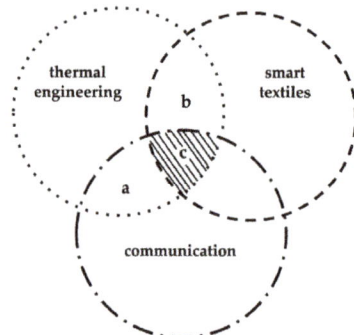

Figure 1. Thermal pixels are at least an intersection between three different technological areas.

1.1. Thermal Communication

The skin is our most expansive organ; it has a complex structure and is involved in a vast number of bodily functions. Apart from the obvious task of defining a bodily inside and a bodily outside and handling moderate mechanical impact, the skin is also involved in maintaining body temperature, mediating sensation, and protecting us from UV-radiation, among other functions [13].

Sensations are mediated through cutaneous sensory systems that receive tactile, thermal painful, and pruritic information and forward them to the central nervous system [14]. Human skin also has two kinds of thermoreceptors, one of which reacts to heat and the other to cold. Thermoreceptor density varies in different parts of the body, and there are many more thermoreceptors that respond to cold than those responding to heat [14]. Temperature exposure can cause acute pain when the skin reaches temperatures above 45 °C or below 5 °C. The temperature at which pain is initiated varies; some examples indicate that being exposed to temperatures below 17 °C can trigger pain sensations [15]. Skin temperature is distinct from body temperature; it normally remains within 32–35 °C but can vary from 20–40 °C [16].

The threshold for thermal perception (the least detectable noticeable difference) is lower for cooler temperatures than for warmer temperatures, and the thresholds are larger for lower body parts than for the upper body [17].

1.2. Thermal Expression

Lee and Lim performed a study on heat as a medium for expression in interpersonal communication [11]. In order to better describe heat in the context of communication, they introduced the term thermal expression, which denotes 'the activity of controlling a particular amount of heat with intention' [11]. Based on this definition, they described four expressive elements as representations of thermal expression, namely temperature, duration, temperature change rate, and location [11].

The thermal modality of human sensation is susceptible to illusions, such as the phenomenon known as the *thermal grill illusion* [17], which refers to a sensation of heating or cooling that occurs in absence of any change in actual skin temperature. Research using this phenomenon include experiments conducted by Manasrah et al. [18], Hojatmadani and Reed [19], and Oron-Gilad et al. [20]. The latter used the thermal grill illusion in the efforts to develop the foundation for a tactile language [20].

It seems rational to use thermal illusions for communicating signals, which greatly reduce the inconvenience experienced by the subject, as the receiver's skin temperature does not have to actually reach the given temperatures every time that a signal is sent.

1.3. Defining Thermal Textile Pixel

In order to create a thermal illusion, adjacent devices need to be kept separated such that they nonetheless maintain different temperatures. A thermal contrast is needed whereby one can perceive changes in temperature over a sufficient small distance, over a sufficiently short time period, or both (difference in perception due to varying thermal conductivities of materials is not considered). Thus, the thermal textile pixel is a spatially and temporally defined textile structure that shows spatial and temporal thermal contrast and can be used in thermal communication. It is a device for thermal expression within the repertoire of non-auditory and nonvisual communication (NANV).

Figure 2a,b illustrate the ideal spatial thermal contrast for both heating and active cooling. The edges should be maximally sharp in order to enable pixel distinction and the temperature T_{max} resp. T_{min} should be detectable. Further on, the temperature of the plateaus (T_{max} resp. T_{min}) should be coherent with the temperature generated by the thermal source.

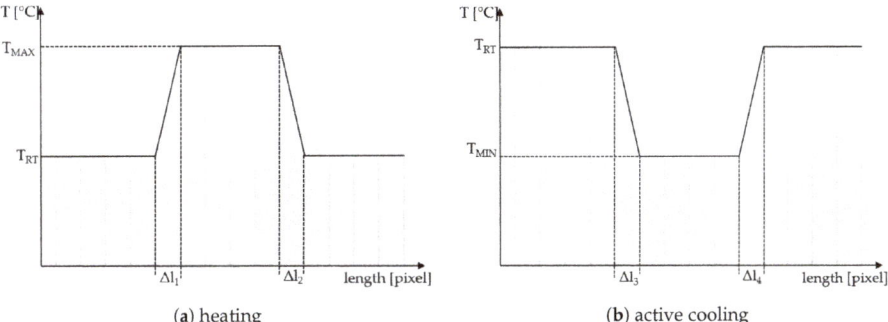

(a) heating   (b) active cooling

Figure 2. Scheme of a function representing the ideal spatial contrast of a thermal textile pixel. T_{RT} is ambient (room) temperature. (**a**) shows the heating signal, whereby T_{MAX} is temperature sent from thermal source, T_{MAX} should be below 45 °C for avoiding pain. (**b**) shows the active cooling signal, T_{MIN} is temperature sent from thermal source, T_{MIN} should be above 5 °C to avoid pain. Δl:s should be minimally short.

Figure 3a,b show the ideal temporal thermal contrasts for heating and active cooling, respectively. The temperatures T_{max} resp. T_{min} should be predicted based on the repeatability of temperatures generated by the thermal source. Compared with spatial thermal development, in which Δl should be very small, for temporal thermal development, we need to control Δt, meaning adherence to thermal source.

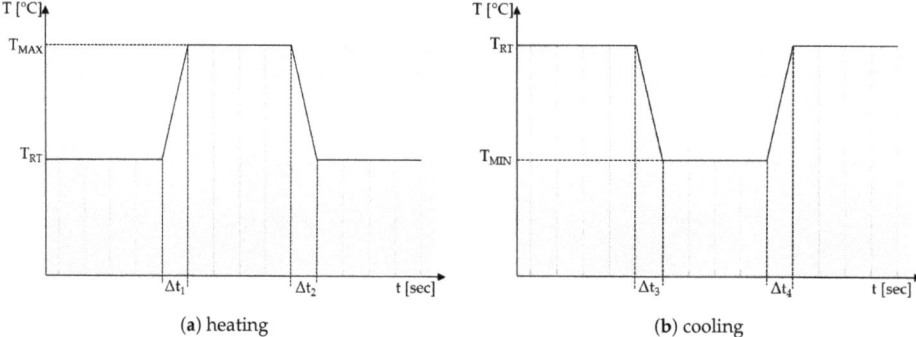

(a) heating (b) cooling

Figure 3. Scheme of a function representing the ideal temporal contrast of a thermal textile pixel. T_{RT} is ambient (room) temperature. (**a**) shows the heating signal, whereby T_{MAX} is temperature sent from thermal source, T_{MAX} should be below 45 °C to avoid pain. (**b**) shows the active cooling signal, T_{MIN} is temperature sent from thermal source, T_{MIN} should be above 5 °C to avoid pain. Depending on what is being communicated Δt:s could be either short or long. In any case it is important that they are controllable.

This study aimed to investigate the effect of textiles on (1) the thermal expression by spatial thermal spread in x-, y- and z-direction, and (2) temporal thermal spread in- and out-of-plane (Figure 4). This was done by measuring the heat spread in- and out-of-plane with an IR camera and measuring the temperature change rate of the textile interface with a thermocouple.

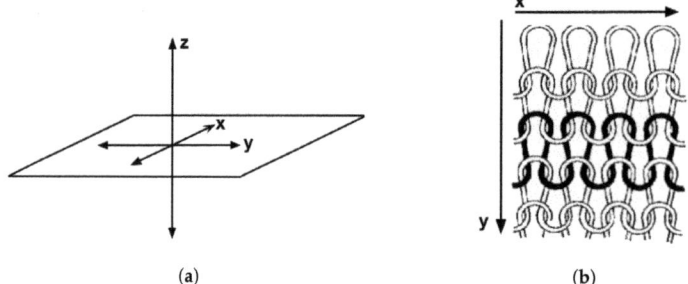

(a) (b)

Figure 4. Overview of the in-plane and out-of-plane directions. (**a**) z-direction out of plane, x- and y-direction in plane. (**b**) y denotes the wale direction, x represents the course direction. Figure from Peterson [21].

Results show that textile interfaces can be used to convey thermal communication signals and there is an anisotropic effect in the temperature spread in the x-y- and z-direction.

2. Results

This section presents the thermal conductivity and temporal and spatial thermal development of six different textile samples. Three samples comprised a plain knit structure consisting of 100% wool (WO), polyamide (PA) and a silver-plated polyamide yarn (Shieldex), the other three samples constituted a terry knit structure consisting of 50/50 WO/PA, 50/50 PA/Shieldex and 100% PA.

Focusing on the material and textile behaviours, a simple realisation of a thermal pixel was used with a Peltier element as the thermal source. A Peltier element was chosen due to its ability to generate both heating and cooling, which is central to mapping how the temperature develops and follows the source.

2.1. Thermal Conductivity

The results of measuring the thermal conductivity of the different samples is presented in Table 1. As anticipated, wool showed the greatest thermal resistance, as it was the most insulating fibre used for testing. The plain knit Shieldex exhibited the highest thermal conductivity.

Table 1. Measured parameters and thermal conductivity calculated with Equation (1) for each of the six textile samples.

Structure	Material	Mass per Unit Area [g/m^2]	Thickness (20 Pa) [mm]	StDv	Air Flow Resistivity (Face) Pa·s/m^2	StDv	Thermal Resistance [m^{2}K/W]	Thermal Conductivity [W/(m·K)]
Plain	Shieldex	239.83	1.56	±0.07	49.82	±1.62	0.015	0.10
	PA	261.96	2.01	±0.18	171.4	±1.14	0.024	0.08
	WO	524.05	3.41	±0.13	79.24	±2.67	0.104	0.03
Terry	Shieldex/PA	273.58	3.03	±0.13	78.8	±1.22	0.044	0.07
	PA	354.22	3.02	±0.08	165.4	±0.89	0.035	0.09
	WO/PA	441.52	4.47	±0.03	71.34	±1.39	0.119	0.04

2.2. Mapping Temporal Thermal Development, Out-of-Plane

Tables 2 and 3 present the thermal curves of the interface temperatures of the six different samples. Each curve represents a cycle, which for heating consisted of 300 seconds of heating followed by 300 seconds of thermal dissipation. For active cooling, the cycle consisted of 150 seconds of heating and 150 seconds of thermal dissipation. A prerequisite for the thermal textile pixel is to be able to follow (with possible some lag) the thermal source. Tables 2 and 3 show the out-of-plane temporal thermal development for the six fabric samples.

Mapping temporal thermal development indicates how quickly the thermal pixel device reacts to the thermal source. In an ideal situation, all thermal energy (heat flux) should be instantly transferred through the textiles, and the textile interface should not reduce the thermal detection percentage. In Tables 2 and 3, the orange curve should overlap with the grey curve. However, the temperature measured at the textile surface (indicated by orange curves) did not reach the intensity given by the thermal impulse/given temperature from the thermal source (indicated by the grey curves). For heating (Tables 2 and 3 (a), (c), (e)), PA performs best with plain knit surface ((c)), whereas Shieldex performs best with terry fabric ((a)). Wool is the least effective for heating both knitting structures. For active cooling ((b), (d), (f)), Shieldex performed best ((b)), followed by PA. Again, wool was the least effective material for cooling both plain and terry structures.

Another observation is that the temperature measured at the textile/Peltier element interface in some of the heating processes ((a), (c) in Table 2, (c) in Table 3), unexpectedly increased above the given temperature (compare blue and grey curves), which is untenable from an energy perspective. We interpreted the reason as being that the thermal system did not reach equilibrium because the temperature was measured at 300 seconds after the Peltier element was turned on. The thermal flux was trapped by the textiles and did not complete the process of thermal energy exchange (heat transfer) with the surroundings. The difference between the two curves (blue and grey) is more significant in the case of wool than the other materials, which indicates that the thermal system reaches the equilibrium state much slower in the presence of that material.

2.3. Mapping Thermal Spatial Development, In-Plane

Infrared thermography was used to record the temperature every 60 seconds during a 300 second test for heating and at 60 seconds and 150 seconds for active cooling. The results of the spatial development are shown in Table 4 for plain knit and Table 5 for terry knit.

Table 2. Temporal thermal development, out-of-plane, for plain knit samples. Measured with thermocouples according to setup in Figure 5. Room temperature (RT) is denoted in black, the orange curve represents the temperature of the textile, and the blue curve represents the temperature of the Peltier element (PE). The light grey curve was measured prior to testing and is equivalent to the thermal energy put into the system, hence representing the ideal outcome for the textile. Temporal thermal development for heating given in (a) plain knit Shieldex sample, (c) plain knit wool sample, (e) plain knit polyamide sample. Temporal thermal development for active cooling given for (b) plain knit Shieldex sample (d) plain knit wool sample, (f) plain knit polyamide sample.

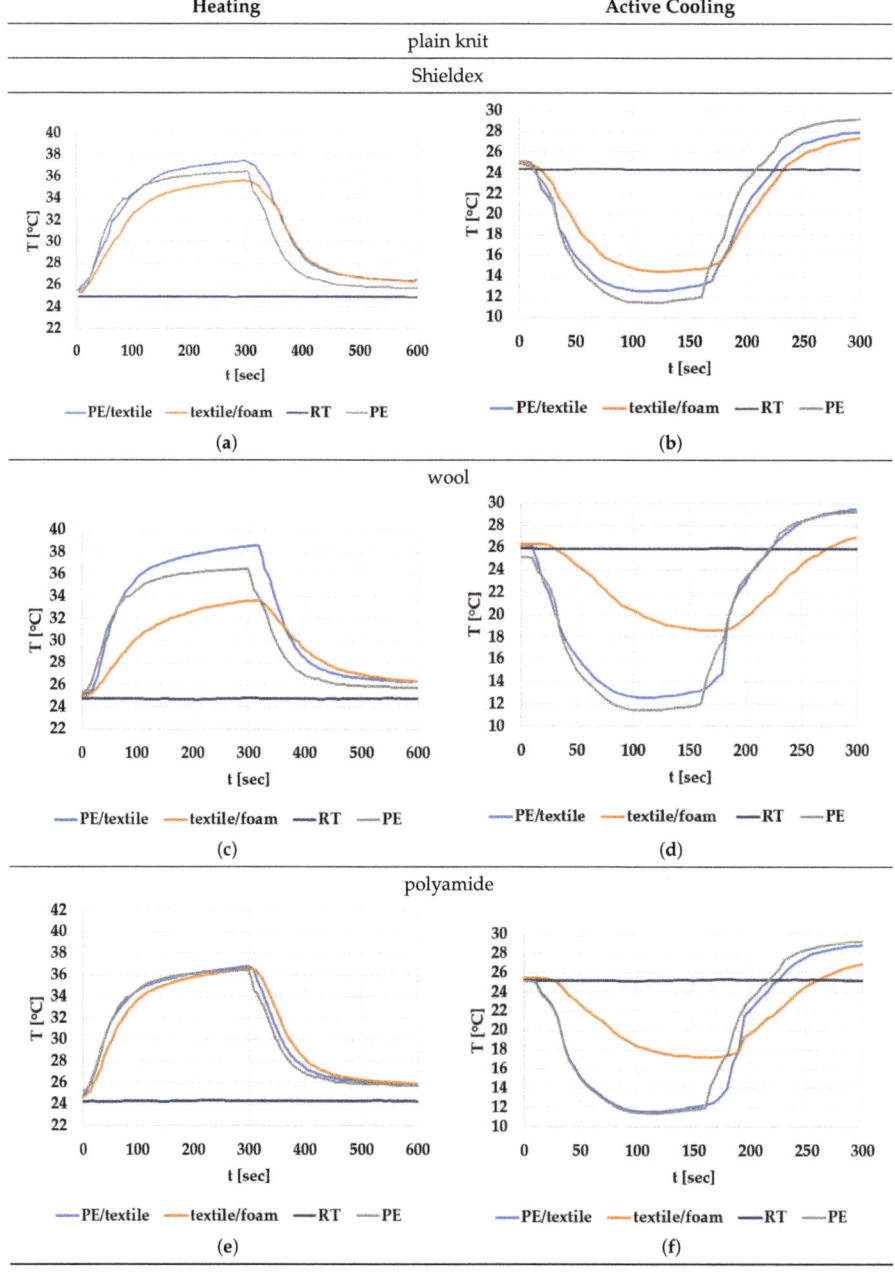

Table 3. Temporal thermal development, out-of-plane, for terry knit samples. Measured with thermocouples according to setup in Figure 5. Room temperature (RT) is denoted in black, the orange curve represents the temperature of the textile, and the blue curve represents the temperature of the Peltier element (PE). The light grey curve was measured prior to the testing and is equivalent to the thermal energy put into the system, hence representing the ideal outcome for the textile. Temporal thermal development for heating given in (a) terry knit Shieldex sample, (c) terry knit wool sample, (e) terry knit polyamide sample. Temporal thermal development for active cooling given for (b) terry knit Shieldex sample (d) terry knit wool sample, (f) terry knit polyamide sample.

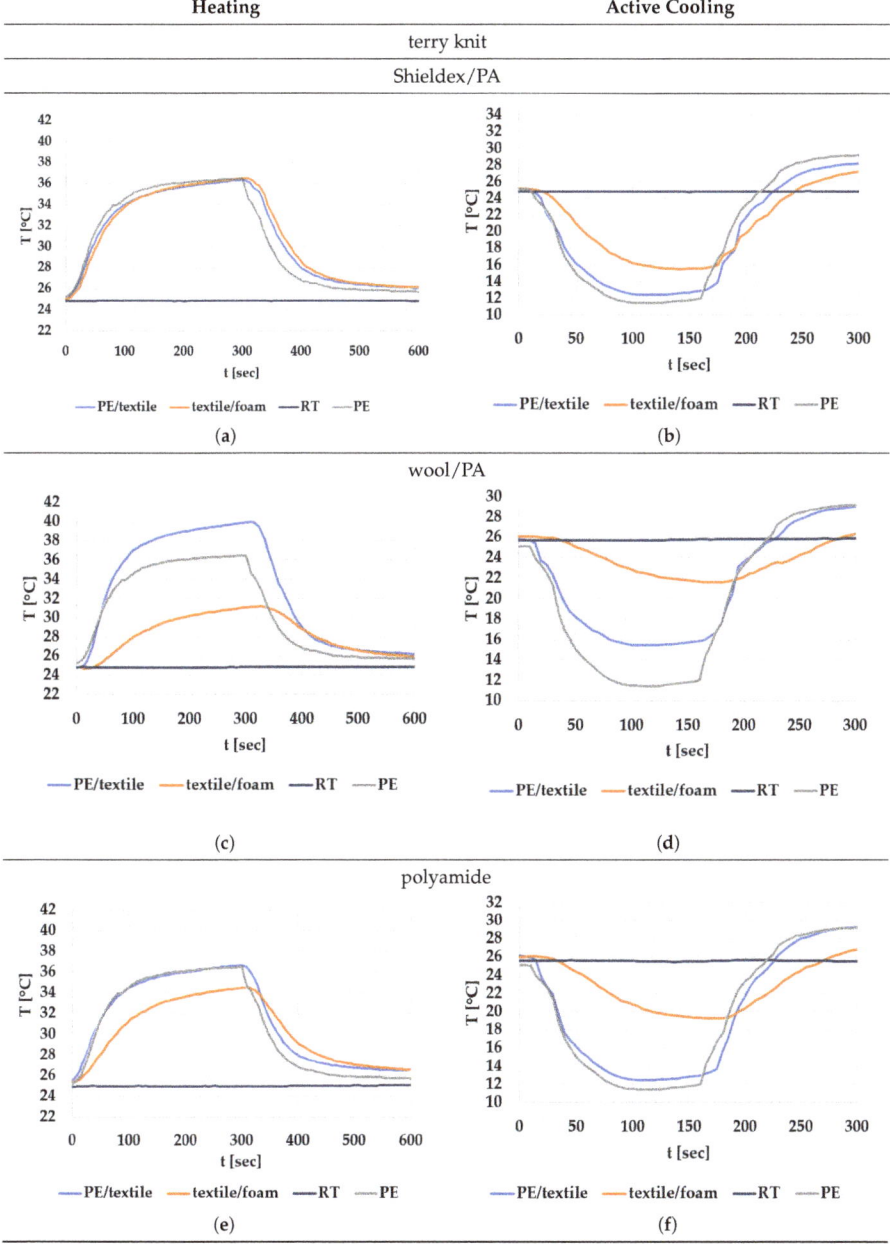

Table 4. Spatial thermal development for plain knit samples, results from thermography. (**a**,**b**) plain knit Shieldex, (**c**,**d**) plain knit polyamide, (**e**,**f**) plain knit wool.

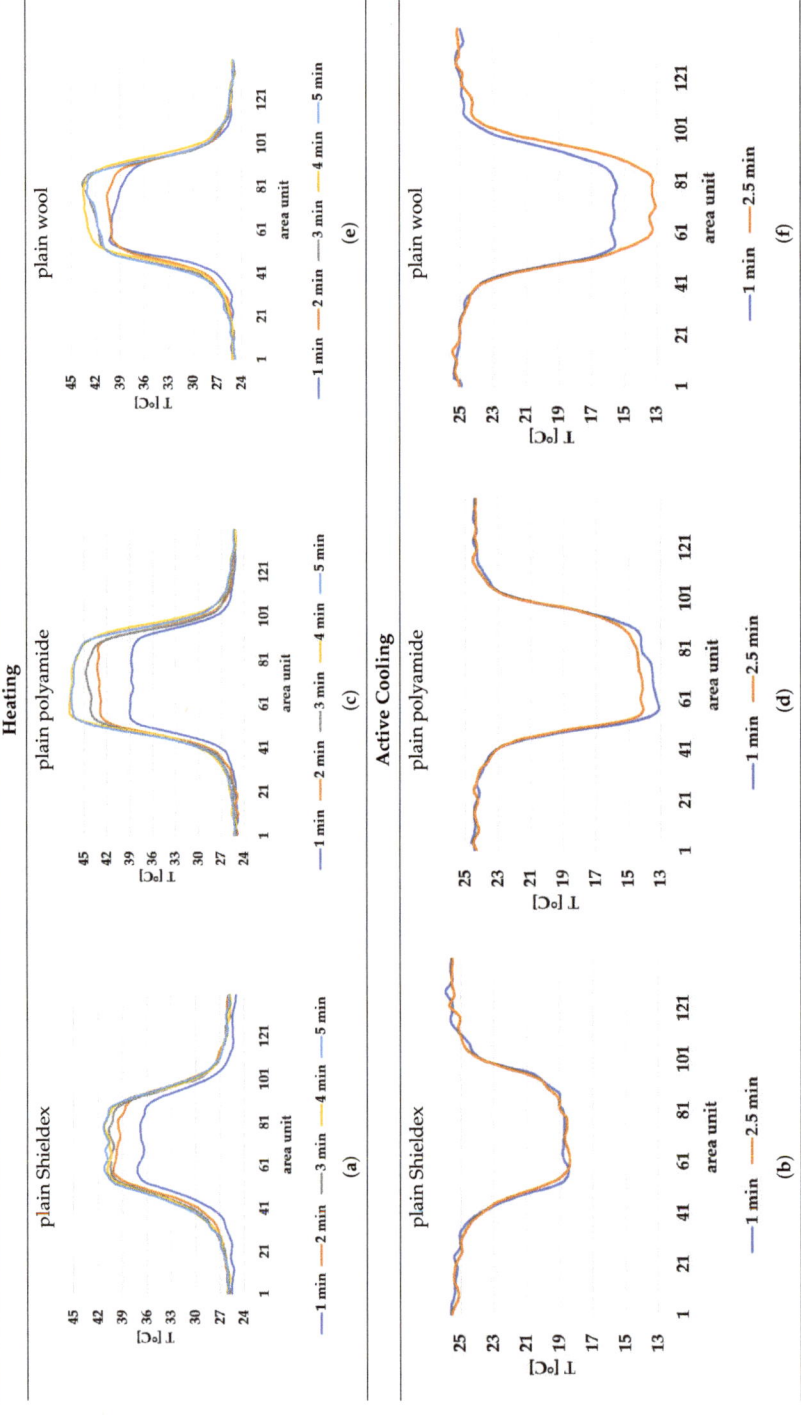

Table 5. Spatial thermal development in-plane for terry knit samples, results from thermography. (**a,b**) terry knit shieldex, (**c,d**) terry knit polyamide, (**e,f**) terry knit wool.

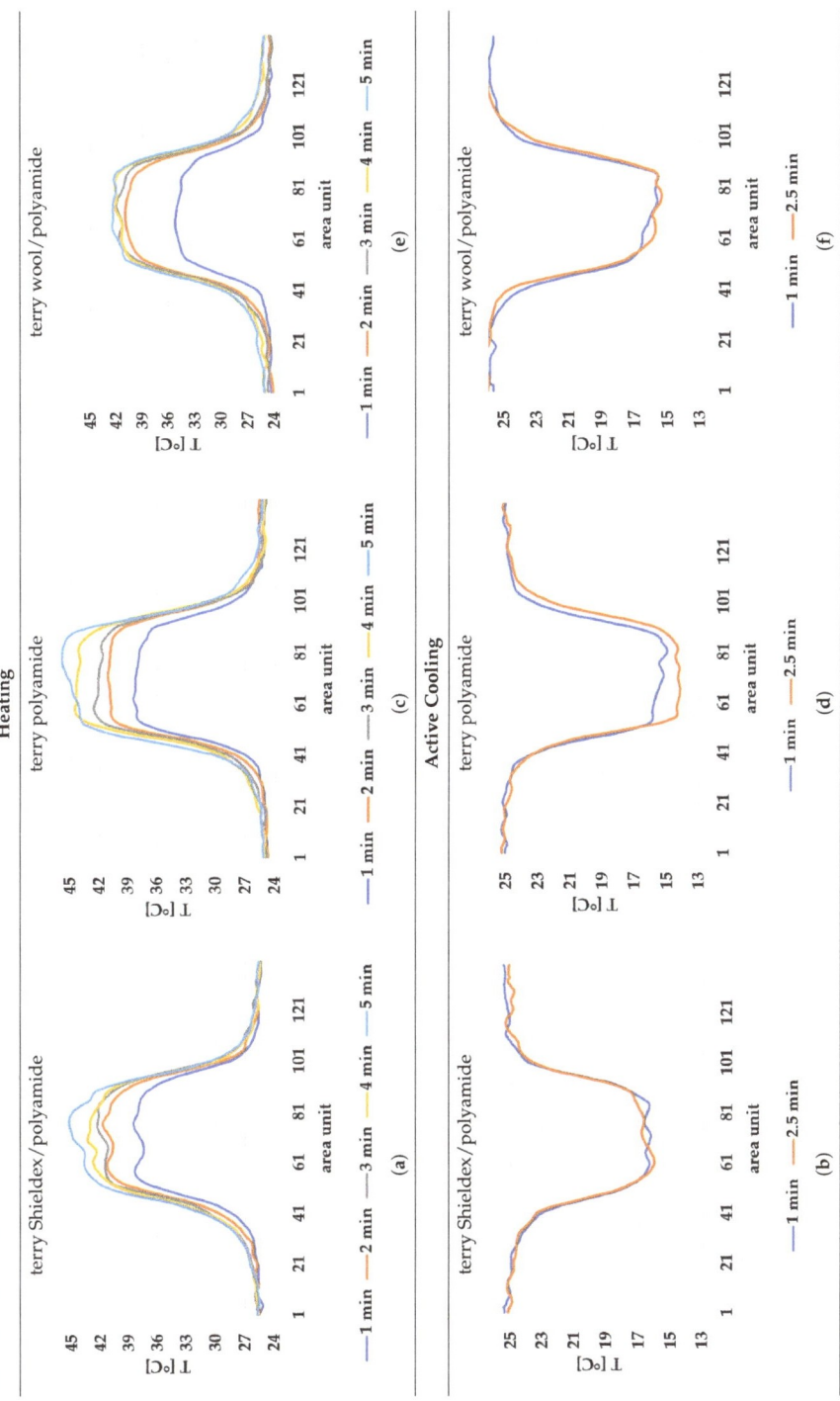

All samples show a bump-like behaviour with relative sharp symmetric ramp up and ramp down. A difference is observable between time to reach T_{max} (T_{min}) as well as in the evenness of the plateaus.

The Shieldex curves ((a), (b), Tables 4 and 5) have a wider plateau, and the temperature spreads more in-plane than for the other two non-conductive materials, thus indicating that conductive materials are not favourable for thermal pixels. The same phenomenon can be seen in the case of the terry fabrics (Table 5). Thermal development is less significant for active heating than for heating.

An asymmetry was observed in how the temperature distributed in x- and y- directions. Table 6 shows an example of the typical behaviour. Both x and y showed bump-like behaviour; however, y was typically more blurred.

Table 6. Directional dependency of spatial thermal development for Shieldex terry knit. (a) directional dependency for heating terry knit Shieldex/polyamide sample. (b) directional dependency for active cooling terry knit Shieldex/polyamide sample.

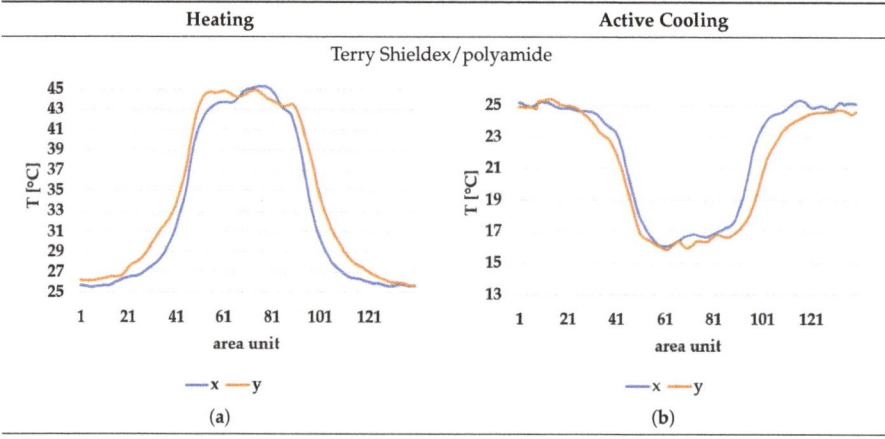

3. Discussion

This work served as an exploratory rather than a comparative study and focused exclusively on investigating heat transfer via textile thermal pixels (therefore excluding psychophysical measurements). Thermal pixels are defined within the context of this work as thermal feedback (both heating and cooling) providing systems consisting of a thermal source paired with a textile component. The thermal output was generated by a Peltier element as a model.

The results provide insight on the impact the textile would have on the elements of thermal expression defined by Lee and Lim [11]: (a) temperature; (b) duration; (c) temperature change rate; and (d) location. Temperatures (a) above detection limits and below pain limits (T_{MAX} and T_{MIN}) could clearly be reached. Durability has been shown to be achievable for a heating device, whereas an arrangement with a heat sink is necessary for active cooling if a thermoelectrical mechanism is used. In order to convey the 'correct' signal, the textile must follow temperature rises and drops with minimal delay (c) of the thermal source or be a consistent function of the thermal source, which also seems feasible. The present study worked with knitted samples rather than readymade garments, and knitting (rather than weaving) was the process of choice. Garments worn close to the body are typically knitted and can be used at almost any body location (d) except the face.

For plain knitting, the use of conductive materials such as Shieldex is unnecessary if only heating is required; however, the use of conductive material significantly improves the performance in the case of terry fabrics. The results presented in Tables 2 and 3 indicate that signals conveyed with active cooling tend to be less distinct than those conveyed with heating; this is most evident in the case of wool and polyamide. Temporal thermal development results for the terry knit samples (Table 3) confirm that such textile structures with more air entrapment are worse conductors. Textile thermal pixel designs should therefore consider both the thermal properties of the fibre and the textile surface structure.

It is plausible that thermal illusions will serve as the basis for thermal communication, as they enable sending signals without altering the skin temperature of the receiver, which could cause discomfort if occurring over a lengthy duration. Constructing such displays for thermal illusions demands the ability to create adjacent but distinguishable areas that can be kept at different temperatures, such as the thermal grill illusion used by Oron-Gilad et al. [20]. As such, thermal spatial resolution is important. Table 6 shows the presence of a certain blurriness of approximately 20%, which is still below what is needed for the Grill effect. It might be interesting to create displays of several textile pixels for sending more complex messages without any use of thermal illusions, in which case the thermal resolutions of both the material and the human should be considered.

The spatial development shows a discrepancy for heat transfer in the x- and y-directions, with a less conductive tendency in the wale direction of the fabric. This could be explained by the structure of the knitted surface and that there is more air entrapment in the wale directions, thus causing an isolating effect.

Time constants seem long for heating and especially for active cooling. Even the quickest plain Shieldex samples did not reach 90% of the max temperature until 100 seconds, and heat dissipation took 200 seconds (Table 2), which resulted in a total of 300 seconds before the pixel is off again. This indicates a very low updating frequency far below Hz, which is not easily overcome. We conclude that thermal communication should be adapted to the urgency of the message.

4. Materials and Methods

This section presents the method, materials, and setups used to measure the thermal conductivity of the textile samples and the measurement of temporal and spatial thermal development.

4.1. Textile Materials

Wool is classified as the most thermally insulating fibre [22], whereas polyamides have the best thermal conductivity performance compared to other traditional non-electroconductive fibres. Metals are also known for their outstanding thermal conductivity. Therefore, one conductive yarn was chosen to represent a highly thermal conductivity option. The scattered thermal features of the fibres were expected to yield a wide spectrum of distinguished behaviours.

Two types of textile constructions were similarly chosen for their great differences. The plain knit has a uniform flat face, as opposed to the terry knit, which is more voluminous and porous due to loop formations on the face of the fabric.

Three plain knit samples were made using 100% Shieldex, 100% polyamide and 100% wool. The polyamide was used as a binding thread to achieve a well-functioning fabric in the terry structure, which resulted in a more stable, higher density type of terry fabric. Therefore, the three terry knit samples were made in 50/50 Shieldex/PA, 50/50 WO/PA, and 100% PA. All samples were made with equal stitch lengths of 11.5 on the back side, and the loop was formed with 9.5 stitching on the front and stitch length 8 in the back.

Samples were manufactured on a flat knitting machine: the Stoll type CMS 330 TC multi gauge. Yarn tension was kept stable and a gauge of 12 was maintained throughout the manufacturing procedure. Samples cut with a diameter of 330 mm were used to ensure flat, even fabric during testing.

4.2. Characterisation of Thermal Conductivity

Thermal resistance (R) was measured with a two-plate tog-test, in compliance with ISO 5085-1:2004 [23]. The samples' thickness (d) was measured with a 'Shirley' thickness gauge by Shirley Development Limited, Manchester according to ISO 5084:1996 [24]. These measurements were necessary in order to 223 calculate the thermal conductivity (k) (Equation (1)) for each sample according to ISO 5085-1:2004 [23].

$$k = \frac{d}{R} \quad [\text{W}/(\text{m} \cdot \text{K})] \tag{1}$$

4.3. Characterisation of Heat Source

Prior to the thermal development study, a reference for the 'ideal' result characterizing the behaviour of the PE only by the mean of thermal outcome from five measurements. All samples were tested at constant room temperature 25 °C. The test samples were exposed to heat and active cooling by an insulated Peltier TEC Module (4), purchased from elektrokit.com [25]. The PE module was provided with power by a dual-tracking DC power supply (9), model TPS-4000 by Topward Electric Instruments Co., Ltd (Hukou, Hsin Chu, Taiwan).

4.4. Characterisation of Temporal Thermal Development, Out-of-Plane

Prior to the thermal development study, a reference was made characterising the 'ideal' behaviour of the PE based on the mean of thermal outcomes from five measurements. All samples were tested at constant room temperature 25 °C. The test samples were exposed to heat and active cooling by an insulated Peltier TEC Module ((4) Figure 5) purchased from elektrokit.com [25]. The PE module was supplied with power by a dual-tracking DC source ((9) Figure 5): model TPS-4000 by Topward Electric Instruments Co., Ltd.

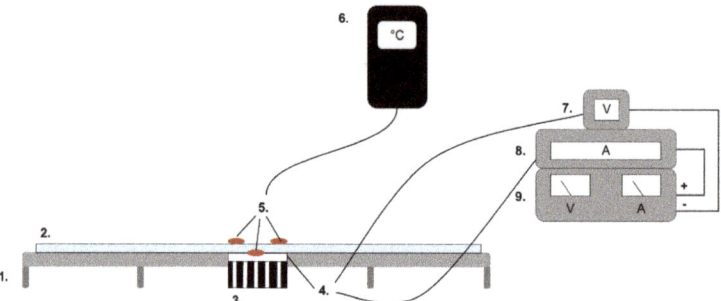

Figure 5. The test setup for the out of plane heat transfer experiment consisted of (1) an elevated board for the sample placement, (2) the textile sample covered by a Styrofoam plate, (3) a heat sink, (4) a Peltier element as thermal source, (5) thermocouples, (6) a data logger thermometer, (7) a voltmeter, (8) a multimeter and (9) a power source.

Figure 6 illustrates the instalment of the samples for testing and the placement of the thermocouples. One thermocouple was placed between the Peltier element and textile sample to give insight on the thermal development of the PE/textile interface. The second device was placed on top of the textile sample in the centre of the area directly affected by the Peltier element. Additionally, a Styrofoam plate was placed on top of the setup to keep the thermocouple in place. The third thermocouple recorded the room temperature (RT). The collected data were recorded by a three-channel data logging thermometer (6): model SD 200 by Extech Instruments.

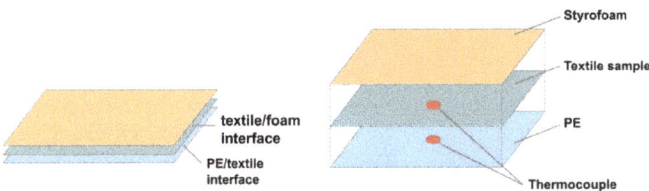

Figure 6. Setup of test specimen for temporal resolution test.

The heating cycle began by exposing the PE to 300 mA of current for 300 seconds before turning off the power supply and letting the system cool down for another 300 seconds. The thermal development after the heat source was turned off is herein referred to as 'thermal dissipation'. A shorter temporal interval in combination with higher current was necessary for the active cooling cycle, as a preliminary study had showed that the heatsink was not sufficient to thermally balance the high amount of heat generated if a cooling effect was demanded over a longer period of time. Hence, the active cooling generated by changing the polarity of the PE was implemented by applying 500 mA for 150 seconds to cause active cooling, and the thermal development was monitored for another 150 seconds after turning the current off.

The mean data of five test rounds was used throughout the study and plotted to generate time dependent thermal curves of both the PE and the textiles measured with the thermocouple.

4.5. Characterisation of Spatial Thermal Development, In-Plane

The in-plane heat transfer was investigated via infrared thermatography. The same setup was used as for the temporal resolution except for the addition of an infrared camera E4 by FLIR, which was fixed at a distance of 25 cm above the test setup; see (10) in Figure 7.

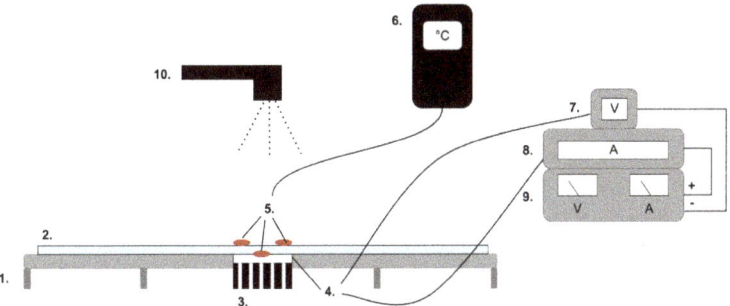

Figure 7. The test setup for the in-plane heat transfer is based on setup in Figure 5 with an the addition of (10) an IR camera (10).

The heat spread was recorded every 60 seconds over a period of 300 seconds for heating. For the active cooling cycle, the heat spread was recorded at 60 and 150 seconds.

By using the FLIR plus tool [26] provided by the IR camera manufacturer, the images taken by the IR camera could be analysed using thermal measurements for every pixel of the picture created by the camera. For clarity and to avoid confusion with the thermal textile pixel, the IR camera pixels will hereafter be referred to as area units. Each dataset from the IR camera consisted of 140 area units and their corresponding temperature measurements. By defining two data lines, x- and y-direction corresponding to the course and whale direction of the fabric, it was possible to compare the thermal spread in each direction (Figure 8).

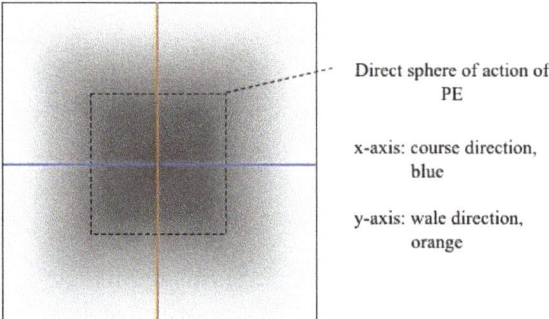

Figure 8. Scheme of IR image, heat spread was made visual by colour gradient and analyszed in the FLIR plus tool.

Author Contributions: Conceptualization, N.-K.P.; methodology, A.S.; validation, A.S., L.G. and N.-K.P.; formal analysis, A.S., E.L.; investigation, A.S.; resources, N.-K.P.; writing–original draft preparation, A.S., E.L.; writing-review and editing, E.L., L.G., N.-K.P; visualization, A.S., E.L.

Funding: This research was funded by European Horizon 2020 research and innovation program grant number 780814, project SUITCEYES and the Smart Textiles initiative (Vinnova).

Conflicts of Interest: The authors declare no conflict of interest.

References

1. Berger, X.; Sari, H. A new dynamic clothing model. Part 1: Heat and mass transfers. *Int. J. Therm. Sci.* **2000**, *39*, 673–683. [CrossRef]
2. Karaca, E.; Kahraman, N.; Omeroglu, S.; Becerir, B. Effects of Fiber Cross Sectional Shape and Weave Pattern on Thermal Comfort Properties of Polyester Woven Fabrics. *Fibres. Text. East. Eur.* **2012**, *92*, 67–72.
3. Marolleau, A.; Salaun, F.; Dupont, D.; Gidik, H.; Ducept, S. Influence of textile properties on thermal comfort. In *IOP Conference Series: Materials Science and Engineering 254, Proceedings of the 17th World Textile Conference AUTEX 2017-Textiles-Shaping the Future, Corfu, Greece, 29–31 May 2017*; IOP Publishing: Bristol, UK, 2017.
4. Song, G. 2—Thermal insulation properties of textiles and clothing. In *Textiles for Cold Weather Apparel*; Williams, J.T., Ed.; Woodhead Publishing: Cambridge, UK, 2009; pp. 19–32.
5. Lenz. Available online: https://www.lenzproducts.com/de/heating-wear.html (accessed on 10 October 2018).
6. Inuheat. 2018. Available online: https://inuheat.com (accessed on 9 October 2018).
7. Embr Labs. Available online: https://embrlabs.com (accessed on 10 October 2018).
8. Fujita, H.; Nishimoto, K. Lovelet: A Heartwarming Communication Tool for Intimate People by Constantly Conveying Situation Data. In Proceedings of the CHI 2004, Vienna, Austria, 24–29 April 2004.
9. SUITCEYES. Available online: https://suitceyes.eu/ (accessed on 12 September 2019).
10. Wilson, G.; Halvey, M.; Brewster, S.; Hughes, S. Some like it hot? Thermal feedback for mobile devices. In Proceedings of the CHI '11 CHI Conference on Human Factors in Computing Systems, Vancouver, BC, Canada, 7–12 May 2011; pp. 2555–2564.
11. Lee, W.; Lim, Y. Explorative research on the heat as an expression medium: Focused on interpersonal communication. *Pers. Ubiquit. Comput.* **2012**, *16*, 1039–1049. [CrossRef]
12. Lee, W.; Lim, Y. Thermo-Message: Exploring the potential of heat as a modality of peripheral expression. In Proceedings of the CHI 2010, Atlanta, GA, USA, 14–15 April 2010.
13. Hossler, F. *Ultrastructure Atlas of Human Tissues*; John Wiley and Sons, Incorporated: Somerset, UK, 2014.
14. McGlone, F.; Reilly, D. The cutaneous sensory system. *Neurosci. Biobehav. Rev.* **2010**, *34*, 148–159. [CrossRef] [PubMed]
15. Møller, A.R. Physiological Pain. In *Sensory Systems: Anatomy and Physiology,* 2nd ed.; Aage R Møller Publishing: Dallas, TX, USA, 2012; pp. 342–343

16. Jones, L.A.; Berris, M. The psychophysics of temperature perception and thermal-interface design. In Proceedings of the 10th Symposium on Haptic Interfaces for Virtual Environment and Teleoperator Systems, Orlando, FL, USA, 24–25 March 2002; pp. 137–142.
17. Kappers, A.M.L.; Plaisier, M. Thermal Perception and Thermal Devices used on Body Parts other than Hand or Face. *IEE Trans. Haptics* **2019**. [CrossRef] [PubMed]
18. Manasrah, A.; Crane, N.; Guldiken, R.; Reed, K.B. Perceived Cooling Using Asymmetrically-Applied Hot and Cold Stimuli. *IEEE Trans. Haptics* **2017**, *10*, 75–83. [CrossRef] [PubMed]
19. Hojatmadani, M.; Reed, K. *Asymmetric Cooling and Heating Perception*; Springer: Cham, Switzerland, 2018; pp. 221–233.
20. Oron-Gilad, T.; Salzer, Y.; Ronen, A. Thermoelectric Tactile Display Based on the Thermal Grill Illusion. In *Haptics: Perception, Devices and Scenarios*; Springer: Berlin/Heidelberg, Germany, 2008; pp. 343–348.
21. Peterson, J. *Trikåteknik, 6:e upplagan*; Textile support Scandinavia: Borås, Sweden, 2013; p. 1.
22. Baxter, S. The thermal conductivity of textiles. *Proc. Phys. Soc.* **1946**, *58*, 105–118. [CrossRef]
23. International Organization of Standardization. *Textiles—Determination of Thermal Resistance (ISO 5085-1:2004)*; International Organisation for Standardization: Geneva, Switzerland, 2004.
24. International Organization of Standardization. *Textiles—Determination of Thickness of Textiles and Textile Products (ISO 5084:1996)*; International Organisation for Standardization: Geneva, Switzerland, 1996.
25. Elektrokit.com. Available online: https://www.electrokit.com/en/ (accessed on 20 April 2018)
26. FLIR Tools+ | FLIR Systems. Available online: https://www.flir.de/products/flir-tools-plus/ (accessed on 24 April 2018).

© 2019 by the authors. Licensee MDPI, Basel, Switzerland. This article is an open access article distributed under the terms and conditions of the Creative Commons Attribution (CC BY) license (http://creativecommons.org/licenses/by/4.0/).

Article

Investigation of the Mechanical and Electrical Properties of Elastic Textile/Polymer Composites for Stretchable Electronics at Quasi-Static or Cyclic Mechanical Loads

Christian Dils [1,*], Lukas Werft [1], Hans Walter [1], Michael Zwanzig [1], Malte von Krshiwoblozki [1] and Martin Schneider-Ramelow [2]

1. Fraunhofer IZM (Institute for Reliability and Microintegration), 13355 Berlin, Germany; Lukas.Werft@izm.fraunhofer.de (L.W.); Hans.Walter@izm.fraunhofer.de (H.W.); Michael.Zwanzig@izm.fraunhofer.de (M.Z.); Malte.von.Krshiwoblozki@izm.fraunhofer.de (M.v.K.)
2. Microperipheric Center, Technical University Berlin, 10623 Berlin, Germany; Martin.Schneider-Ramelow@izm.fraunhofer.de
* Correspondence: christian.dils@izm.fraunhofer.de

Received: 24 September 2019; Accepted: 30 October 2019; Published: 1 November 2019

Abstract: In the last decade, interest in stretchable electronic systems that can be bent or shaped three-dimensionally has increased. The application of these systems is that they differentiate between two states and derive there from the requirements for the materials used: once formed, but static or permanently flexible. For this purpose, new materials that exceed the limited mechanical properties of thin metal layers as the typical printed circuit board conductor materials have recently gained the interest of research. In this work, novel electrically conductive textiles were used as conductor materials for stretchable circuit boards. Three different fabrics (woven, knitted and nonwoven) made of silver-plated polyamide fibers were investigated for their mechanical and electrical behavior under quasi-static and cyclic mechanical loads with simultaneous monitoring of the electrical resistance. Thereto, the electrically conductive textiles were embedded into a thermoplastic polyurethane dielectric matrix and structured by laser cutting into stretchable conductors. Based on the characterization of the mechanical and electrical material behavior, a life expectancy was derived. The results are compared with previously investigated stretchable circuit boards based on thermoplastic elastomer and meander-shaped conductor tracks made of copper foils. The microstructural changes in the material caused by the applied mechanical loads were analyzed and are discussed in detail to provide a deep understanding of failure mechanisms.

Keywords: textile/polymer composite; stretchable electronics; smart textiles; mechanical and electrical properties; quasi-static and cyclic mechanical loading; life-time expectancy

1. Introduction

Stretchable circuit boards are required for applications where an applied single or repeated mechanical load results in deformation of the system without loss of the desired functionality. In the first case, electronic components are integrated onto two-dimensional substrates that are easier to manufacture than complex three-dimensional ones, before the final shaping. In these forming processes, one-time loads act on the circuit carriers [1]. In the second case, electronics are integrated into systems that are repeatedly stretched. These include, for example, smart clothing for safety, work, sport or healthcare applications, conductors in soft robotics, and wearable sensors in medical bandages or skin-adhesive patches. All applications require non-destructive single or multiple deformation of the circuit board [2]. Current commercial solutions are based on a non-linear design of copper-based conductor tracks, with limited mechanical reliability. In this work, therefore, the copper foil-based conductors are substituted with conductive fabrics.

As is known, commonly used textile fibers show no intrinsic electrically conductive properties. However, the raw fiber textile can be processed into an electrically conductive material by plating a thin metal layer onto their surface, in which gold, silver or copper is usually applied [3]. These fibers are further twisted into electrically conductive yarns that can be embroidered into conductive tracks and even used to contact electronic modules [4]. Tensile tests have shown that such metallized yarns have an elongation at a break of 20%, with the initial electrical resistance quadrupling before breakage [5]. It has also been reported that a thin silver plating has no influence on the tear strength of the fiber; however, the thermoelectric properties of conductive filaments should be considered for the respective processing and application [6]. Conductor tracks based on metallized yarns can be stretched to nearly 50% by embroidering them as a zigzag structure onto elastic fabrics [7]. Yet, in wearable applications, elastic loads can occur from 5% (back) to 60% (along the elbow). Embroidered conductive yarns therefore do not meet the mechanical requirements for reliable use, especially for tight, body-worn garments that are used to measure vital signs and are therefore subject to high dynamic mechanical loads. Even elastic fabric-embedded and preconditioned stainless steel filaments do not exhibit sufficient mechanical and electrical properties under cyclic loading [8].

Thus, we propose the use of metallized polymer-based fabrics, which, due to textile processing techniques, produce a material with higher ductility than a single yarn- or fabric-based material on metal wires. The investigation of the mechanical and electrical properties of these materials, structured into conductor tracks and embedded in an elastic matrix, is the object of this article. In Figure 1, the schematic drawings and SEM images of three different types of selected conductive fabrics are shown.

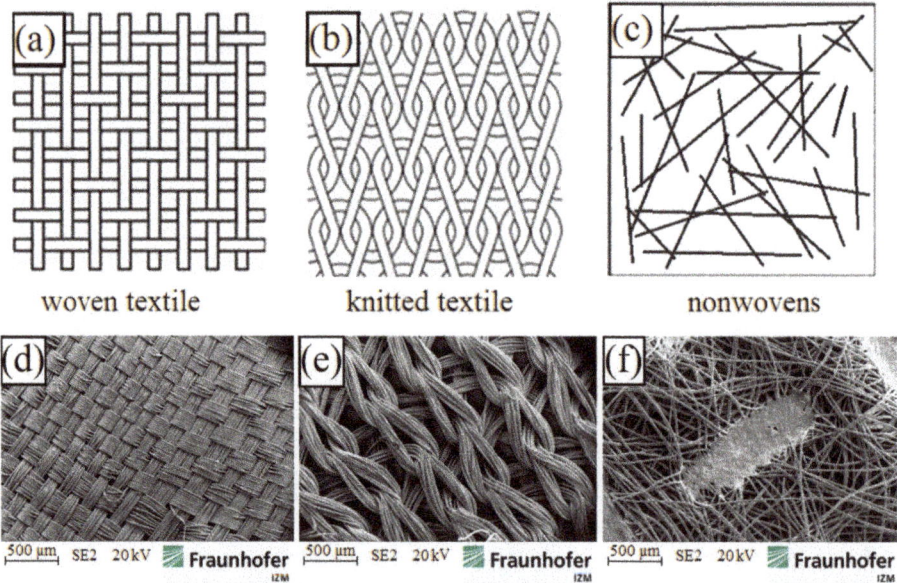

Figure 1. Schematic drawings above of: (**a**) woven; (**b**) knitted and (**c**) nonwoven fabrics and SEM images below of: (**d**) woven; (**e**) knitted and (**f**) nonwoven fabrics.

2. Materials and Methods

All selected fabrics are made of silver-plated polyamide (PA 6.6) fibers, which are embedded in thermoplastic polyurethane (TPU). Important material properties are summarized in Table 1.

Table 1. Mechanical and electrical properties of the used materials [9–12].

Property	Unit	TPU	PA 6.6	Ag
melting temperature T_m	[°C]	160	260	962
glass transition temperature T_G	[°C]	−40	70	/
electrical conductivity (at 20 °C) κ	[Ωm]$^{-1}$	10^{-9}	10^{-12}	61.4×10^6
strain at break (at 20 °C) A	[%]	650	20	50
tensile strength (at 20 °C) R_m	[MPa]	50	85	125–195

2.1. Thermoplastic Polyurethane (TPU)

Thermoplastic polyurethane in film form is used as a substrate material. TPU combines the properties of thermoplastics and elastomers. At temperatures above the glass transition temperature (at −40 °C), TPU has an elastic behavior. This feature is required for later application as a stretchable circuit board. In contrast to elastomers, the selected TPU film also has a melting point (at 160 °C). Above this temperature, it can be thermoformed [13].

2.2. Polyamide 6.6 (PA 6.6)

In the textiles industry, polyamide is the second most used man-made fiber [14]. It is a flexible, thermoplastic polymer that is characterized by high strength. Polyamide filaments (single fibers) are made by melt spinning [14]. The molten polymer is pressed through spin dosage plates, frozen by cooling it down under airflow, and extended to a multiple of its original length. While stretching, the crystalline areas in the polyamide slide off each other. Thereby, they are aligned in the pulling direction. The polyamide becomes stronger and stiffer and the elasticity decreases. The stretching of the polyamide can be partially undone by a heat treatment, at which the required temperature is between the glass transition temperature and the melting temperature. The crystalline areas take on a more energetically favorable form. Under mechanical load, this tempered polyamide filament is more elastic than the non-tempered one [9].

2.3. Thin Silver Layer (On PA 6.6 Filaments)

The fabrics, made from polyamide fibers, are plated with a thin silver layer. Silver is chosen because of its high electrical conductivity and chemical resistance [15]. In addition, Krshiwoblozki [16] described silver as a preferred surface metallization for contacting electronic modules on textile conductors by means of a non-conductive adhesive bonding process. The silver layer on the selected fabrics is around 50–200 nm thick and consists of many silver crystals smaller than 100 nm. It is not a uniform layer as it is known from foil.

2.4. Manufacturing of Elastic Textile/Polymer Composites for Stretchable Circuit Boards

For electrical insulation and protection against external agents, the conductor tracks are embedded in a matrix of TPU. Figure 2 illustrates the developed process flow in the manufacturing of stretchable circuit boards made of textiles and elastomer films.

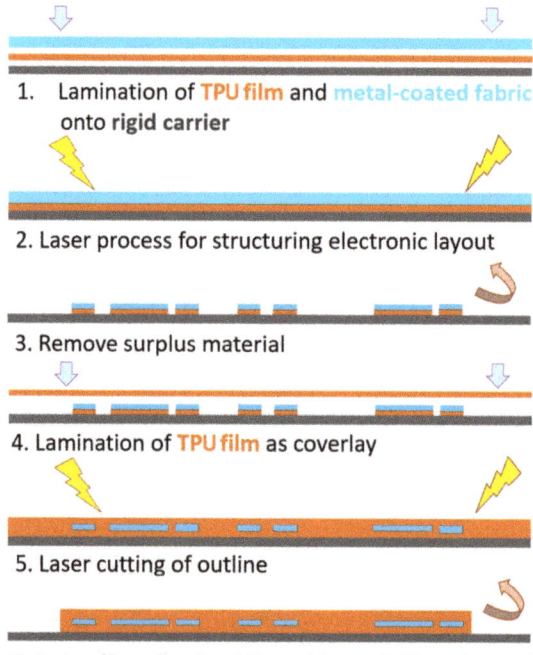

Figure 2. Process flow of stretchable circuit boards based on elastic textile/polymer composites.

A rigid carrier is used to improve handling during manufacturing. A TPU film and a conductive fabric are fixed and bonded at an elevated temperature and pressure onto the carrier (see Figure 2, step 1). The electrical conductor track is structured using a CO_2-laser (see step 2). The excess material is removed, so only the structure of the conductive pattern remains on the carrier (see step 3). Afterwards, a second TPU film is laminated as a cover layer (see step 4). Then the sample outline is laser-cut. Finally, the stretchable composite can be lifted from the rigid carrier (see steps 5 and 6). In Figure 3 the design and the realized test pieces are shown.

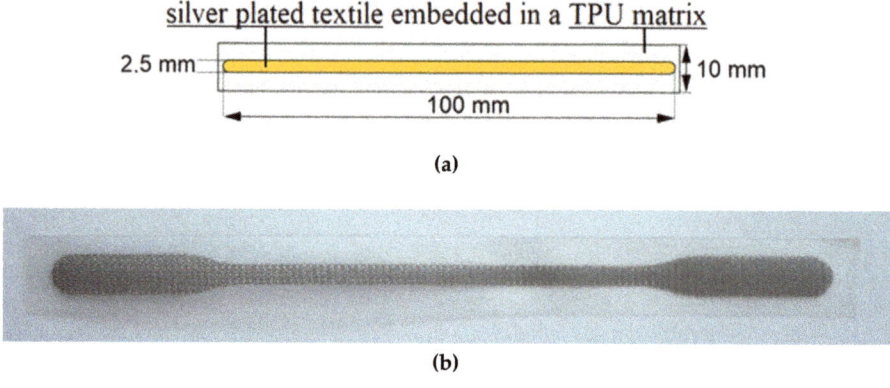

Figure 3. Design of test piece: (**a**) schematic design; (**b**) realized test sample.

It has been observed that the lamination of conductive fabrics has a significant impact on the electrical and mechanical properties of the material. In general, the metallized polyamide filaments are contacted to each other selectively, not over their entire surface. The number of contact points per surface again differs between the woven, the knitted and nonwoven fabric, depending on the voids and type of the thread intersections. Depending on the textile processing, the resulting contact pressure between the filaments plays an important role in the electrical resistance according to Holm's contact theory [17]. Furthermore, the silver on the PA 6.6 filaments does not consist of a continuous layer. Rather, it is made up of many tiny silver nanoparticles. The silver-plating has an uneven thickness of 50 nm to 200 nm, so the electrical conductivity is limited by the thinnest area.

Temperatures far below the melting temperature of silver, which also occur while laminating the samples, have a great influence on the silver layer. The individual silver nanoparticles sinter together to form a more compact layer (see Figure 4), whereby the contact area is increased, and the electrical conductivity is improved.

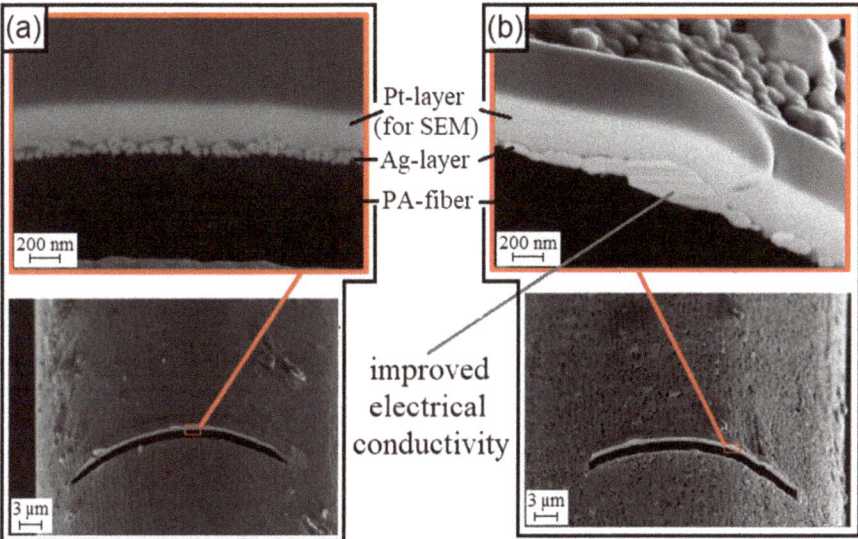

Figure 4. SEM-image of a cut with a focused ion beam (FIB) on a thermally: (**a**) untreated; (**b**) tempered silver-plated PA 6.6 [18].

The sheet resistance is used to compare the electrical resistances of two-dimensional conductors with different width and lengths, such as electrically conductive fabrics. It is calculated from the measured resistance and the quotient of conductor width and length ($R_{square} = R \cdot \frac{width}{length}$). The sheet resistances of the individual fabrics in different stages of processing are shown in Figure 5. The starting sheet resistances of the textile structures before lamination (without heat and pressure influence, marked with A) are compared to the resistances of the textile structures after lamination (with heat and pressure influence, for sheet resistance measurement without embedding in TPU, marked with B).

The test pattern with conductor tracks made of knitted textiles shows the lowest electrical resistance of the three tested fabrics, followed by the woven and the nonwoven structure. In all fabrics, the electrical resistance has dropped by about one third due to the lamination.

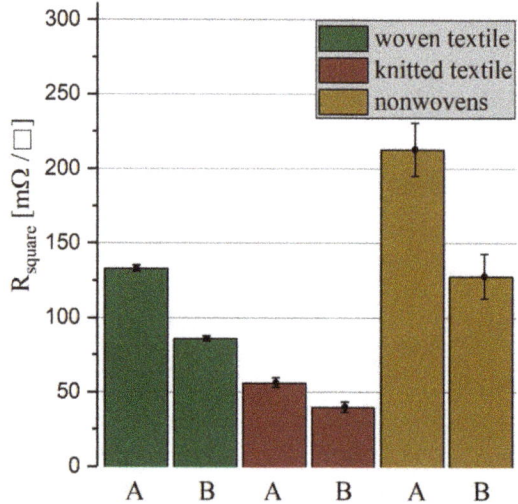

Figure 5. Sheet resistances of the three different fabrics before (**A**) and after (**B**) the influence of pressure and heat during lamination.

2.5. Quasi-Static Mechanical Load

For simulating the quasi-static mechanical load to the samples, the material testing machine TIRAtest 28025 (TIRA GmbH, Schalkau, Germany) is used for the uniaxial tensile test. Simultaneously, the electrical resistance of the conductor tracks is measured. For this purpose, the test samples are contacted with crimp connectors and connected to a digital multimeter Fluke 175 (Fluke Corporation, Everett, WA, USA). The boundary conditions for these tests are listed in Table 2.

Table 2. Boundary conditions of the tensile test.

Parameter	Unit	Value
strain rate	$[\frac{\%}{min}]$	1
entire sample length	[mm]	100
test length l_0	[mm]	31

2.6. Cyclic Mechanical Load

The cyclic mechanical load, which acts on the stretchable circuit boards in many dynamic applications, is simulated by repeated uniaxial loading and relaxation. In order to characterize the change of the electrical behavior during the cyclic mechanical load test, the stretch tester developed and described by Born [19] is used. This measures the initial resistance of the clamped samples, pulls them to a preset stretch, further measures the electrical resistance in the stretched state, and then returns to the home position to measure the electrical resistance again. The stretch tester repeats these cycles until the conductor breaks or the test is terminated manually. The boundary conditions during these cyclic mechanical load tests are listed in Table 3 and Figure 6, the time-progress of the strain-controlled cyclic tests is shown.

Table 3. Parameters of the cyclic tests.

Parameter	Unit	Value
test frequency f	[Hz]	0.04
entire sample length	[mm]	200
test length l_0	[mm]	100

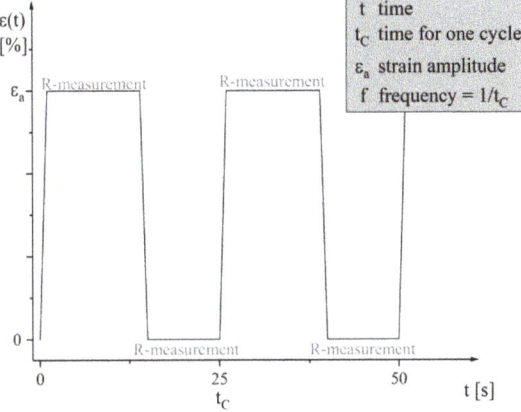

Figure 6. Time-progress of the strain-controlled cyclic tests.

3. Results

3.1. Quasi-Static Mechanical Load

The results of the tested mechanical and electrical properties of the three elastic textile/polymer composite samples are listed in Table 4 and shown in Figure 7. The following formula is used to calculate the initial sheet resistance in Table 4:

$$R_{0,\,square} = R_0 \cdot \frac{b}{l} = R_0 \cdot \frac{2.5\ mm}{100\ mm} = \frac{R_0}{40} \qquad (1)$$

Table 4. Mechanical and electrical properties of tested elastic textile/polymer composite-based stretchable circuit boards.

Property	Unit	Woven Textile	Knitted Textile	Nonwoven Textile
max. force F_{max}	[N]	21–22.5	19–21	11–13
strain at break A	[%]	31–40	65–74	70–75
initial resistance R_0	[Ω]	4.8–6.4	2.4–2.8	7.8–20.1
initial sheet resistance $R_{0,\,square}$	[mΩ/□]	120–160	60–70	195–502.5
resistance at break R_{break}	[Ω]	11.5–16.4	7.4–8.9	61–100
max. resistance change $\Delta R_{max} = \frac{R_{break}}{R_0}$	[-]	2.5–3.5	2.6–3.7	5–8

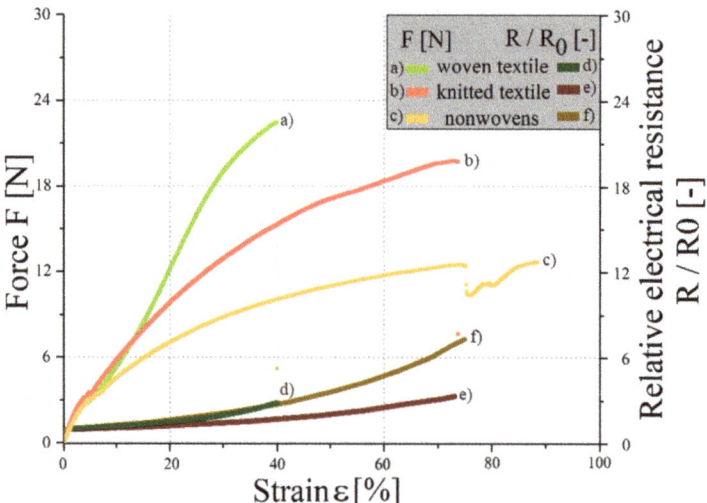

Figure 7. Force-strain diagram and relative electrical resistance changes at a quasi-static load.

In the case of woven and knitted structures, the mechanical damage causes the electrical failure. When the elastic textile/polymer composite breaks at about 38% and 72%, respectively, the silver structure also breaks. For nonwovens, the loosely connected filaments are pulled apart. Only a small amount of mechanical load acts on the fabrics. From an elongation of approx. 75%, the sample is no longer electrically conductive, as the filaments have completely separated from each other. Furthermore, the TPU matrix holds the composite together. The samples with conductor tracks made of nonwovens have a high initial resistance with great fluctuations. While quasi-static mechanical loading, they show the highest relative increase in resistance of the investigated structures. Due to their inconstant mechanical and electrical properties, the nonwoven fabrics enable no reliable use of them, so consequently they are excluded from further investigations. The knitted fabric shows the lowest initial and breaking resistance, as well as the lowest relative increase in resistance under quasi-static mechanical load. Likewise, it has the highest elongation at break of the tested structures. It is, regarding the mechanical and electrical properties, the most suitable structure for single forming.

As expected, the elongation at break of the threads (A = 20%, [5]) is lower than those of the textile structures. The increased stretchability is attributable to, inter alia, the processing into fabrics. The tensile load of the fabrics is partly modified by the fact that the thread intersections strengthen under bending loads, so that a higher elasticity is achieved. In addition to processing into fabrics, the threads are also subjected to heat and pressure during lamination in TPU. In order to find out what influences the individual process steps have on the samples, further tests are conducted. Figure 8 illustrates the elongation at break point of all steps during processing, exemplary for samples made of knitted textiles.

The stretchable circuit boards have an average elongation at break point of about 72%. Knitted fabric samples that have undergone the same temperature and pressure conditions as the processed circuit boards, but which have not been embedded in TPU, have an elongation at break point of 68%. The non-tempered knit fabric is stretchable up to 55%. The increase in stretchability by heat and pressure treatment is due to a reduction of the degree of stretching (compare to Section 2.2). This leads to reduced strength and increased ductility.

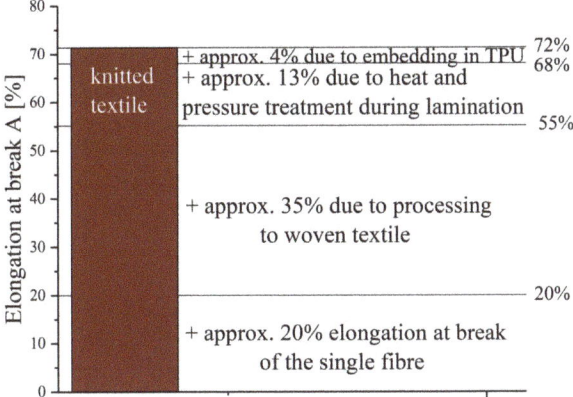

Figure 8. Influence of the process steps on the elongation at break point of the knitted fabric samples (absolute indication of the elongation at break point).

All in all, it can be concluded that the processing of the threads into textiles increases the elongation at break point of the knitted structure by 35% (in absolute percentage). An equivalent evaluation of the results has shown that processing the fibers to woven structures increases the stretchability only by 2%. With 38% (woven-) to 72% (knitted fabric), the stretchable conductors made of textile structures have a higher elongation at break point than the silver-plated polyamide yarns (A = 20%, [5]). Other approaches to stretchable circuit boards, such as those with meander-shaped tracks made of copper foil, can be nonrecurring and elongated up to A = 300% [20].

3.2. Cyclic Mechanical Load

With cyclic mechanical loading of the samples, no classical conductor break was observed, but no or only a very late electrical failure occurred. Experiments demonstrate that the cyclic load at silver-plated fabrics causes a resistance increase up to 10,000 times the initial value, whereby significant differences between loaded and unloaded measurements can be seen. The improved conductance in loaded measurements could be due to the increased number of contact points due to cross-sectional contraction (see Figure 9).

Thus, the tests are evaluated on a specific application. A body conformable, textile-based electrode is used for the wearable measurement of vital parameters, for example for ECG or EMG measurements. Recent publications describe promising results for capacitively-coupled textile electrodes compared to skin contact dry electrodes [21,22]. High electrode impedance may increase noise, but this effect can be reduced by using appropriate (digital) signal processing techniques. In our example, therefore, an absolute electrical resistance increase including supply line of the textile electrode up to $R_{max} = 1\ k\Omega$ can be used as a failure criterion.

A statistical evaluation according to Rossow [23] allows one to determine a survival probability (90%) of the samples. According to Coffin [24] and Manson [25], an evaluation of the measurement results is possible (see Figure 10), so the life expectations of various stretchable circuit boards can be compared to each other.

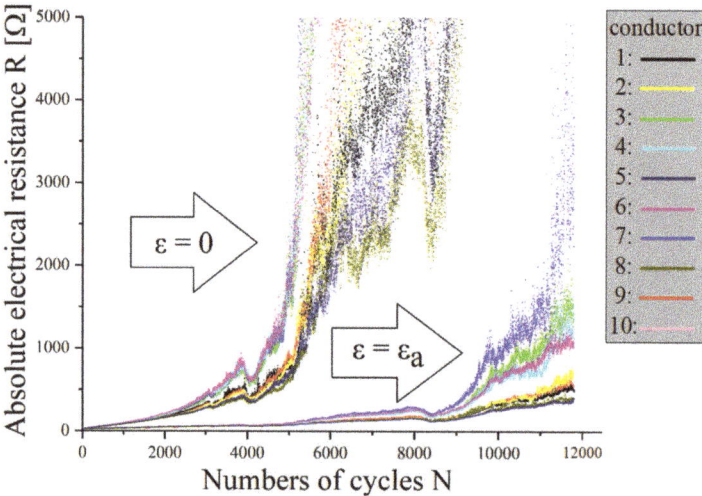

Figure 9. Electrical resistance changes of elastic textile/polymer composite-based stretchable electronics with conductors made of silver-plated knitted textiles under cyclic mechanical load at a strain amplitude of 22.1%. For definition of measurements, see Figure 6.

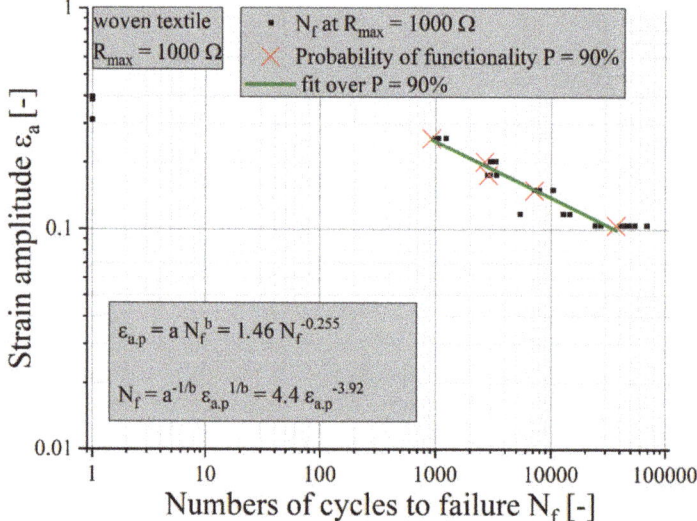

Figure 10. Life expectancy for elastic textile/polymer composite-based stretchable electronics with conductors made of silver-plated woven textiles under a cyclic mechanical load with $R_{max} = 1$ kΩ.

The results of the Coffin–Manson equations for samples with textile structures with the failure criterion of Rmax = 1 kΩ are shown in Figure 11. To compare the lifetime, the life expectancy of a sample with meandering copper tracks is added [26].

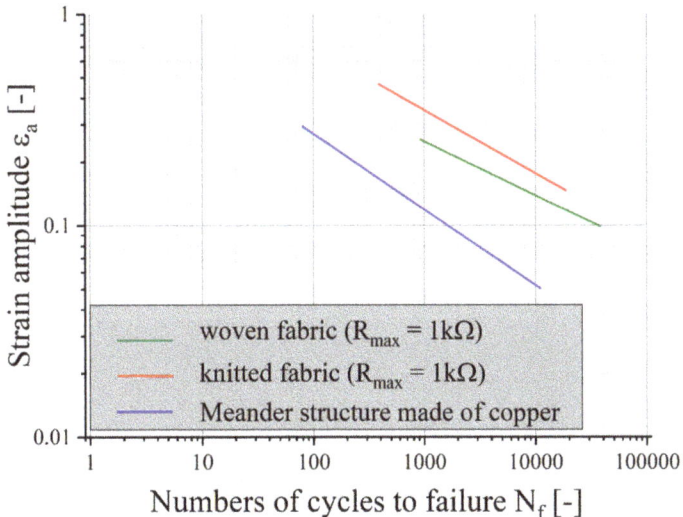

Figure 11. Life expectancy for elastic textile/polymer composites with conductors made of silver-plated woven or knitted textiles (R_{max} = 1 kΩ) compared with samples made of meandering copper tracks [26].

In the chosen application example, the life expectancy of samples with textile conductor materials is much higher than that of copper foil. The lifetime of the knitted structure is the highest among the samples tested. However, if another application requires less resistance, life expectancy will decrease for textile materials only. For applications where higher electrical resistance is sufficient, for example as electrodes, antennas or sensors, stretchable circuit boards with conductors made of textiles are very well suited.

3.3. Analysis of Failure Mechanisms

The microscopic analysis compares a heavily stressed sample with knitted conductor tracks (300,000 cycles at 25% strain amplitude) with an untested sample.

3.3.1. Transversal Microsection

In order to observe the internal structure of the untested samples, transversal microsections were made and analyzed with a visible light microscope (VLM) (see Figure 12).

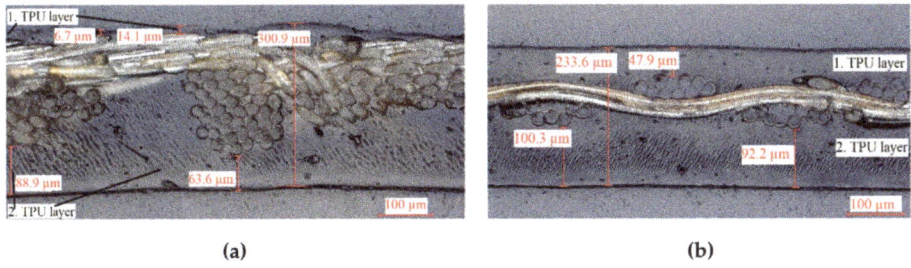

Figure 12. Micrographs of the elastic textile/polymer composites: (**a**) knitted conductive textile; (**b**) woven conductive textile.

Unlike the composite with conductive woven fabric, the knitted sample is not in the center of the TPU matrix and thus not in the neutral fiber. During the first lamination, the TPU layer melts and the knitted material is pressed into the TPU. In the second lamination step, TPU flows into the remaining gaps and bonds to the first TPU layer. However, the knit is pressed until the edge of the material compound. As a result, some polyamide threads are exposed, and they are no longer isolated and can react with the environment.

3.3.2. Surface Analysis

The exposed polyamide threads of the knit-based composite in the TPU layer can also be seen on the SEM images (see Figure 13, left).

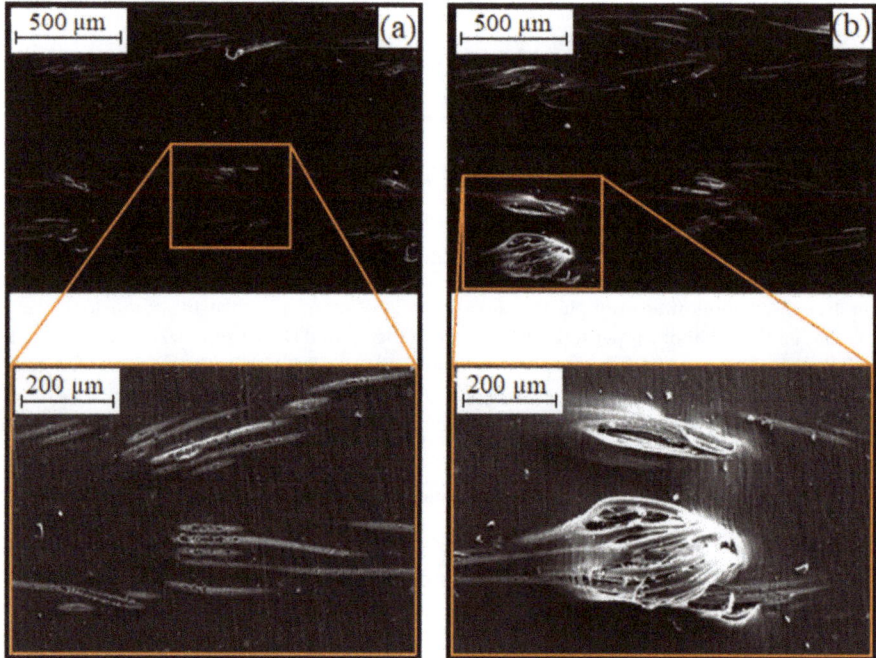

Figure 13. SEM images of samples with conductors made of knitted textiles: (**a**) untested; (**b**) loaded (300,000 cycles at a strain amplitude of 25%).

Energy-dispersive X-ray spectroscopy shows that the silver layer on the polyamide filaments is still intact (see Figure 14, left). If the number of cycles increases, the defects in the TPU matrix become larger, so that the polyamide threads can move further out of the composite material (see Figure 13, right). Due to the repeated stretching and bending of the threads, the silver layer breaks (see Figure 14, right). This results in an increase in electrical resistance.

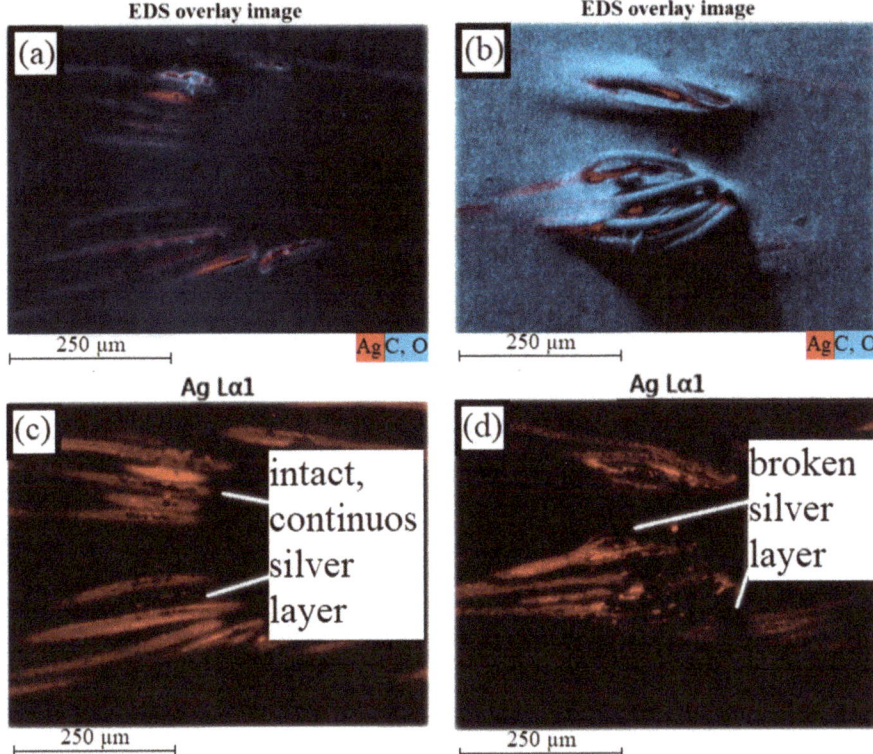

Figure 14. EDS overlay image of knitted sample (silver in red, polymer in blue): (**a**) untested; (**b**) loaded; EDS overlay image of knitted sample (only silver): (**c**) untested; (**d**) loaded.

3.3.3. Analysis of the Inner Structure

In order to investigate the internal structure of the composite material, micro computed tomography (micro CT) is used. In the areas between the loops, even in the parts of their cross-connections, the untested fiber surfaces are still largely smooth. In the tested sample it can be seen that in the area of the loops, the previously smooth surfaces become rougher (see Figure 15).

Knowing that the greatest damage to the silver layer occurred in the contact areas of the loops, transverse microsections of an untested and a tested sample were prepared and analyzed using a focused ion beam (FIB) and SEM in the contact area. Figure 16 shows the polyamide threads of the untested sample.

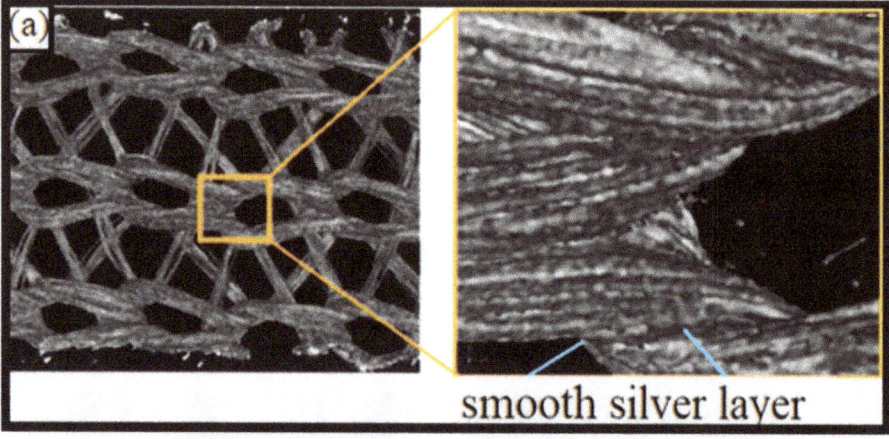
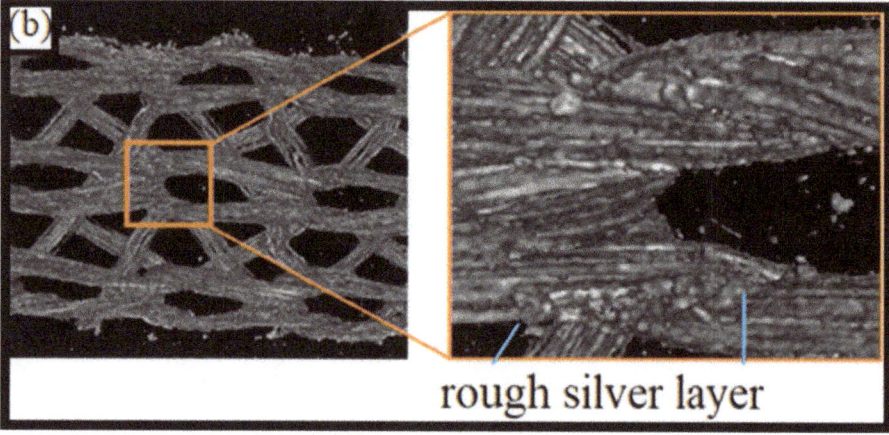

Figure 15. Micro CT as a 3D view of a sample with conductors of knitted textiles: (**a**) untested reference sample; (**b**) tested sample after 300,000 cycles at a strain amplitude of 25%.

The silver nanoparticles on the filaments are sintered together to form a continuous, no uniform layer. This layer is around 50 nm to 200 nm thick (see Figure 16, bottom, right). The individual crystallites are still clearly visible. Complete sintering does not occur. The sintering takes place beyond the boundaries of a silver layer. If two silver-plated polyamide filaments are contacted to each other, the layers partially sinter together during lamination (see Figure 16, top, right). The gap between the layers shows that the FIB cut does not exactly pass through the contact plane at this point. The cut is either a little before or behind the contact plane.

The FIB cut on a tested sample (see Figure 17) shows a discontinuous silver layer on the polyamide filaments (see bottom, left). In the right images, several fibers meet so that the silver layer is thicker than that of a single fiber. Partially, whole silver-flakes dissolved and shifted within the material composite (see right).

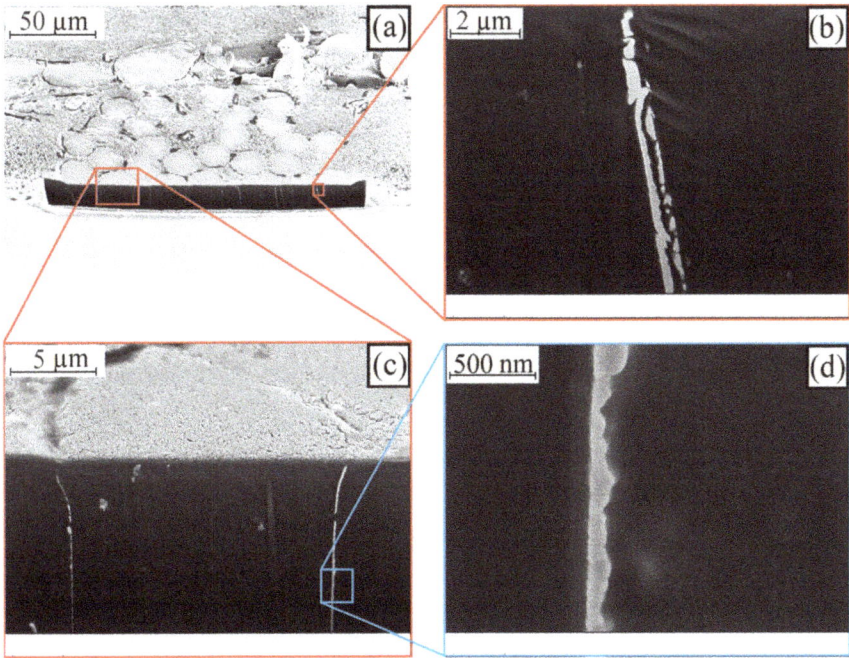

Figure 16. SEM image of a FIB cut on an untested sample with conductors made of knitted textiles at magnification of: (**a**) 500×; (**b**) 10,000× (**c**) 4000× and (**d**) 50,000×.

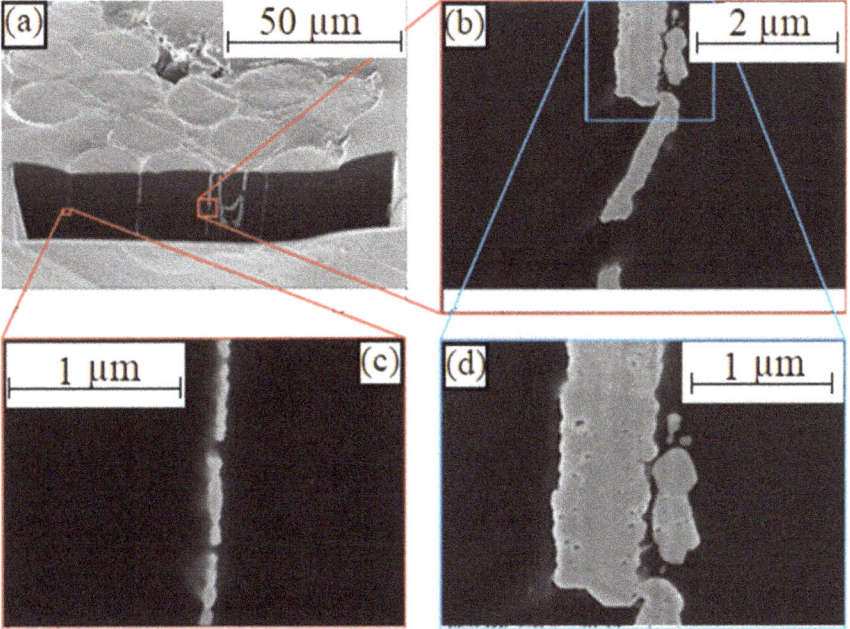

Figure 17. SEM image of a FIB cut on a tested sample with conductors made of knitted textiles at magnification of: (**a**) 1000×; (**b**) 20,000×; (**c**) 50,000× and (**d**) 40,000×.

4. Discussion

In this work, the mechanical and electrical properties of elastic textile/polymer composites for stretchable circuit boards under quasi-static or cyclic mechanical load were investigated on three conducting fabrics (knitted, woven, and nonwoven) embedded in a thermoplastic elastomer matrix.

It was observed that after lamination, the sheet resistances of all tested fabrics were reduced due to a sintering of silver nanoparticles and an increase in the contact points of the conductive fibers with each other. The silver-plated knit shows the best properties in the results of this investigation. It has the lowest initial and the least increase in sheet resistance under quasi-static mechanical load. In particular, the metallized knit fabric has the highest life expectancy under a cyclic mechanical load and thus exceeds the life expectancy of commercially available stretchable circuit boards based on thermoplastic polyurethane and meander-shaped conductor tracks made of copper foil.

With a cyclic mechanical tensile load of the textile–elastomer composite, no classical conductor cracks could be observed. Instead, the silver-plated polyamide fibers rub against each other at the loop of courses and wales intersections, so that over time the thin silver layer partially flakes off from the filament, thereby decreasing the electrical conductivity in the loop running direction. However, due to the presence of secondary filaments that transverse the loop running direction, which are subjected to lower linear mechanical load and friction, the electrical conductivity is maintained but decreases with mechanical load over time.

Due to the relatively high sheet resistances of conductive fabrics and the described electrical behavior under a mechanical load, an application, in particular, in the field of body-near textile sensors and electrodes is conceivable. Therefore, a life expectancy of a typical textile-integrated electrode with high mechanical strain as it occurs for example in knitted shirts was presented.

5. Conclusions

In the next step, the lamination of the conductive textiles into the elastomer matrix has to be optimized in order to prevent out-of-plane displacement and thus substantial damage to the surface metallization of the filaments. A further optimization of the mechanical load capacity is expected by a meandering design of the tracks, analogous to the currently used concepts for copper foil based stretchable conductor structures. Due to the mechanical failure behavior of the tested materials, no conductor breaks occur up to a high load, and we expect a good washing resistance of the textile/ elastomer-based circuit board. In conjunction with conductive textiles with a thicker surface metallization, the novel composite and manufacturing concept presented here can be used for most e-textile and smart clothing applications as well as multi-layer textile circuit boards for the realization of more complex textile-integrated electronic systems. Further fields of application are conceivable in the field of soft robotics. Pneumatically controlled, active morphing soft elastomer surfaces can be extended with conductive textiles to an electronic skin with sensor functions.

Author Contributions: Conceptualization, C.D., L.W. and M.S.-R.; methodology, C.D., L.W. and M.S.-R.; software, L.W.; validation, C.D., L.W. and M.S.-R.; formal analysis, L.W.; investigation, L.W., H.W. and M.Z.; data curation, L.W.; writing—original draft preparation, C.D. and L.W.; writing—review and editing, M.v.K., H.W. and M.S.-R.; visualization, L.W. and M.Z.; supervision, M.S.-R.; project administration, C.D.; funding acquisition, C.D.

Funding: This research was partly funded by the European Unions Horizon 2020 research and innovation program under Grant Agreement No. 825647 and the Federal Ministry of Education and Research (BMBF) under Grant Agreements No. 16FMD01K, 16FMD02 and 16FMD03.

Acknowledgments: The authors would like to thank Martin Haubenreißer for the manufacturing of the test samples, Jörg Gwiasda for the realization of micrographs and Renè Vieroth for the definition of the electronic application fields for the selected metallized textiles.

Conflicts of Interest: The authors declare no conflict of interest.

References

1. Kallmayer, C.; Schaller, F.; Löher, T.; Haberland, J.; Kayatz, F.; Schult, A. Optimized Thermoforming Process for Conformable Electronics. In Proceedings of the 2018 13th International Congress Molded Interconnect Devices (MID), Würzburg, Germany, 25–26 September 2018; pp. 1–6. [CrossRef]
2. Wang, C.F.; Wang, C.H.; Huang, Z.L.; Xu, S. Materials and Structures toward Soft Electronics. *Adv. Mater.* **2018**, *30*, 1801368. [CrossRef]
3. Steinmann, W.; Schwarz, A.; Jungbecker, N.; Gries, T. *Fibre-Table—Electrically Conductive Fibres*; Shaker: Aachen, Germany, 2014.
4. Linz, T. Analysis of Failure Mechanisms of Machine Embroidered Electrical Contacts and Solutions for Improved Reliability. Ph.D. Thesis, University Ghent, Ghent, Belgium, 2011.
5. Simon, E.P. Analysis of Contact Resistance Change of Embroidered Interconnections. Bachelor's Thesis, Technical University Berlin, Berlin, Germany, Bedey Media GmbH, Hamburg, Germany, 2009.
6. Breckenfelder, C.; Dils, C.; Seliger, H.W. Electrical properties of metal coated polyamide yarns. In Proceedings of the 4th International Forum on Applied Wearable Computing 2007, Tel Aviv, Israel, 12–13 March 2007; pp. 1–8.
7. Tangsirinaruenart, O.; Stylios, G. A Novel Textile Stitch-Based Strain Sensor for Wearable End Users. *Materials* **2019**, *12*, 1469. [CrossRef]
8. Isaia, C.; McNally, D.S.; McMaster, S.A.; Branson, D.T. Effect of mechanical preconditioning on the electrical properties of knitted conductive textiles during cyclic loading. *Text. Res. J.* **2019**, *89*, 445–460. [CrossRef]
9. Eyerer, P.; Schüle, P. *Polymer Engineering 1*; Springer: Berlin, Germany, 2019.
10. Hornbogen, E.; Warlimont, H. *Metalle—Struktur und Eigenschaften der Metalle und Legierungen*; Springer: Berlin, Germany, 2016.
11. Product Information Platilon U. Available online: https://solutions.covestro.com/-/media/covestro/solution-center/brands/downloads/imported/1561582168.pdf (accessed on 20 September 2019).
12. Elastollan—Material Properties. Available online: http://www.polyurethanes.basf.de/pu/solutions/elastollan/en/function/conversions:/publish/content/group/Arbeitsgebiete_und_Produkte/Thermoplastische_Spezialelastomere/Infomaterial/elastollan_material_uk.pdf (accessed on 20 September 2019).
13. Mark, J.E. *Physical Properties of Polymers Handbook*; Springer: New York, NY, USA, 2011.
14. Tobler-Rohr, M.I. The supply chain of textiles. In *Chapter 2 Handbook of Sustainable Textile Production*; Woodhead: Manchester, UK, 2011.
15. Adams, D.; Alford, T.L.; Mayer, J.W. *Silver Metallization—Stability and Reliability*; Springer: London, UK, 2008.
16. Von Krshiwoblozki, M.; Linz, T.; Neudeck, A.; Kallmayer, C. Electronics in Textiles—Adhesive Bonding Technology for Reliably Embedding Electronic Modules into Textile Circuits. *Adv. Sci. Technol.* **2013**, *85*, 1–10. [CrossRef]
17. Holm, R. *Electric Contacts, Theory and Applications*, 4th ed.; Springer: New York, NY, USA, 1967.
18. Foerster, P. Untersuchungen zu Eigenschaften von Nanosilberschichten auf Polyamidfasern. In *Student Research Project as Part of the Degree Programm Electrical Engineering*; Technical University Berlin: Berlin, Germany, 2010.
19. Born, S. *Konzeptionierung* und *Realisierung* eines *Prüfstandes* zur *Qualifizierung* von *Dehnbaren Schaltungsträgern*. Bachelor's Thesis, Technical University Berlin, Berlin, Germany, 2010.
20. Vieroth, R.; Löher, T.; Seckel, M.; Dils, C.; Kallmayer, C.; Ostmann, A.; Reichl, H. Stretchable Circuit Board Technology and Application. In Proceedings of the 2009 International Symposium on Wearable Computers, Linz, Austria, 4–7 September 2009; pp. 33–36. [CrossRef]
21. Taelman, J.; Adriaensen, T.; van der Horst, C.; Linz, T.; Spaepen, A. Textile Integrated Contactless EMG Sensing for Stress Analysis. In Proceedings of the 2007 29th Annual International Conference of the IEEE Engineering in Medicine and Biology Society, Lyon, France, 22–26 August 2007; pp. 3966–3969. [CrossRef]
22. Fuhrhop, S.; Lamparth, S.; Heuer, S. A textile integrated long-term ECG monitor with capacitively coupled electrodes. In Proceedings of the 2009 IEEE Biomedical Circuits and Systems Conference, Beijing, China, 26–28 November 2009; pp. 21–24. [CrossRef]
23. Rossow, E. *Introduction to the Principles and Application of Sampling Plans by Attributes*; EOQC Monograph: Rotterdam, The Netherlands, 1968.
24. Coffin, L. A study of cyclic thermal stress on a ductile metal. *Trans. Am. Soc. Mech. Eng.* **1954**, *76*, 931–950.

25. Manson, S. *Behaviour of Materials under Conditions of Thermal Stress*; NACA Report No.1170; National Advisory Committee for Aeronautics: Washington, DC, USA, 1954.
26. Grams, A.; Kuttler, S.; Löher, T.; Walter, H.; Wittler, O.; Lang, K.-D. Lifetime modelling and geometry optimization of meander tracks in stretchable electronics. In Proceedings of the 2018 19th International Conference on Thermal, Mechanical and Multi-Physics Simulation and Experiments in Microelectronics and Microsystems (EuroSimE), Toulouse, France, 15–18 April 2018; pp. 1–5. [CrossRef]

© 2019 by the authors. Licensee MDPI, Basel, Switzerland. This article is an open access article distributed under the terms and conditions of the Creative Commons Attribution (CC BY) license (http://creativecommons.org/licenses/by/4.0/).

Article

Characterisation of Electrical and Stiffness Properties of Conductive Textile Coatings with Metal Flake-Shaped Fillers

Veronica Malm [1,*], Fernando Seoane [1,2,3] and Vincent Nierstrasz [1]

1. Textile Materials Technology, Department of Textile Technology, Faculty of Textiles, Engineering and Business, University of Borås, 50190 Borås, Sweden
2. Institute for Clinical Science, Intervention and Technology, Karolinska Institutet, 171 77 Solna Stockholm, Sweden
3. Department of Medical Care Technology, Karolinska University hospital, 141 57 Huddinge, Sweden
* Correspondence: veronica.malm@hb.se; Tel.: +46-33-435-4560

Received: 30 September 2019; Accepted: 24 October 2019; Published: 29 October 2019

Abstract: Two conductive formulations containing different types of micron-sized metal flakes (silver-coated copper (Cu) and pure silver (Ag)) were characterised and used to form highly electrically conductive coatings (conductors) on plain and base-coated woven fabrics, the latter in an encapsulated construction. With e-textiles as the intended application, the fabric stiffness, in terms of flexural stiffness and sheet resistance (R_{sh}), after durability testing (laundering and abrasion) was investigated and related to user friendliness and long-term performance. Bare and encapsulated conductors with increasing amounts of deposited solids were fabricated by adjusting the knife coating parameters, such as the coating gap height (5, 20, 50, and 200 µm), which reduced the R_{sh}, as determined by four-point probe (4PP) measurements; however, this improvement was at the expense of increased flexural stiffness of the coated fabrics. The addition of a melamine derivative (MF) as a cross-linker to the Cu formulation and the encapsulation of both conductor types gave the best trade-off between durability and R_{sh}, as confirmed by 4PP measurements. However, the infrared camera images revealed the formation of hotspots within the bare conductor matrix, although low resistances (determined by 4PP) and no microstructural defects (determined by SEM) were detected. These results stress the importance of thorough investigation to assure the design of reliable conductors applied on textiles requiring this type of maintenance.

Keywords: conductivity; metal flake; coating; e-textile; encapsulation; durability; stiffness

---

1. Introduction

Metals have attracted increasing interest as conductive fillers in textile coatings and printing formulations for implementing textile-electronic integration in e-textiles and, in particular, in wearable measurement applications [1–3]. The interest is likely to depend on the superior electrical conductivity characteristics of metals [4], which is also why metal fillers are the focus in this experimental research. Note that carbon black, intrinsically conductive polymers [5] and graphene [6] are the promising alternatives. High and reliable electrical conductivity is often required for power distribution or for interconnecting leads for sensors or transducers, e.g., textile electrodes for current stimulation, biopotential sensing [7,8], etc. Electrical conductivity depends on the ability of electrons and ions as charge carriers to move through the material—the deposited film matrix in this case; therefore, solid-to-solid contacts between metal fillers forming a three-dimensional (3D) conductive network embedded in the insulating polymeric matrix is of paramount importance to obtain conductivity in the deposited film.

Compared to smaller metal particles, micron-sized flakes typically have a high aspect ratio, which means that less filler is required to generate enough contacts between flakes to achieve a sufficient conductive pathway [3]. However, the size of the filler also comes with certain drawbacks, such as increased stiffening effects when deposited on flexible fabrics. Less flexible fabrics and brittle conductive networks are prone to cracking upon bending [7,9,10], which adversely affect the comfort of the wearer and the long-term usage (durability) due to the loss of conductivity. Since these types of conductive fillers possess superior conductive properties, micron-sized flakes as metal fillers still remain a current research topic [2], specifically in terms of overcoming the associated challenges [11–13].

The encapsulation of conductive films (conductors) on textile substrates is a promising approach to protect from material fatigue caused by mechanical stress [11,12,14] and chemical attacks, and for some applications, this is the only alternative to prevent electrical short circuits between conductors and surface oxide film formation. In terms of environmental impact, encapsulation should also prevent the release of metal particulates from functionalised metal fabrics into sweat [15] or washing effluents [16,17], which may have a potential risk to humans [18] and other biological organisms [19]. Komolafe et al. [14] supposed that the encapsulation technique may be helpful for protecting the conductor from abrasion, although it does not structurally optimise the stress response of the conductor. Stress developed during processing may lead to a change in the layer curvature and misalignment between layers, which can cause severe reliability problems in microelectronic devices [20]. Therefore, the performance of encapsulated conductors applied on flexible substrates is important, which should be addressed when working with conductive formulations and their application on textiles, as done in this study.

Furthermore, for certain applications, exposed conductors can be desirable, e.g., when skin-electrode interaction is desired, and the addition of a cross-linker to a conductive formulation to strengthen the film matrix is a possible approach to improving the durability. Melamine cross-linkers as additives for waterborne coating formulations have been extensively studied for their ability to improve the mechanical and chemical resistance of industrial coatings [21]. However, to the best of the authors' knowledge, the addition of a melamine derivate to a waterborne polyurethane formulation containing metal flakes has not been addressed in any recent paper.

Though encapsulation or cross-linking may be beneficial for the durability in terms of the conductive performance, there is also the risk of counteracting effects, such as increased stiffness, weight, and brittleness, which influence the comfort and user friendliness. These effects can be caused by a substantial increase in the amount of applied material or excessive cross-links tying the polymer chains together, resulting in increased stiffness. Therefore, a trade-off between fabric stiffness and flexibility must be addressed to improve the durability of the conductive network while retaining the flexibility of the fabric.

In this experimental research, conductive formulations containing metallic flake-shaped fillers coated on fabrics in a bare and an encapsulated construction are studied. First, the different coating formulation compositions, coating conditions and parameters to adjust the shear rate and viscosities are confirmed. These factors are then linked to the physical properties of the conductors and further related to the electrical and mechanical properties. Second, the morphological and electrical properties correlating to the factors and their relation to the durability during washing and abrasion are analysed. Finally, infrared images of the electrical conductors are obtained, and the optimisation of reliable conductors is discussed in this work. This work provides a guideline for how to characterise bare and encapsulated coated, printed or alternatively manufactured conductive fabrics using conventional textile test methods and a straightforward electrical resistance measurement method.

2. Materials and Methods

2.1. Materials

2.1.1. Substrate

A plain-woven fabric (Almedahl-Kinna AB, Kinna, Sweden) made from poly(ethylene terephthalate) (PET) fibres, scoured and heat-set by the supplier, was used as the substrate. The fabric was composed of a continuous multifilament yarn, with a thread count of 30.5;24 and a mass per unit area of 110 g/m^2.

2.1.2. Coating Formulation

This study includes three formulations containing metal flakes as the conductive filler component. One is a water-based polyurethane (PU)-compound containing an emulsifier agent, thickening agents and silver-coated copper flakes with an average diameter of <12 µm (denoted Cu). The Cu formulation had a solids content of approximately 66 wt% and a density of 1 g/cm^3 according to the supplier (Product U1005, Bozzetto Group, Italy). The Cu formulation was investigated with the addition of a cross-linker (3 wt%) based on a melamine derivative (MF) with a minimum content of free formaldehyde (Reacel MFC, Bozzetto Gmbh Krefeld, Germany) (denoted Cu$_{MF}$). Third, a stretchable formulation containing pure silver flakes <5 µm (PE872, DuPont Electronic Materials Ltd., Bristol, UK) (denoted Ag) was included in the study. The Ag formulation had a solids content of 61 wt%, as determined by the author by differentially weighing the formulation before and after drying according to the conditions in Table 1.

For the interfacial layer between the fabric substrate and the conductor, an aqueous aliphatic polyether polyurethane dispersion (RUCO-COAT PU 1110, Rudolf GmbH, Geretsried, Germany) thickened with a rheology modifier (Borchi Gel L75N, OMG Borchers GmbH, Langenfeld, Germany) was used as a basecoat. To encapsulate the conductor, an elastic thermoplastic polyurethane film (TPU) (LB700-100, Loxy AS, Halden, Norway) was used.

2.1.3. Sample Manufacturing

Figure 1a shows and illustration of coated samples with the conductor applied directly on top of the fabric (series: Cu_05, Cu_20, Cu_50, and Cu_200; Cu$_{MF}$_05, Cu$_{MF}$_20, Cu$_{MF}$_50, and Cu$_{MF}$_200; and Ag_05, Ag_20, Ag_50, and Ag_200), sample notations are further explained in Table 1. Fabric substrates of 20 × 30 cm^2 were attached to an even surface, and the formulation was manually applied using a film applicator (ZUA 2000, Zehntner GmbH Testing Instruments, Sissach, Switzerland) using a procedure similar to the knife-over-blanket technique. Samples with four different gap heights (h) of 5, 20, 50, and 200 µm were prepared with a coating speed (v) of 0.6 m/min. The corresponding shear rates, drying and annealing conditions in a low convection heat oven (D-36-03-EcoCell, Rycobel NV, Deerlijk, Belgium) are shown in Table 1. The shear rate $\dot{\gamma}$ was calculated according to:

$$\dot{\gamma} = \frac{v}{h} \tag{1}$$

Encapsulated conductive samples (Series: B_Cu$_{MF}$05_T, B_Cu$_{MF}$20_T, B_Cu$_{MF}$50_T and B_Cu$_{MF}$200_T; and B_Ag05_T, B_Ag20_T, B_Ag50_T, and B_Ag200_T) were prepared in the four-layer construction shown in Figure 1b. A basecoat was first applied to the fabric with the knife-over-blanket technique, as described above, with an applicator gap height of 100 µm. The basecoat was then dried at 90 °C for 20 min without an annealing step to increase the adhesion to the coated conductive layer applied in the next step. The conductive layer was dried and annealed before the top layer was applied, see Table 1. To expose the conductors for electrical resistance characterization, a polytetrafluoroethylene (PTFE) mask with rectangular stripes was applied on the conductive layer (see Figure 1b) before the top-layer TPU film was laminated using a hot-press plate set at a temperature of 130 °C for 1 min and 40 s.

Table 1. Coating conditions.

Sample	Shear Rate (Formulation)	Solids Deposit (Conductor)	Drying Temp:Time	Annealing Temp:Time
	(s^{-1})	(g/m^2)	(°C:min)	(°C:min)
Cu_05	2000	64	80:5	150:2
Cu_20	500	72	80:10	150:3
Cu_50	200	85	80:15	150:4
Cu_200	50	180	80:20	150:5
CuCL_05	2000	64	80:5	150:2
CuCL_20	500	73	80:10	150:3
CuCL_50	200	85	80:15	150:4
CuCL_200	50	185	80:20	150:5
B_Cu$_{MF}$05_T	2000	56	80:5	150:2
B_Cu$_{MF}$20_T	500	67	80:10	150:3
B_Cu$_{MF}$50_T	200	80	80:15	150:4
B_Cu$_{MF}$200_T	50	166	80:20	150:5
Ag_05	2000	47	-	130:20
Ag_20	500	65	-	130:30
Ag_50	200	72	-	130:50
Ag_200	50	160	-	150:40
B_Ag05_T	2000	44	-	130:20
B_Ag20_T	500	64	-	130:30
B_Ag50_T	200	81	-	130:50
B_Ag200_T	50	149	-	150:40

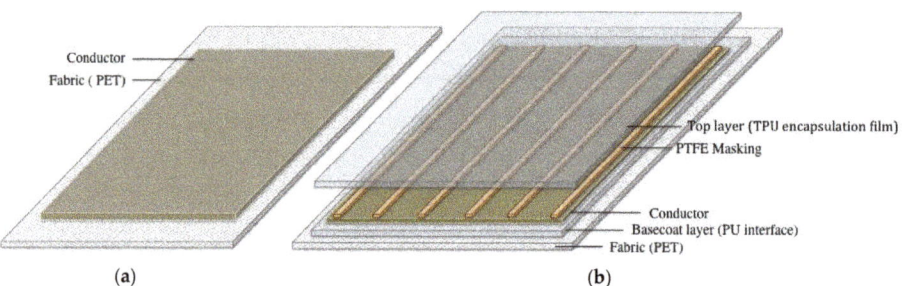

Figure 1. Illustration of the coated samples with different constructions: (**a**) conductor on plain fabric and (**b**) layered construction of an encapsulated conductor with PTFE masking prepared for washing treatments.

2.2. Characterisation

2.2.1. Viscosity

The viscosities of the conductive and basecoat formulas were measured using a Paar Physica MCR 500 rheometer (Anton Paar, Ostfildern, Germany) at 23 °C with a cone and plate setup at a distance of 50 µm. Three measurements of each formulation were taken at shear rates between 0.01 and 2000 s^{-1}, under simulated process conditions during the coating procedure. The shear rates are listed in Table 1.

2.2.2. Surface Topography and Morphology

The surface topography (or surface roughness) of coated samples was evaluated by a non-contact (optical) high-resolution surface profiler (Wyko NT 1100, Veeco Instruments Inc. Tucson, AZ, USA).

The arithmetical mean deviation of the profile, (R_a), with the dimensional unit of micrometre, was obtained from 2D roughness measurements in the vertical z direction. The surface morphology (visual appearance) was studied by means of scanning electron microscopy (SEM-JEOL JSM 6301F, JEOL Ltd., Tokyo, Japan) of the sputtered samples.

2.2.3. Flexural Stiffness

The conventional flexural stiffness of the coated samples was characterised by a fixed-angle flexometre according to the SS-ISO 4604:2011. The test was performed by sliding a conditioned (12 h, 22 °C, 65% RH) sample (25 × 200 mm^2) over the edge of a horizontal surface until the tip bends under its own weight and makes a 41.5° angle to the horizontal by touching the inclined surface. The length of the overhang was measured, and the tests were made in both weft and warp direction and were repeated on both ends in face up and down orientations with a total of four readings per sample. The conventional flexural stiffness (G) in millinewton metres was calculated according to the standard, and the average flexural stiffness from four replicates was used.

2.2.4. Electrical Resistance

The electroconductive properties were evaluated based on an electrical resistance technique, which involved the inline 4PP arrangement using a probe with four rectangular electrodes arranged as shown in Figure 2a and further described elsewhere [22]. The electrode surface was coated with a textured silver conductive epoxy layer, and a weight of 2 kg was placed on top of the probe before each measurement to obtain better contact with the coated textile surface. For plain conductors, the probe was consequently placed contacting the surface of the conductor at five positions on each rectangular sample in both the weft (x) and warp (y) direction, as shown in Figure 2b,c. To enable measurement on the encapsulated samples and because no significant resistance anisotropy was found when using the 4PP arrangement, measurements on the encapsulated samples were taken only in the weft direction.

Electrodes were connected to a multimetre (Agilent 34401A) set in four-wire resistance mode with a low test current flow of 1 mA and an accuracy of ±4 mOhm. The values were recorded after 1 min to circumvent initial fluctuations. The outer electrodes were current (I) carrying, and the inner electrodes (forming a 20 × 20 mm^2 square) were voltage (V) sensing. The values are presented as sheet resistance (R_{sh}) according to:

$$R = R_{sh} \frac{L}{W} = \frac{\rho}{t} \frac{L}{W} \qquad (2)$$

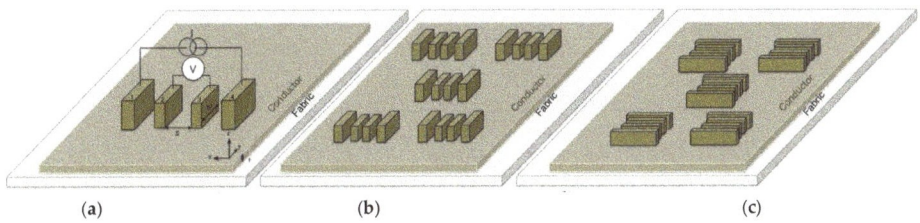

Figure 2. (a) Four-point probe setup and the placement of the probe at five different positions in the (b) weft and (c) warp directions in the 4PP arrangement.

To validate the ohmic linearity for all coated samples current–voltage (IV) curves were obtained using the described arrangement with an additional power supply and by varying the current from 100 up to 1000 milliamperes, see Figure A1 in Appendix A.

2.3. Durability Test

2.3.1. Washing

Durability to laundering was studied according to the standard EN ISO 6330:2012 method 4G with a ballast load of 1 kg (Type III) using a 40 °C hand washing procedure with detergent in a laboratory washing machine (Electrolux Wascator FOM71 MP, Stockholm, Sweden). To avoid samples from getting stuck in the drum, the samples were attached to ballasts using tag fasteners with the coated surface facing outward. Five replicates were characterised for their electrical resistance before and after washing using the four-point probe method. The influence of the drying procedure between washing cycles was studied, and line drying at room temperature (24 °C) for 24 h and line drying in a heat cabinet at 50 °C for one hour were compared.

2.3.2. Abrasion

The impact of abrasion on the coated samples with one conductive layer was tested with a Martindale 2000, using the standard SS-EN ISO 5470-2 method 2 and dry tests. The influence of abrasion on the electrical resistance was evaluated by performing 4PP measurements in the warp and weft direction on three replicates. For both plain and encapsulated samples 4PP measurements were made before and in between each inspection stage, that is, after 1600, 3200, 6400, 12,800, 25,600, and 51,200 revolutions corresponding to 100, 200, 400, 800, 1600, and 3200 abrasion cycles. For the encapsulated samples, electrical measurements were only made in the weft direction and on three different replicates from the same sample series in between each inspection stage.

3. Results and Discussion

3.1. Influence of the Formulation Type and Coating Parameters on the Physical, Electrical and Mechanical Properties

3.1.1. Viscosity Behaviour of the Formulations

Figure 3 shows the viscosity versus shear rate for the conductive formulations listed in Table 1, including the formulation used for the interface layer between the textile substrate and the conductor layer. For all formulations, the viscosity decreased with increasing shear rate related to the coating conditions in Table 1, demonstrating pseudoplastic behaviour. At shear rates below $10\ s^{-1}$, the addition of the MF had a slight dilution effect on the Cu_{MF} formulation, producing a lower viscosity than that of the plain Cu formulation. However, with increasing shear rates (50, 200, 500 and $2000\ s^{-1}$) by tuning the application conditions, the differences were suppressed, resulting in overlapped curves. The differences at the lower shear rates are more relevant for the film-forming process after application, where a lower viscosity offers increased levelling conditions, whereas viscosity differences at higher shear rates can theoretically influence the amount of solids deposited.

At lower shear rates, the viscosity of the Ag formulation was higher than that of the Cu and Cu_{MF} formulations, and with increased shear rates up to $300\ s^{-1}$, the shear thinning behaviour was less steep, showing a more Newtonian profile, i.e., the viscosity was less influenced by the shear rate. This indicates that the binder polymers within the Ag formulation are of high molecular weight and long enough that polymer entanglement restricts their alignment in the flow direction with increased shear rates up to $300\ s^{-1}$. However, at increase shear rates, a more pseudoplastic behaviour was pronounced.

3.1.2. Influence of Solids Deposit on Surface Topography and Morphology

In Figure 4a the surface roughness (R_a) values were observed to decrease with an increased amount of solids (because of the gap height). This was because the increased amount of deposited solids levelled out the surface topography of the fabric. The phenomenon is also what caused the significant difference in R_a between the plain-coated and base-coated sample series, the value of the latter decreasing to half of the value of the former. This is supported by the SEM images in Figure 4b showing the surface morphology differences of Cu- and Ag-coated samples produced with the lowest (5 µm) and the highest (200 µm) coating gaps, including the encapsulated samples with the highest coating gap.

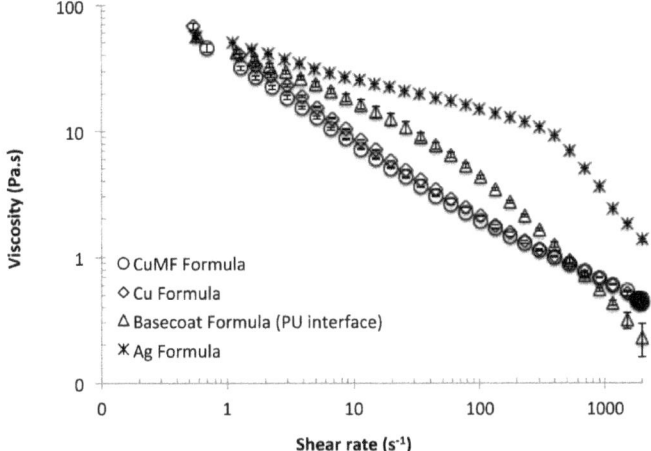

Figure 3. Viscosity versus shear rate for the three conductive formulations and the formula used at the interface.

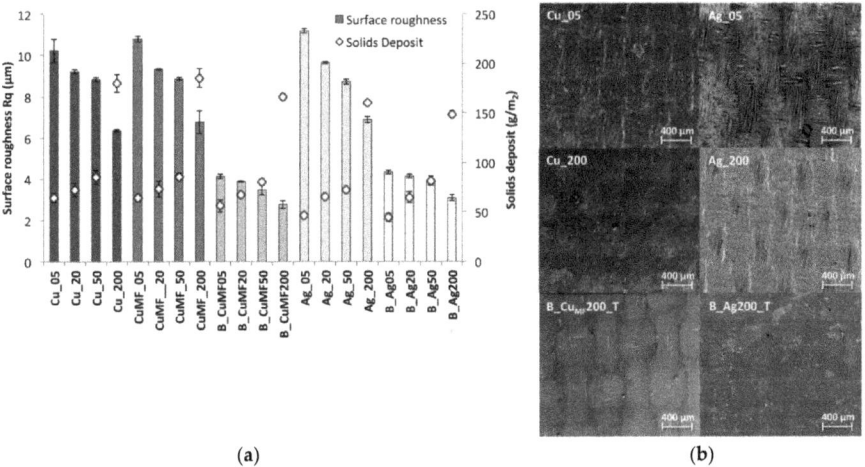

Figure 4. (**a**) Surface roughness of the conductor layers and solids deposition depending on the formulation and gap height. Individual standard deviations were used to calculate the intervals, and the plot displays a 95% confidence interval (CI) for the mean. (**b**) Scanning electron microscopy (SEM) images of samples coated with the Cu and Ag formulations showing the mass per unit area in brackets Cu_05 (64 g/m^2), Cu_200 (180 g/m^2), B_Cu$_{MF}$200_T (166 g/m^2), Ag_05 (47 g/m^2), Ag_200 (160 g/m^2) and B_Ag200_T (149 g/m^2).

The basecoat has a levelling effect on the fabric texture before the conductor is applied, which also hinders the formulation from penetrating the cavities between the yarns. The total decreases in R_a with increased solids deposition is therefore not as pronounced (<1.3 µm) for the base-coated sample series compared to the plain-coated samples, which decreased by over 4 μm. Thus, a more homogenous morphology can be expected for base-coated conductor layers with less variation in the surface topography. However, an undulated microstructure conductive layer is still visible on the surface of the encapsulation film, even on the base-coated samples, as seen in Figure 4b.

It is important to note that surface roughness is a material property that may influence electrical resistance measurements since the contact between the conductor layer and the electrodes of the measurement probe likely contribute to the electrical contact resistance.

No significant difference in R_a was seen between the Cu and Cu_{MF} samples, since there was also no difference in the amount of deposited solids, see Figure 4a. This correlates well with the equivalent pseudoplastic behaviour of the Cu and Cu_{MF} formulations, as seen in Figure 3. Accordingly, equivalent penetration of the formulation into the woven structure can also be assumed. When comparing the plain-coated and encapsulated Cu_{MF} sample series with the same gap heights, a trend towards less solids deposition for base-coated fabrics was observed. This is the result of the preventative effect of the formulation flowing into the fabric, as discussed above.

A lower amount of solids was deposited on the Ag-coated samples than the samples coated with the Cu and Cu_{MF} formulations. The plain coated and base-coated Ag samples showed the same trends. However, the Ag formulation showed a higher viscosity at the relevant shear rates depending on the processing conditions (see Table 1), which generally leads to a larger amount of applied material. The reason for this is most likely the lower solids content in the Ag formulation (61 wt%) than in the Cu formulation (66 wt%). Furthermore, the concentration of metal flakes in each formula should be taken into account, although this information is unknown to the authors. Though the filaments in the yarn appear on the surface of Cu_05 show a rather coarse texture, a coherent film without cracks or blisters was achieved. However, for Ag_05, the deposited amount of 47 g/m^2 was not enough to achieve a coherent film, as the filament yarn was only partly covered with the material. On the other hand, with an increased amount of deposited material, as for Ag_200, the film was more properly covered.

3.1.3. Influence of Solids Deposit on Electrical Resistance

Figure 5 shows the R_{sh} as a function of solids deposit. For all samples, the R_{sh} values are below 1.5 Ω/sq, indicating that all of the conductor layers possess high electroconductive properties. A significant decrease in R_{sh} with increased solids deposit (because of increased coating gap) from 44 to 185 g/m^2 was observed and attributed to the increased number of flakes within the polymer matrix, creating more contact points between flakes and more possible routes for electron conduction through the 3D network. A similar decrease in R_{sh} is seen for the Cu and Cu_{MF} curves on plain fabrics, which can be correlated to the similarities in increased solids deposition. The decreasing trend is, however, not as pronounced for the B_Cu_{MF}_T curve, and no significant difference in R_{sh} between the plain-coated and encapsulated conductor was noticed, even though there was a clear difference in solids deposition between the samples, especially those with the largest amount (Cu: 180, Cu_{MF}: 185 and B_CuMF_T: 166 g/m^2). The observed phenomena may be related to the more uniform microstructure of the conductor applied on the base-coated samples, which facilitates electron transport pathways, resulting in a low R_{sh} with a lower amount of deposition. On the other hand, conductors on plain fabric conform to the substrate texture, resulting in an undulated film with increased electron path lengths, and the equal R_{sh} values were possibly compensated by the increased deposition.

Compared to the Cu sample, the Ag sample showed a more pronounced decrease in R_{sh} with increased deposition, with a total decrease of more than one order of magnitude from 1.1 to 0.03 Ω/sq. The explanation for this is that silver possesses a lower intrinsic resistivity than copper, 16 and 17 nΩ m respectively [4]. It was interesting to see that Ag_05 had a mean R_{sh} value of 1.2 Ω/sq, although the SEM images in Figure 4b revealed a non-coherent topography. This result indicates that the Ag formulation penetrates the fabric and forms a conductive network between the interstices of the yarn filaments.

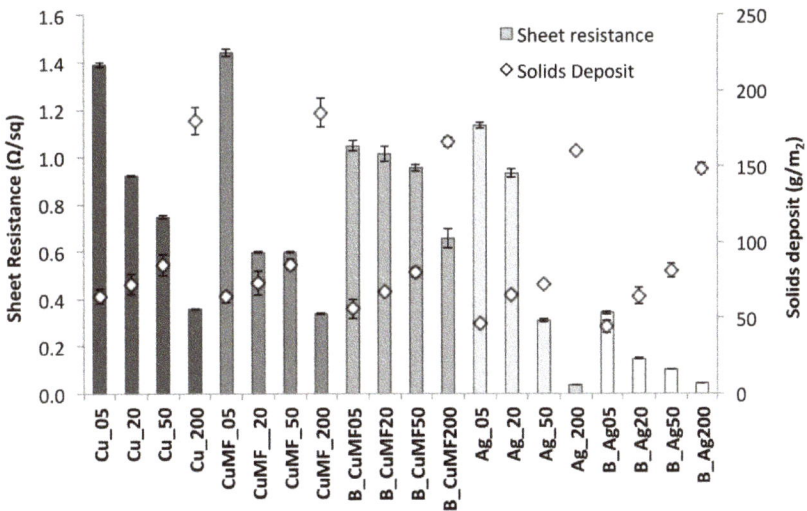

Figure 5. Sheet resistance and solids deposition for all samples using the inline four-point probe configuration. In this diagram, the thickness of the encapsulated conductor is not taken into account; thus, the R_{sh} instead of the bulk resistivity is displayed.

3.1.4. Influence of Solids Deposit and Construction on Fabric Flexibility

The fabric stiffness in terms of conventional flexural stiffness of the plain and coated fabrics with the lowest and highest amount of solids deposition (generated from a gap height of 5 and 200 μm) is presented in Figure 6, showing average values in the warp and weft direction, as well as in the face up and down orientation. For the plain fabric, no significant difference in the flexural stiffness in the warp and weft direction or the face up or down orientation of the samples was observed. This means that fabric anisotropy because of differences in thread count did not influence the fabric stiffness. Furthermore, it shows that the addition of the MF did not impact the fabric stiffness, as the confidence intervals overlapped for the Cu and Cu$_{MF}$ samples within the same gap span. For all samples, increased deposition lead to a substantial increase in flexural stiffness, as shown by the flexural stiffness values starting from ~0.01 mN m for plain fabric and increasing by a factor of 10 for the samples with the least deposition and almost doubling for the samples with the most deposition. The layered B_Ag200_T samples had the highest flexural stiffness values of ~0.38 mN m. Most certainly, this stiffening behaviour would decrease the drapeability and comfort, even when these types of formulations are printed instead of coated as in this study, provided that the same amount of deposition material is used.

For this study, it was interesting to differentiate between the faces of the fabric, especially since fabric flexibility leads to perturbations to the electrical potential within a conductor. Flexibility in the context of conductors is a counterproductive property, where higher flexural stiffness will reduce the dimensional changes upon bending, having less impact on the conductive network. This is to reduce perturbations within the conductor, whereas lower flexural stiffness is preferred to enhance the drape and comfort for the wearer. Therefore, for closer examination, the flexural stiffness of the fabric samples in both the face up and face down orientations are plotted in Figure 6. A significant difference depending on fabric face orientation was observed for samples with a larger amount of coating (except for Ag_200), and the differences further increased for samples with a layered construction. The higher values of the face up orientation reflect the enhanced resistance to bending and are relative to the larger amount of applied material, forming a stiffer fabric surface. Lower values are seen for fabrics in the face down orientation, and with an increased amount of applied material, it is evident that these samples bend more easily under their own weight, resulting in a lower stiffness. This trend applies when the majority of the coating stays on the fabric surface and does not severely penetrate the fabric construction, as in the case of Ag_200, of which the flexural stiffness is equal regardless of whether the coated fabric is face up or down. The reason for this is likely that the polymer diffused inside the fabric construction, resulting in the mechanical interlocking of the woven yarn construction because of the penetration of the applied viscous formulation. For the Ag_200 sample and in the face down orientation, the resistance to bending was even slightly larger than that of the layered B_Ag200_T sample, despite the fact that the latter had an increased amount of applied material.

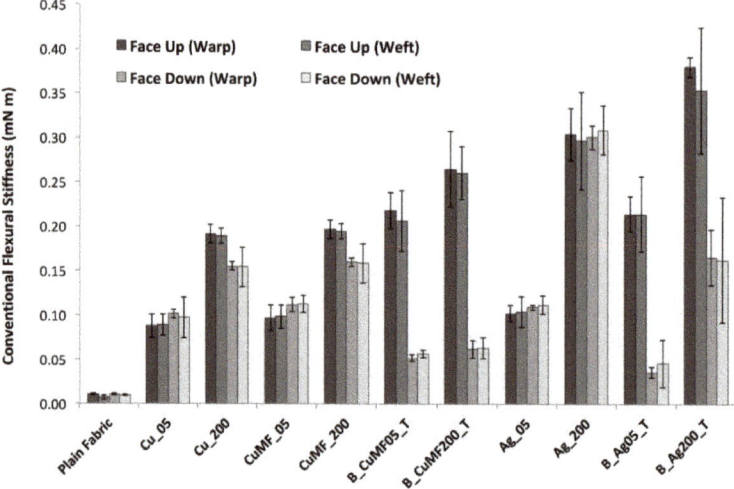

Figure 6. Conventional flexural stiffness of plain and coated fabrics with the lowest and highest amount of deposited solids (received from gap heights of 5 and 200 m) in the warp and weft direction and whether the sample is in the face up or down orientation. Individual standard deviations were used to calculate the intervals, and the plot displays a 95% CI for the mean.

3.2. Influence of Durability Testing on the Morphology and Electrical Properties

3.2.1. Influence of Washing on the Morphology

In Figure 7, the SEM images show the surface appearance of plain Cu- and Ag-coated samples after five washing cycles. These samples were produced with the lowest (5 μm) and the highest (200 μm) coating gaps and showed major differences in appearance. The encapsulated samples did not visually show any significant differences before or after the washing procedure and are therefore not included in this section. In general, the washing procedure affected the surface appearance of the plain-coated fabrics to a large extent. The Cu_05 samples with the lowest deposition displayed a film topography with a larger degree of roughness compared to the untreated sample (for comparison, see Figure 4). The surface showed an embossed character where the boundaries between the filament yarns are constituted. Apparently, this is a combined effect from the harsh mechanical treatment in the drum and the chemical and wetting processes during the washing cycles. A similar topography was seen for samples with the lowest deposition and containing the cross-linker; thus, the addition of a MF did not visually influence the conductor resistance to washing. However, with increased deposition, as seen for washed Cu_200, the topography was unaffected and was similar to the untreated sample. One can argue that thicker conductive coatings with an increased amount of deposited solids and a smoother topography generally withstands the mechanical abrasions in the drum to a larger extent.

The SEM images of the Ag-coated topography show clear visible cracks for both Ag_05 and Ag_200. Increased amounts of Ag formulation did not improve the resistance to washing.

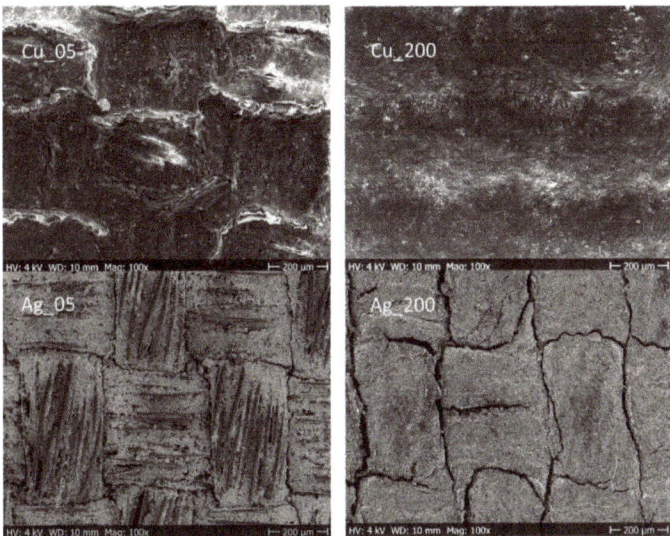

Figure 7. Top-view SEM images showing the microstructures of the coated samples after five washing cycles.

3.2.2. Influence of Washing on the Electrical Resistance

The influence of washing on the R_{sh} for the plain samples is seen in Figure 8. The R_{sh} value for both the Cu and Cu$_{MF}$ conductor layers (see Figure 8a) increased with the number of washing cycles, and after five cycles, these layers resulted in values below 6 and 3.3 Ω/sq, respectively. However, the Cu$_{MF}$ samples were shown to be less affected by washing, and because of the similarities in terms of the amount of deposited solids and the penetration of the formulation in Cu- and Cu$_{MF}$ samples (see Table 1 and the morphology section), this deviation could be related to the different network structures. The Cu$_{MF}$ matrix likely has a larger degree of polymer chain interdiffusion or even chemically cross-linked polymers than the Cu matrix without the addition of a MF. Nevertheless, the infrared camera images in Figure 8a reveal the formation of hotspots within the conductor, even though the resistance of the conductor to washing in terms of the R_{sh} was reasonably good and the microstructure was seemingly intact (based on the SEM images in Figure 7). Most likely, the mechanical stress exerted on the coated textile during washing lead to reduced contact between flakes, resulting in capacitive junctions within the matrix, where a constriction of the charges causes substantial heating and the possible risk of conductor malfunction.

For the plain Ag layers applied using coating gaps from 5 to 50 µm, there was a sharp increase in the R_{sh} with increased washing cycles from one to five cycles, see Figure 8b. This fragile behaviour is clearly the result of cracking of the layers after the washing treatment, as displayed in the SEM images in Figure 7. However, the Ag_200 samples applied with the largest amount, maintained values below 5 Ω/sq after five washing cycles, despite the appearance of visible cracks in the microstructures of these samples. This means that the Ag flake network was maintained at several connection points within the polymer matrix closer to the fabric. Still, the application of increased amounts of Ag formulation obviously did not improve the resistance of the microstructure to washing, which supports the idea that this type of Ag formulation needs to be protected to obtain a stable and coherent conductor, at least when using the coating procedure followed herein.

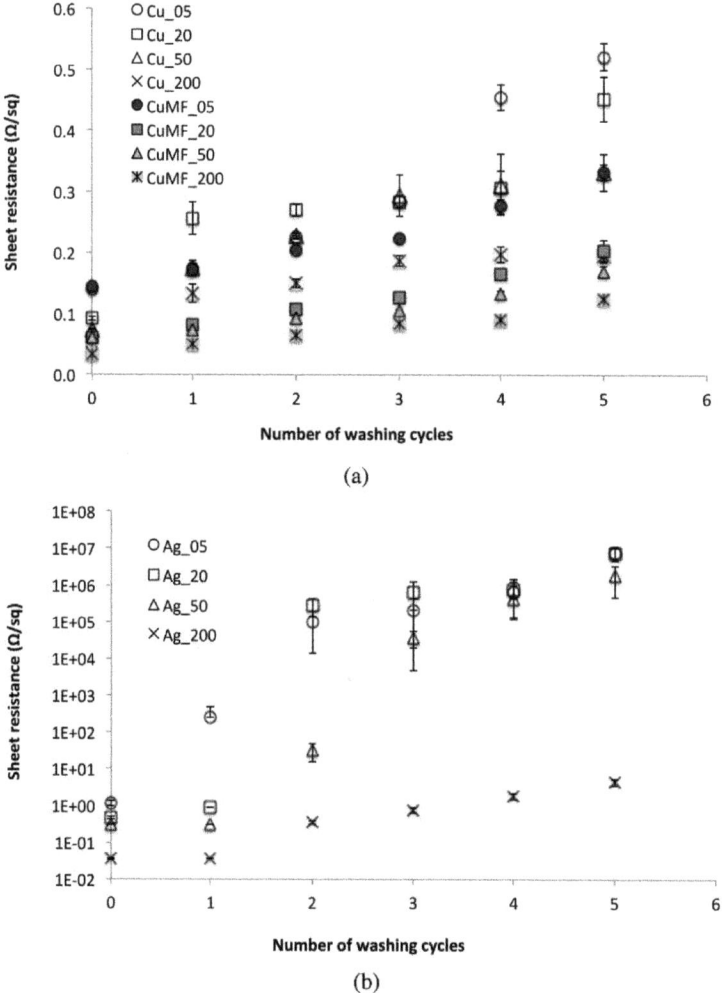

Figure 8. Sheet resistance of (**a**) the plain Cu and Cu$_{MF}$ formulation and (**b**) the Ag-coated samples before and after five washing cycles. Error bars represent 95% CI for the mean, and individual standard deviations were used to calculate the intervals.

3.2.3. Encapsulation and Drying in between Washing Cycles.

Figure 9a shows the R$_{sh}$ of the encapsulated samples before and after five washing cycles using two different drying procedures: line-drying at 24 °C and a RH of 54% and at 50 °C in a heat cabinet. For the B_Cu$_{MF}$_T sample series, a significant decrease in resistance by more than 90% was noticed when increasing the drying temperature from 24 to 50 °C. For the B_Ag_T sample series, the resistance decreased up to 80% regardless of the increasing deposition amount. Possibly, the encapsulation or the interface layer material absorbs water during washing, and the line-drying procedure at 24 °C over 24 h is insufficient. Retained water could lead to dimensional changes and induce stresses in the conductor layer, thereby increasing the resistance, whereas increased drying temperatures (50 °C) promote water evaporation, minimising residual stresses within the conductor and leading to decreased resistance values.

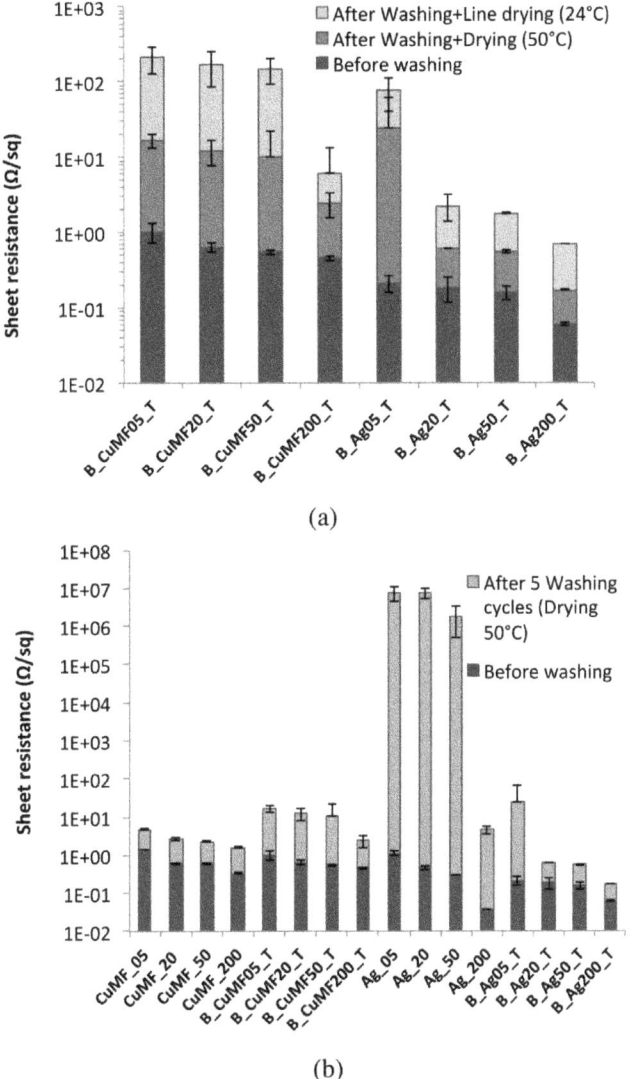

Figure 9. (a) Sheet resistance before and after different drying procedures in between washing cycles for the encapsulated samples and (b) the sheet resistance before and after five washing cycles with a 50 °C drying step for samples with and without encapsulation.

The protective performance of conductor encapsulation was evaluated, and the R_{sh} of samples with and without encapsulation before and after five washing cycles are compiled in Figure 9b. When comparing the Cu_{MF} and $B\_Cu_{MF}\_T$ sample series, there was a significant increase in R_{sh} after washing regardless of the deposited amount of formulation. The samples with the lowest amount of conductor (CU_{MF} 05 and $B\_Cu_{MF}05\_T$) showed the largest difference, with values increasing from 3 to 16 Ω/sq. The difference decreased with an increased amount of formulation, and for CU_{MF} 200 and $B\_Cu_{MF}200\_T$, the difference was only approximately 1 Ω/sq. Despite that encapsulation of the conductors resulted in increased R_{sh} after washing, it is important to show that encapsulation provided protection against mechanical damage of the conductor, as shown in the IR camera images inset in

Figure 9b. Compared to the washed Cu_{MF}_200 sample, which showed uneven heat distribution because of hotspots within the conductor, the B_ $Cu_{MF}200\_T$ sample showed an even heat distribution with no signs of hotspots in the encapsulated conductor. When comparing plain and encapsulated Ag-coated samples, it is clear that encapsulation of the conductor is required for protection during washing, as the results show a decrease in R_{sh} of more than five orders of magnitude for the encapsulated washed samples.

3.2.4. Influence of Abrasion on the Electrical Resistance

The effect of abrasion on the R_{sh} for samples with different formulations and layered constructions is shown in Figure 10. A trend in increasing R_{sh} with an increasing number of abrasion cycles is observed for all samples. For most samples, the total increase in R_{sh} after 3200 abrasion cycles was between two to three orders of magnitude, except for the silver-coated samples with a larger amount of coating (Ag_200 and B_Ag200_T), where the increase was below one order of magnitude, maintaining values below 0.4 Ω/sq. For the plain-coated Cu and Cu_{MF} samples, the R_{sh} values did not significantly differ with increased abrasion cycles depending on the added cross-linker. This means that neither the physically dried nor the possibly cross-linked conductor was sufficiently stable to withstand distortion from repetitive abrasion cycles in a meaningful way. Compared to the plain-coated Cu and Cu_{MF} sample series, the encapsulated samples showed a more gradual increase in R_{sh} with increased amount of conductor, depending on the gap height (5, 20, 50 and 200 µm) and the number of abrasion cycles. Furthermore, the total increase in R_{sh} was approximately 1.2 orders of magnitude, at least for the B_Cu_{MF}_T samples with an increased gap between 20–200 µm compared to the plain-coated samples within the same sample series for which the total increase in R_{sh} was closer to two orders of magnitude. Thus, the encapsulated conductor in the Cu_{MF} formulation with a solids deposition from 67 to 166 g/m^2 (see Table 1) withstands abrasion cycles to a larger extent compared to the plain-coated conductors within similar ranges of solids deposition.

A similar trend and a more gradual increase in R_{sh} with an increased amount of solids and number of abrasion cycles were observed for the encapsulated B_Ag_T samples. Compared to the plain-coated Ag formulation, the encapsulated Ag formulation also maintained R_{sh} values below 10 Ω/sq even after 800 abrasion cycles. Possibly, the encapsulation protects the conductor from abrasion up to a point with a solid deposition of approximately 80 g/cm^2. However, with increased amounts over 150 g/cm^2, there was no significant difference because of encapsulation, as the plain-coated and encapsulated samples with the largest amount of applied material showed similar behaviour. For both samples, the total increase in R_{sh} with increased abrasions was less than one order of magnitude, and the values were below 0.4 Ω/sq.

For the plain-coated samples, no significant weight loss was detected after abrasion, although there was a visual fading and strengthening of the metallic gloss for the Cu- and Ag-coated samples, respectively, and the abrading cloth was slightly coloured. Therefore, the increased R_{sh} with increased abrasion cycles can only be partly explained by the loss of some of the polymer matrix and conductive material. Evidently, other reasons must also underlie the increasing R_{sh} trends, as displayed for the encapsulated samples as well. The results support that a balance must be found between the amount of applied formulation and the encapsulation to improve the resistance to abrasion.

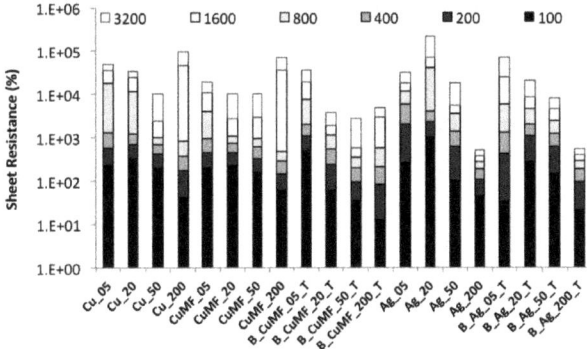

Figure 10. Sheet resistance before and after 100 to 3200 abrasion cycles for the plain and encapsulated conductors.

4. Conclusions

The obtained results confirm that both the addition of a cross-linking additive and encapsulation improved the durability of metallic conductive coatings and that increasing the amount of applied conductive material can reduce the R_{sh}. An improvement in the durability and electrical conductivity was achieved by increasing the amount of deposited solids or by conductor encapsulation, but this was at the expense of increasing the flexural stiffness of the coated fabric. The flexural stiffness of the coated fabric increased by a factor of ten for samples with approximately 60 g/m^2 of deposition, while it redoubled for samples produced with 170 g/m^2 deposition. The flexural stiffness was also higher for samples in which the conductive polymer formulation diffused deeper into the fabric, interlocking the woven yarn construction, and for tested samples with the conductor fabric in the face up orientation. With a more rigid conductor, regardless of the coating method used, a less drapable coated fabric and perhaps also a less comfortable e-textile is obtained. More importantly, a more rigid conductor can be related to enhanced resistance to bending and prevention of dimensional changes, which can deteriorate the conductive network. The results showed that the largest amount of deposition and conductor encapsulation significantly improved the resistance to electrical fatigue after durability testing, which confirms that not only a trade-off between stiffness and flexibility of conductors must be considered but also that for certain conductive coatings, a high degree of stiffness is the only alternative to obtain durable electrical textile coatings. Note that the specific impact on comfort remains unknown and requires future studies targeting specific applications of e-textiles and wearable sensing.

The results indicate that the addition of a cross-linking additive improved the electrical fatigue properties because of the formation of a stronger polymeric network. This was confirmed after excluding possible impacting factors such as viscosity, deposition amount and surface roughness. However, infrared camera images revealed hotspots within all the washed plain-coated conductor matrixes, even those with preserved low R_{sh} and without microstructural defects, as seen in the SEM images. The phenomenon was not seen for encapsulated conductors that had a more homogeneous morphology, because the conductive formulation was hindered from penetrating cavities between the yarn filaments. Given that trapped water within the bare conductor matrix can be excluded, the shearing forces during durability testing are suggested to have a larger impact on the bare conductive matrix than the encapsulated conductors, which were interlocked within the yarn construction, introducing small ruptures within the matrix and between the conductive flakes and leading to capacitor-type junctions, which at worst, can cause the conductor to malfunction.

This paper confirms that conductive formulations applied to flexible substrates, such as textiles, must be extensively studied, specifically using durability tests, to ensure that the applied conductors are properly constructed with high and reliable conductive properties for long-term performance for e-textile applications.

Author Contributions: V.N., F.S. and V.M. conceptualised this project; V.N. and F.S. supervised this project; V.M. carried out this research and wrote the original draft. Writing—review and editing was conducted by V.M. and F.S.

Funding: The work was supported by the R&D Board at the University of Borås.

Conflicts of Interest: The authors declare no conflict of interest.

Appendix A

In Figure A1a–e current–voltage (IV) curves for all samples are shown. The data plotted are obtained from four points probe (4PP) measurements on finite samples and on three replicates. They all show a linear curve. The IV plotted data in all figures show positive slopes for conductive samples ohmic linearity.

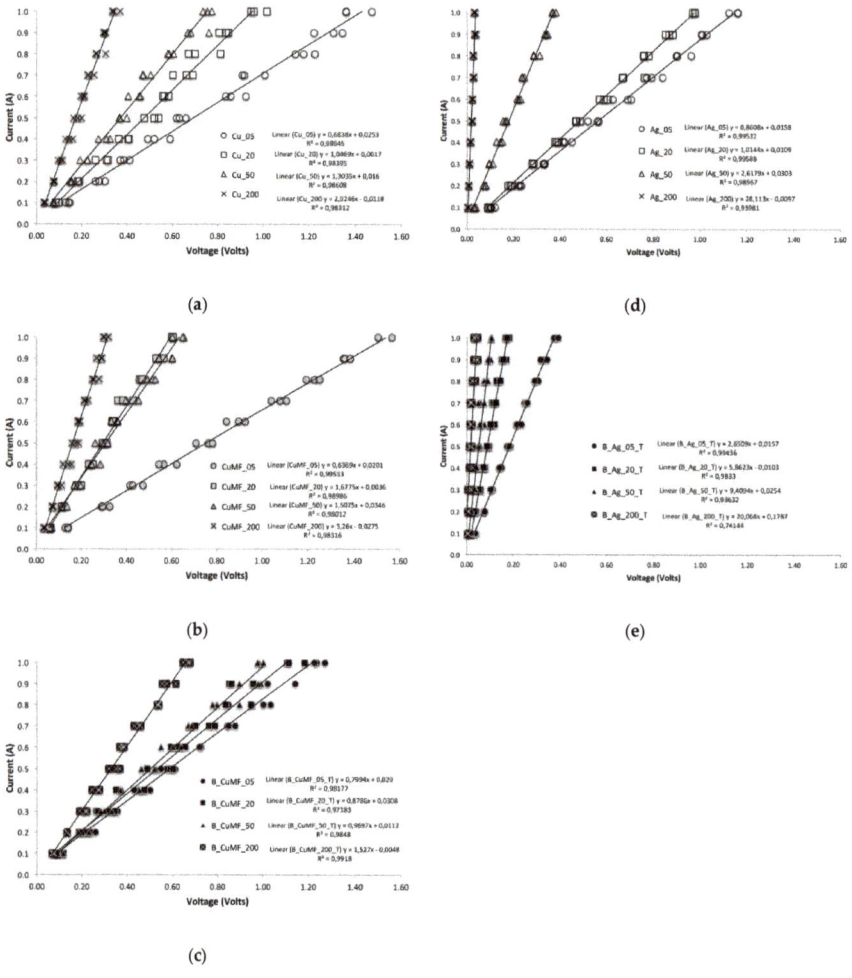

Figure A1. Current versus voltage curves (**a**) for sample series: Cu_05, Cu_20, Cu_50, and Cu_200; (**b**) for samples series: Cu$_{MF}$_05, Cu$_{MF}$_20, Cu$_{MF}$_50, and Cu$_{MF}$_200; (**c**) for samples series: B_Cu$_{MF}$05_T, B_Cu$_{MF}$20_T, B_Cu$_{MF}$50_T and B_Cu$_{MF}$200_T; (**d**) for samples series: Ag_05, Ag_20, Ag_50, and Ag_200; and (**e**) for sample series: B_Ag05_T, B_Ag20_T, B_Ag50_T, and B_Ag200_T.

References

1. Jin, H.; Nayeem, M.O.G.; Lee, S.; Matsuhisa, N.; Inoue, D.; Yokota, T.; Hashizume, D.; Someya, T. Highly Durable Nanofiber-Reinforced Elastic Conductors for Skin-Tight Electronic Textiles. *ACS Nano* **2019**, *13*, 7905–7912. [CrossRef] [PubMed]
2. Stoppa, M.; Chiolerio, A. Wearable Electronics and Smart Textiles: A Critical Review. *Sensors* **2014**, *14*, 11957–11992. [CrossRef] [PubMed]
3. Topp, K.; Haase, H.; Degen, C.; Illing, G.; Mahltig, B. Coatings with metallic effect pigments for antimicrobial and conductive coating of textiles with electromagnetic shielding properties. *J. Coat. Technol. Res.* **2014**, *11*, 943–957. [CrossRef]
4. Kasap, S.; Koughia, C.; Ruda, H.E. Electrical Conduction in Metals and Semiconductors. In *Springer Handbook of Electronic and Photonic Materials*; Kasap, S., Capper, P., Eds.; Springer International Publishing: Cham, Switzerland, 2017; pp. 19–28. [CrossRef]
5. Villanueva, R.; Ganta, D.; Guzman, C. Mechanical, in-situ electrical and thermal properties of wearable conductive textile yarn coated with polypyrrole/carbon black composite. *Mater. Res. Express* **2018**, *6*, 016307. [CrossRef]
6. Yapici, M.K.; Alkhidir, T.; Samad, Y.A.; Liao, K. Graphene-clad textile electrodes for electrocardiogram monitoring. *Sens. Actuators B Chem.* **2015**, *221*, 1469–1474. [CrossRef]
7. Paul, G.; Torah, R.; Beeby, S.; Tudor, J. The development of screen printed conductive networks on textiles for biopotential monitoring applications. *Sens. Actuators A Phys.* **2014**, *206*, 35–41. [CrossRef]
8. Mondal, K. Recent Advances in Soft E-Textiles. *Inventions* **2018**, *3*, 23. [CrossRef]
9. Kazani, I.; Hertleer, C.; De Mey, G.; Guxho, G.; Van Langenhove, L. Dry cleaning of electroconductive layers screen printed on flexible substrates. *Text. Res. J.* **2013**, *83*, 1541–1548. [CrossRef]
10. Merilampi, S.; Laine-Ma, T.; Ruuskanen, P. The characterization of electrically conductive silver ink patterns on flexible substrates. *Microelectron. Reliab.* **2009**, *49*, 782–790. [CrossRef]
11. Yang, K.; Torah, R.; Wei, Y.; Beeby, S.; Tudor, J. Waterproof and durable screen printed silver conductive tracks on textiles. *Text. Res. J.* **2013**, *83*, 2023–2031. [CrossRef]
12. Gordon Paul, R.T.; Kai, Y.; Steve, B.; John, T. An investigation into the durability of screen-printed conductive tracks on textiles. *Meas. Sci. Technol.* **2014**, *25*, 11.
13. Matsuhisa, N.; Kaltenbrunner, M.; Yokota, T.; Jinno, H.; Kuribara, K.; Sekitani, T.; Someya, T. Printable elastic conductors with a high conductivity for electronic textile applications. *Nat. Commun.* **2015**, *6*, 7461. [CrossRef] [PubMed]
14. Komolafe, A.O.; Torah, R.N.; Yang, K.; Tudor, J.; Beeby, S.P. Durability of screen printed electrical interconnections on woven textiles. In Proceedings of the 2015 IEEE 65th Electronic Components and Technology Conference (ECTC), San Diego, CA, USA, 26–29 May 2015; pp. 1142–1147.
15. Wagener, S.; Dommershausen, N.; Jungnickel, H.; Laux, P.; Mitrano, D.; Nowack, B.; Schneider, G.; Luch, A. Textile Functionalization and Its Effects on the Release of Silver Nanoparticles into Artificial Sweat. *Environ. Sci. Technol.* **2016**, *50*, 5927–5934. [CrossRef] [PubMed]
16. Limpiteeprakan, P.; Babel, S.; Lohwacharin, J.; Takizawa, S. Release of silver nanoparticles from fabrics during the course of sequential washing. *Environ. Sci. Pollut. Res.* **2016**, *23*, 22810–22818. [CrossRef] [PubMed]
17. Mitrano, D.M.; Rimmele, E.; Wichser, A.; Erni, R.; Height, M.; Nowack, B. Presence of nanoparticles in wash water from conventional silver and nano-silver textiles. *ACS Nano* **2014**, *8*, 7208–7219. [CrossRef] [PubMed]
18. Gliga, A.R.; Skoglund, S.; Odnevall Wallinder, I.; Fadeel, B.; Karlsson, H.L. Size-dependent cytotoxicity of silver nanoparticles in human lung cells: The role of cellular uptake, agglomeration and Ag release. *Part. Fibre Toxicol.* **2014**, *11*, 11. [CrossRef] [PubMed]
19. Arvidsson, R.; Molander, S.; Sandén, B.A. Assessing the Environmental Risks of Silver from Clothes in an Urban Area. *Hum. Ecol. Risk Assess. Int. J.* **2014**, *20*, 1008–1022. [CrossRef]
20. Gunda, M.; Kumar, P.; Katiyar, M. Review of Mechanical Characterization Techniques for Thin Films Used in Flexible Electronics. *Crit. Rev. Solid State Mater. Sci.* **2017**, *42*, 129–152. [CrossRef]

21. Winnik, M.A.; Pinenq, P.; Krüger, C.; Zhang, J.; Yaneff, P.V. Crosslinking vs. interdiffusion rates in melamine-formaldehyde cured latex coatings: A model for waterborne automotive basecoat. *J. Coat. Technol.* **1999**, *71*, 47–60. [CrossRef]
22. Åkerfeldt, M.; Strååt, M.; Walkenström, P. Influence of coating parameters on textile and electrical properties of a poly(3,4-ethylene dioxythiophene): Poly(styrene sulfonate)/polyurethane-coated textile. *Text. Res. J.* **2013**, *83*, 2164–2176. [CrossRef]

© 2019 by the authors. Licensee MDPI, Basel, Switzerland. This article is an open access article distributed under the terms and conditions of the Creative Commons Attribution (CC BY) license (http://creativecommons.org/licenses/by/4.0/).

Article
A Novel Textile Stitch-Based Strain Sensor for Wearable End Users

Orathai Tangsirinaruenart and George Stylios *

Research Institute for Flexible Materials, Heriot Watt University, Edinburgh EH14 4AS, UK; ot32@hw.ac.uk
* Correspondence: g.stylios@hw.ac.uk

Received: 11 February 2019; Accepted: 22 April 2019; Published: 7 May 2019

Abstract: This research presents an investigation of novel textile-based strain sensors and evaluates their performance. The electrical resistance and mechanical properties of seven different textile sensors were measured. The sensors are made up of a conductive thread, composed of silver plated nylon 117/17 2-ply, 33 tex and 234/34 4-ply, 92 tex and formed in different stitch structures (304, 406, 506, 605), and sewn directly onto a knit fabric substrate (4.44 tex/2 ply, with 2.22, 4.44 and 7.78 tex spandex and 7.78 tex/2 ply, with 2.22 and 4.44 tex spandex). Analysis of the effects of elongation with respect to resistance indicated the ideal configuration for electrical properties, especially electrical sensitivity and repeatability. The optimum linear working range of the sensor with minimal hysteresis was found, and the sensor's gauge factor indicated that the sensitivity of the sensor varied significantly with repeating cycles. The electrical resistance of the various stitch structures changed significantly, while the amount of drift remained negligible. Stitch 304 2-ply was found to be the most suitable for strain movement. This sensor has a wide working range, well past 50%, and linearity (R^2 is 0.984), low hysteresis (6.25% ΔR), good gauge factor (1.61), and baseline resistance (125 Ω), as well as good repeatability (drift in R^2 is −0.0073). The stitch-based sensor developed in this research is expected to find applications in garments as wearables for physiological wellbeing monitoring such as body movement, heart monitoring, and limb articulation measurement.

Keywords: textile-based stretch sensors; stitch structure; wearable stretch sensor; conductive thread

1. Introduction

In the last decade there has been an increasing interest for developing different types of wearable sensors. There are many sensor types that have shown potential as wearable sensors amongst them piezoresistive films which show good change in resistance by simple changes to their geometry and observed microcracking contributing to high gauge factors [1,2]; although their durability and strechability are areas of concern for further development. Capacitive sensors are another common type seeing in touch screens because of their good sensitivity, low energy and adaptability [3]. They have however mainly being used for pressure because they suffer from environmental noise and hence difficult for use in wearable applications. Textile-based sensors are desirable for wearable end users [4,5] because they are comfortable, flexible, and not obstructive to the wearer's everyday activities. A textile sensor can be designed and presented in numerous types and forms. One particularly interesting type is as a strain resistance sensor, which is achieved by the alteration of the mechanical properties of the material under stress/strain deformation, whilst its flexibility allow care of wrapping of the body of the wearer. Textile strain sensors can detect stretch, displacement, and force resulting from large movement of joint bending [6] or small body movements such as breathing [7]. Many studies have investigated theoretical and practical relationships between the electrical resistance and elongation of conductive fabrics [8–10]. Thomson and Kelvin [11] found that the resistance changes of a conductor were affected by stretching. Zhang et al. [12,13] and Li et al. [14] showed that the electrical resistance occurring between overlapping knitted loops is a major factor contributing to the overall resistance. Likewise, Ehrmann et al. [15] found that different yarns, stitch dimensions, and different fabric directions affect the elongation and time-dependent resistance behavior of the sensor.

There are many changes in properties when a knitted fabric-based stretch sensor is extended. Previous researches [16–18] have shown that one interesting effect of stretching a fabric is that it increases its electrical resistance along the conductive thread. This is caused by opening up the stitch and so breaking the parallel contact points, making it necessary for the current to travel in series rather than in parallel. This increase in conductive path leads to greater resistance [17,19,20] as shown in Figure 1.

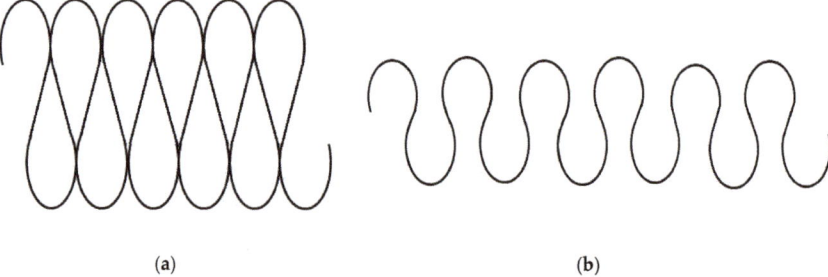

Figure 1. Conductive path knitted fabric model (**a**) in the relaxed and (**b**) stretched position [1,15,20].

Another effect is due to the stretching of the conductive yarn itself [20]. It is well known that when a conductive material is stretched its cross-section is reduced, while its conductive path is increased, leading to greater resistance. The relative dominance of each effect is dependent on the stitch type, the strain and the type of conductive yarn.

In this study, our objective is to explore the prospect of developing a flexible wearable textile strain sensor. The optimum design of such sensor is investigated in different types of stitch structures, having as criteria the ability to measure strain deformations and electromechanical properties. Hence, a study was under taken in which conductive threads were sewn under different configurations, directly onto a knitted fabric and their characteristics investigated.

2. Experimental Materials

In this study, our primary aim is to find a particular stitch assembly that has a good stretch ability as well as electrical performance. The stitch assembly is formed by using conductive thread and is constructed using stretchable single jersey fabrics which allow the stitches to be flexible, unobtrusive, and achieve linearity. The preliminary studies of fabric suitability were undertaken to find the ideal base fabric. Six different fabrics were investigated under the criteria of having high elastic recovery, Table 1. A single jersey of Nylon 4.44 tex/2-ply with Spandex 7.78 tex and weight of 260 g/m^2 was found ideal for use as the fabric substrate base sensor, because it has the highest elastic recovery at 93%, as shown in Figure 2.

Table 1. Specification of single jerseys fabrics.

Nylon Yarn Count (Tex)	Sample	Spandex Yarn Count (Tex)	Content Nylon/Spandex (%)	Weight g/m²	Thickness mm	Yarn Density per cm	
						Courses	Wales
4.44/2-ply	A	2.22	91/9	212	0.60	24	34
	B	4.44	88/12	223	0.60	23	33
	C	7.78	75/25	260	0.55	20	39
7.78/2-ply	D	2.22	94/6	286	0.65	19	33
	E	4.44	93/7	313	0.65	19	36
	F	7.78	83/17	323	0.60	18	33

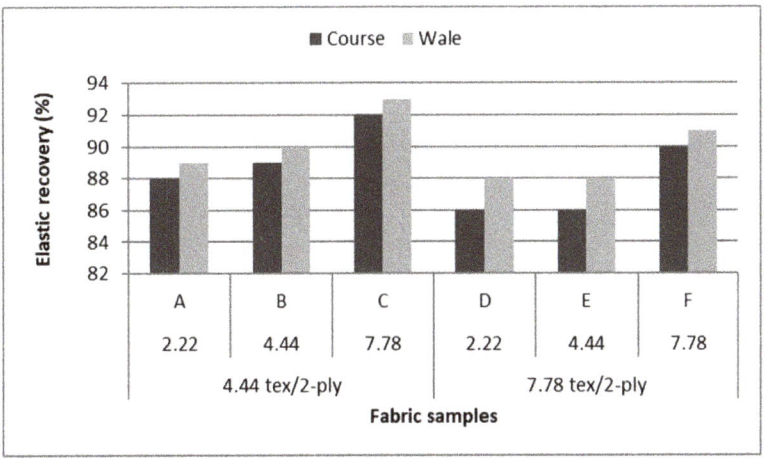

Figure 2. Elastic recovery values of six different nylon/spandex fabrics in course-wise and wale-wise direction.

Trials with different stitching configurations followed. Seven samples 50 × 250 mm² were made by stitching along the wale direction of the fabric. Samples with four stitch types—304 (Zigzag stitch), 406 (2-needle multithread chain stitch: rear side), 506 (4 threads overlock stitch), and 605 (3-needles covering chain stitch)—were constructed. Two types of conductive threads used were silver plated nylon 117/17 2-ply and 235/34 4-ply, Statex Productions & Vertriebs GmbH, Bremen, Germany. Their electrical specification is given in Table 2, and their properties in Table 3. Both threads are commercial, and they have also been used by other researchers [16,17,21–24].

Table 2. Specification of conductive threads.

Conductive Thread	Thread Size (Tex)	Linear Resistance (Ω)
Silver plated nylon 117/17 2-ply	33	500 Ω/meter
Silver plated nylon 235/34 4-ply	92	50 Ω/meter

Table 3. Properties of conductive threads.

Conductive Thread	Maximum Load (N)	Energy at Break (N)	Extension at Break (mm)	Elongation (%)	Tenacity (N/Tex)
Silver plated nylon 117/17 2-ply	10.78	10.77	65.25	26.1	0.4
Silver plated nylon 234/34 4-ply	46.27	45.88	113.17	45.27	0.365

These conductive sewing threads are lightweight, flexible, soft, durable, strong, and do not suffer from permanent distortion after being bent. Therefore they are ideal for machine sewing for typical garment operations. The electrical resistance of the two threads used are given in Figure 3. After imposing the threads to five-cycle tensile testing shown in Figure 4, the resistance response of both threads increases as they are stretched (loading) and decreases upon relaxation (unloading). This good overall electrical behavior renders them suitable for wearable end-uses. The results also show that the threads have a small delay in becoming fully relaxed.

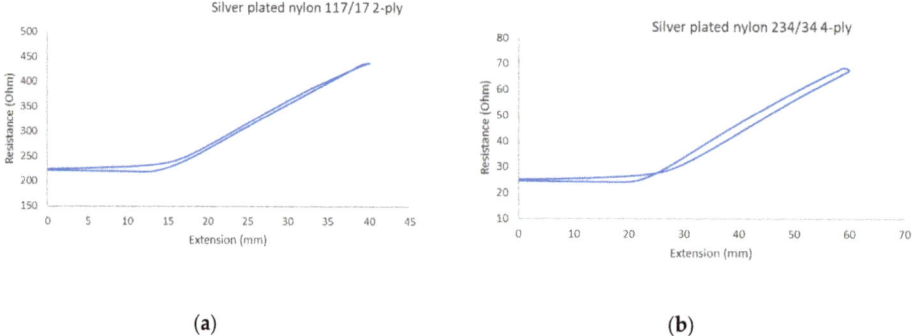

Figure 3. Resistance against extension and resistance against time in different type of silver plated nylon conductive thread: (**a**) 117/17 2-ply and (**b**) 234/34 4-ply.

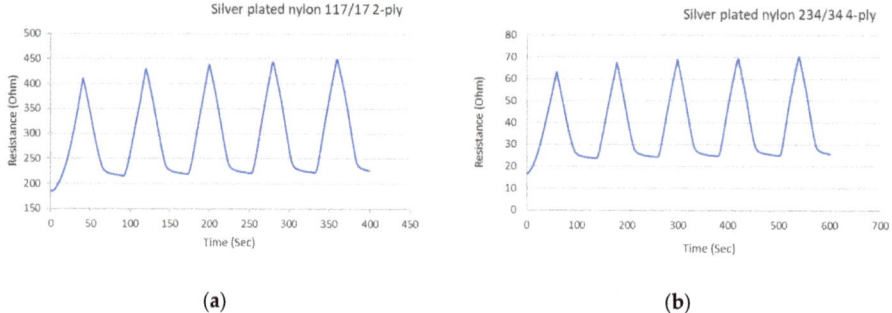

Figure 4. Sewing thread tensile performance during 5 cycles. (**a**) 117/17 2-ply; (**b**) 234/34 4-ply.

For testing consistency, three of the four stitch types have the same characteristics (5 stitches per cm and 7 mm wide). However, stitch type 304 by virtue of its design needs to have different dimensions: 11 stitches per cm and 3 mm wide, as shown in Table 4 where the conductive thread is represented by the black line in the stitch structure.

Table 4. The experimental stitch structure sensors.

Stitch Type	Stitch Structure	Stitch per 10 mm	Stitch Width (mm)	Close-Up of the Stitched Samples
304		11	3	

Table 4. *Cont.*

Stitch Type	Stitch Structure	Stitch per 10 mm	Stitch Width (mm)	Close-Up of the Stitched Samples
406		5	7	
506		5	7	
605		5	7	

2.1. Experimental Methods

2.1.1. Measured Properties and Characteristics

In order to investigate the suitability of the different stitch types, various properties and characteristics of each stitch were measured, which will briefly be defined here with reference to Figure 5.

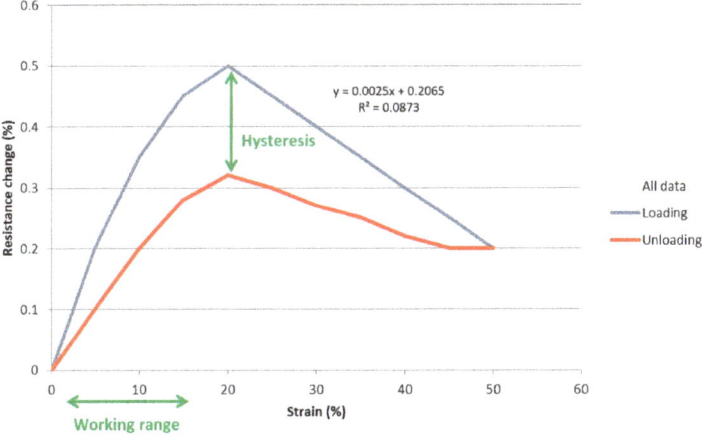

Figure 5. A typical graph example of resistance change (%) vs. strain (%), for defining sensor characteristics.

Working Range

The working range of a sensor determines its end-use suitability. A sensor with a wider working range will be more adaptable to a wider variety of uses. The working range is measured as a percentage of strain, that being the change in length divided by (the unstretched original) length. A sensor which has a working range starting from its unstretched condition (rest position, i.e., zero) will be more useful than a sensor with a working range between two stretched states. It is clear that the sensor in Figure 5 has a working range from 0 to 15%, but above that it becomes unsuitable due to nonlinearity and nonmonotonicity.

Monotonicity

A sensor will be more useful if it is monotonic. This means that the resistance changes in a constant incremental value upward or downward as the strain increases or decreases, respectively. The resistance value of the sensor in Figure 5 does not alter in its constant increment; instead it is reaching to a peak at 20% strain and then falling immediately after. This means that, when in use, it would be difficult to determine if the sensor is giving a measurement of resistance changes of say 0.25% because its strain can be in the 8 to 12% or in 35 to 45% region. So we say that this sensor has no monotonicity.

Gauge Factor

The sensitivity of sensor resistance changes to strain, is given by

$$Gauge\ Factor = \frac{\left(\frac{\Delta R}{R}\right)}{\left(\frac{\Delta l}{l}\right)} = \frac{\left(\frac{\Delta R}{R}\right)}{\varepsilon}$$

where R and l are the unstretched resistance and length, respectively; ΔR and Δl are the changes in resistance and length due to stretching, respectively; and ε is the strain. The gauge factor is unitless and is calculated by taking the gradient of lines of best fit on a graph of resistance change vs. strain. For example, the line of best fit in Figure 5 follows equation y = 0.0025x + 0.2065, so its gauge factor is 0.0025. The ideal sensor would have a high gauge factor (high sensitivity), so that changes in resistance are large in relation to strain changes. This would make it easy to detect a small change in strain because these would show as a large change in resistance. If the sensor shows very small changes in resistance in relation to changes in strain (i.e., a resistance vs. strain graph showing a horizontal strain line), it makes it very difficult to accurately measure small changes in strain, hence leading to poor sensor performance.

Linearity

Linearity is the proportion of change in resistance in relation to the proportion of change in strain. A perfect sensor would have a high and uniform resistance change over the whole of the strain range, so making it easy to calculate the strain from any given resistance measurement. There are various metrics for linearity. The metrics used here are visual inspection of the graphs along with assessment of the R^2 value (coefficient of determination), which indicates how well a line of best fit represents the data it is fitted to. A straight line cannot fit well to a highly curved relationship between resistance and strain, so giving low R^2 values, as shown in Figure 5. It should be noted that low R^2 values could also be caused by random data scatter or large hysteresis.

Hysteresis

The hysteresis of a sensor is the difference between the resistance at any given strain on the loading cycle and the resistance at that same strain on the unloading cycle. Ideally, the sensor would have the same resistance at that strain point regardless of whether it is loading or unloading, so that one strain value can always be measured with one resistance value, without the need to know which cycle the sensor is in, which is impractical. Figure 5 shows a high hysteresis. If a resistance change of say 0.2% was measured, it could correspond to a strain of 5% or 10%, depending on the direction of loading.

Repeatability

Drift is another characteristic, which is determined by repeating cycles, measured by any changing characteristic divided by the number of cycles. In this work, the change of the unstretched resistance, and the change in the gauge factor, between the second and 99th cycles (for reasons explained later) was taken as an indicator of repeatability.

Other Properties/Characteristics Not Investigated

There are many other factors that determine the suitability of a wearable sensor and its end-use, including response/time delay, creep, sensitivity to other types of deformations (e.g., twisting), sensitivity to environmental conditions, stiffness, ease of manufacture, comfort, and aesthetics. Although these other properties may be important they are not essential for the purpose of this investigation.

2.2. Experimental Apparatus and Procedure

Figure 6 shows the experimental apparatus and set up. The sensor samples are clamped between the jaws of an Instron 3345 Tensile Tester and the two ends of the sensor are connected to a digital millimeter (DMM Agilent U1273A/U1273AX, Agilent Technologies Inc., Santa Clara, CA, USA). Installing the Agilent GUI Data Logger software on a PC is used to record the electrical resistance response to extension and recovery cycles, at a sampling rate of approximately 1 Hz. The jaws were electrically isolated from the fabric sensor with a layer of synthetic polymer rubber, so that only the resistance of the sensor itself would be measured by the millimeter. The jaws were set 150 mm apart and 250 mm long in the sensor direction (50 mm is needed at either side of the sample for clamping).

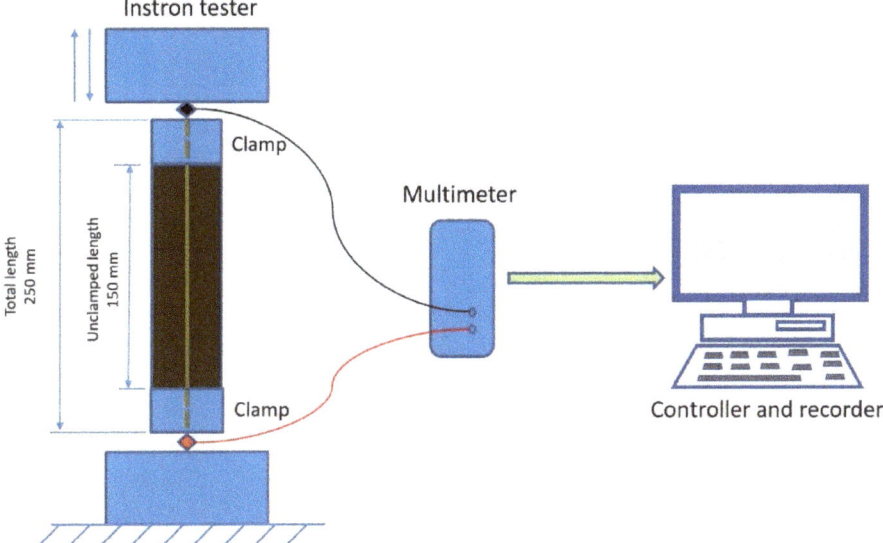

Figure 6. Experimental setup.

Data from the tensile tester and millimeter were subsequently aligned and overlapped using digital timestamps. Each cyclic test is performed at the rate of 200 mm/min and at 50% extension in 10 cycles. In order to observe the repeatability of the samples further tests of 100 cycles were also performed.

Standard atmospheric conditions were used for the experiment i.e.; temperature at 20 ± 2 °C and 65% ± 2% R.H, and all samples were allowed 24 h in the lab for conditioning, prior to testing.

2.3. Data Analysis Procedure

This section will describe and analyze the way in which the various sensor characteristics were derived from the stress/strain required data. Extension and resistance were converted into strain (%) and percentage resistance change (%), as below.

$$\varepsilon = \frac{\Delta l}{l}$$
$$\text{Resistance Change } (\%) = \frac{\Delta R}{R}$$

where Δl is the extension value from the tensometer and l is the unstretched gauge length, Figure 6 (150 mm). Calculating the percentage resistance change was more complicated because in this study we are interested in the percentage resistance change of the stretching portion of the sensor only and not including the portion within the jaws. The portion in the jaws gives a constant additional resistance, herein called "resistance bias". In the unstretched state, the sensor was assumed to have an approximately uniform resistance per mm length. Therefore the resistance of the clamped portion (the resistance bias), is given by

$$R_{bias} = (\textit{proportion of length in clamps}) \times (\text{total unstretched resistance})$$

The proportion of length in the clamps is taken from Figure 6 as 0.4 (100/250 mm), while the total unstretched resistance is different for each stitch and was taken as the resistance at time = 0 (the first data sample). The resistance bias is subtracted from all resistance data samples to give the resistance of the unclamped portion at any one time. To calculate the percentage resistance change, an original or baseline resistance is needed. Figure 7 shows a typical plot of resistance vs. time.

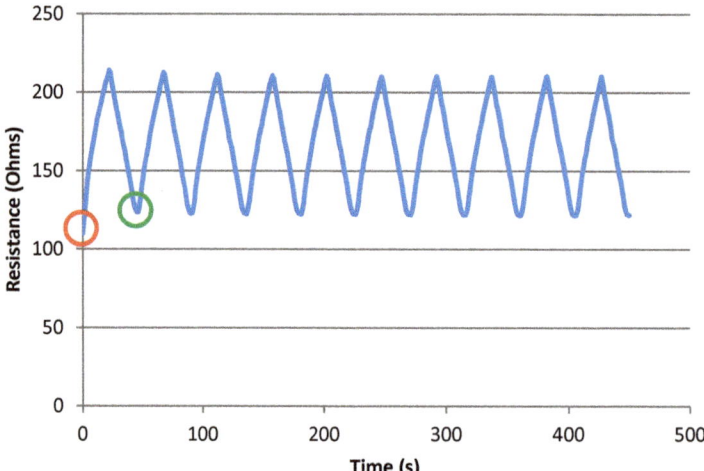

Figure 7. Example of resistance vs. time plot.

Note that the unstretched resistance in the first cycle (circled in red) is significantly lower than that of subsequent cycles, where it becomes stable. Therefore, the first cycle of each sample is not included in the data. Hence, for end use measurements it is advisable that sensors are pre-stretched in order to normalize the data, this was also observed by others [17,19]. In our experiments from now on, only data from the second cycle onwards is being used (circled in green).

Our aim is to find the values of working range and gauge factor for each stitch type. However, there is likely to be some amount of variation between cycles, therefore data for a number of cycles were laid on top of each other and averaged to find an assembly average. Ten cycles were used starting from the second cycle, as discussed. For example, the resistance at the start of the cycle was taken to be the average of the resistances at the start of cycles 2 through to 11. The second value of resistance was the average of the second sample of resistance for cycles 2 through to 11, and so on, until a complete averaged cycle for each stitch type was established. The strain data was processed similarly. An example of the percentage resistance change vs. strain for a sample average is given in Figure 8.

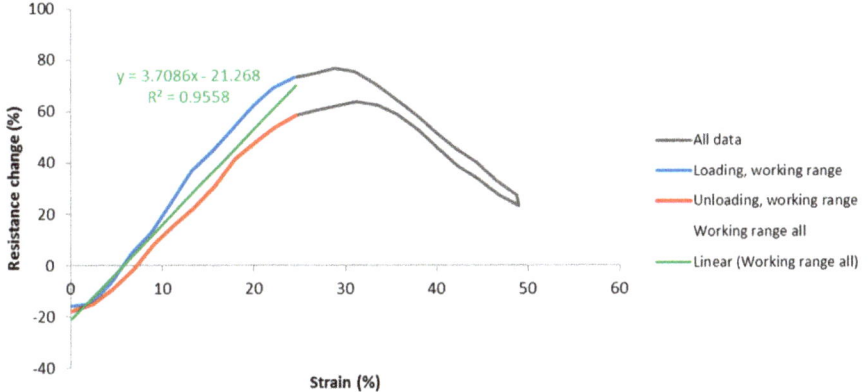

Figure 8. Example of assembly average over 10 cycles.

The grey curve is the assembly average over 10 cycles for this particular stitch, and it is reaching approximately 25% of strain beyond where the response becomes highly nonlinear and nonmonotonic. Therefore the working range of this sensor is considered to be 0 to 25% strain. This is marked on the graph by the blue and red lines, denoting loading and unloading directions respectively. A green line of best fit was then fitted to the data within the working range, revealing the gauge factor of this sensor (in this case 3.71), and a coefficient of determination (R^2, in this case 0.95). As stated, R^2 is a measure of how well the data fits the line of best fit, indicating nonlinearity, scatter, and spread due to hysteresis.

Finally, to assess repeatability, the resistance change vs. strain graphs of the 2nd and 99th cycles were compared, as shown in Figure 9. The second cycle was used for reasons explained earlier, and the 99th cycle was used because the 100th cycle showed some discontinuity due to machine stopping. This graph revealed changes in characteristics such as gauge factor, hysteresis and unstretched resistance. During experiments stitch structure 304 could only be realized with 2-ply 33 tex sewing thread as the 4-ply 92 tex thread was distorting the fabric (bouncing), due to the tensions and geometry of this particular stitch type.

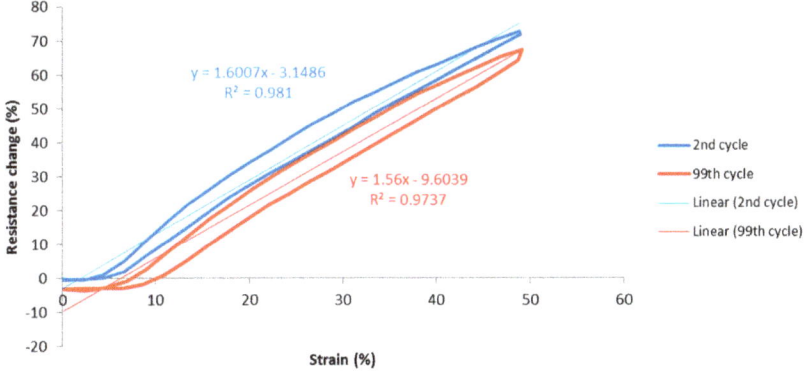

Figure 9. Example repeatability graph.

3. Experimental Results, Analysis and Discussion

To help explain early sensor characteristics, close-up photos were also taken for visual inspection of the stitch deformation. Table 5 shows close-up photographs of one typical sample of each stitch type, relaxed, and stretched, which will be referred to as the data is analyzed further.

Table 5. Close-up photographs of one sample of each stitch type, relaxed, and stretched.

Stitch Type	Conductive Thread Tex	Relaxed State	Stretched State
304	2-ply 33		
406	2-ply 33		
	4-ply 92		
506	2-ply 33		
	4-ply 92		
605	2-ply 33		
	4-ply 92		

Figure 10 shows the percentage resistance change vs. strain for the various stitch types, averaged over all samples of each stitch type, and over the 2nd to 11th cycles. Data derived from these graphs are tabulated in Table 6.

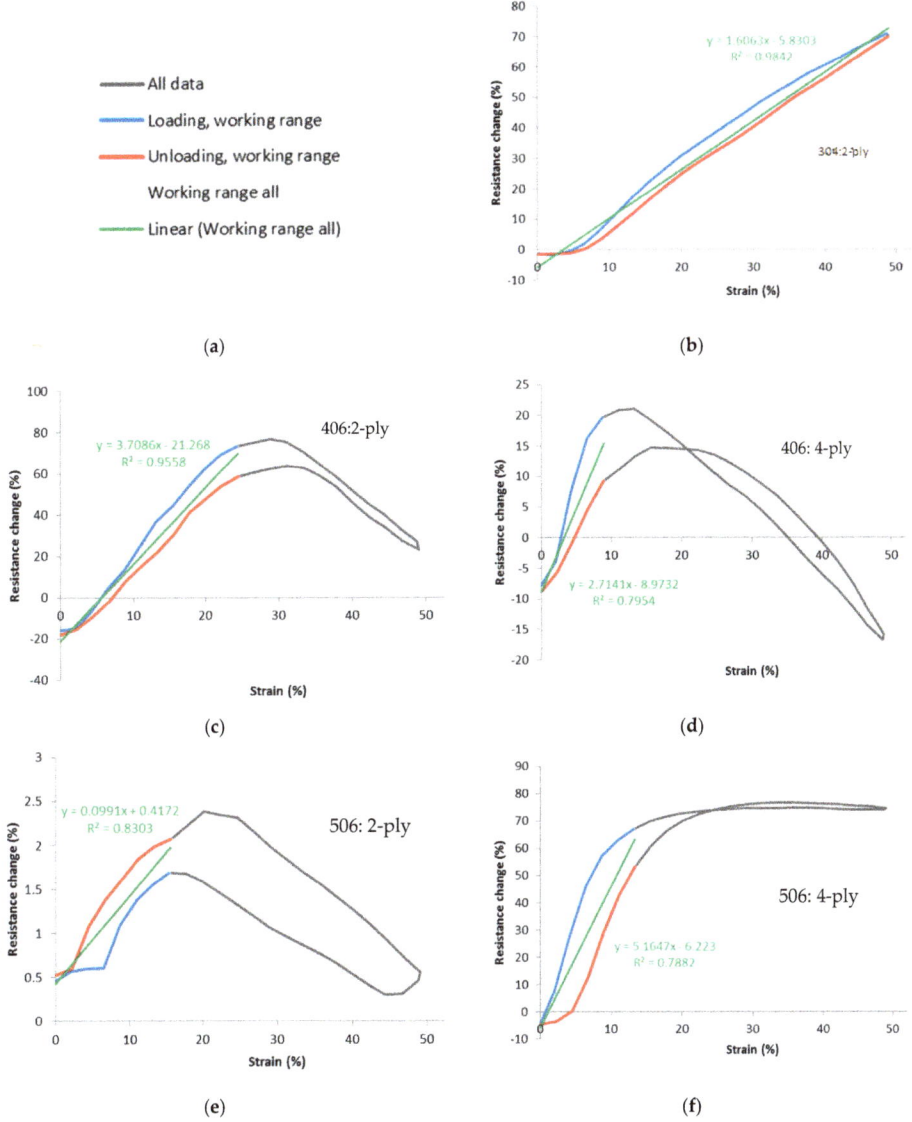

Figure 10. *Cont.*

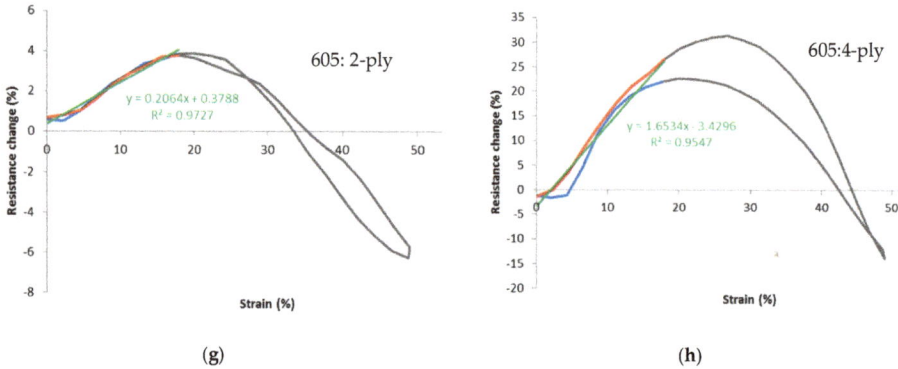

Figure 10. Graphs of percentage resistance change vs. strain, assembly averages over the 2nd to 11th cycles. (**a**) Legend for all graphs; (**b**) 304: 2-ply; (**c**) 406: 2-ply; (**d**) 406: 4-ply; (**e**) 506: 2-ply; (**f**) 506: 4-ply; (**g**) 605: 2-ply; and (**h**) 605: 4-ply. Note the different vertical axis scales. 2-ply is thread 117/17, 33 tex and 4-ply is thread 235/34, 92 tex.

Table 6. Tabulated data from percentage resistance change vs. strain of the assembly from graph averages. Stitch 2-ply uses 117/17 33 tex thread and 4-ply uses 235/34 92 tex.

Stitch Thread Type Tex	Working Range (% strain)	Gauge Factor Over Working Range	Baseline Resistance (Ω)	R^2 Overall Range	R^2 Over Working Range	Max Hysteresis Overall Range (% ΔR)
304 2-ply	0–50	1.61	125	0.984	0.984	6.25
406 2-ply	0–25	3.71	71.5	0.350	0.956	15.1
406 4-ply	0–8	2.71	55.6	0.255	0.795	11.4
506 2-ply	0–16	0.0991	649	0.0419	0.830	0.975
506 4-ply	0–12	5.16	46.7	0.601	0.788	33.6
605 2-ply	0–18	0.206	240	0.499	0.973	2.02
605 4-ply	0–18	1.65	39.3	0.0254	0.955	11.1

The most striking aspect of the percentage resistance change vs. strain of the graphs is the performance of the 304 2-ply stitch. This is highly linear, monotonic, and has a working range that appears to potentially pass 50% extension. It also features low hysteresis and a reasonable high gauge factor, corroborating good linearity. Stitch 506, 4-ply also shows some good attributes, whilst all other stitches followed a trend of increasing resistance followed by decreasing resistance, as the fabric is stretched. This limited their working range to a maximum of 8 to 25% (406 4-ply and 2-ply respectively), which suggests that there may be competing effects in their resistance from extension.

From theory previously described, it is thought that one effect of extending the fabric is to open up contacts between adjacent lengths of conductive thread, thus increasing the conductive path by moving the resistive lengths into series rather than parallel and thus increasing resistance. This might account for the rise in resistance. In all stitches, the initial sensitivity of resistance to strain is low, before increasing rapidly. The likely cause of this is that few contacts are opened up before a certain level of strain is reached.

Something, which might account for the decrease in resistance is the change in dimensions of the conductive thread itself, is altering the resistance per unit length locally. It was previously stated that a conductor increases resistance as it is stretched, due to the longer conduction path and reduction in cross-sectional area. Although one might expect the total length of the conductive thread to increase as the fabric is stretched, it may be the case that it decreases, in certain stitch configurations, hence reducing the sensor resistance. Figure 11 shows an example of fabric stretching with a stitch geometry that might lead to a reduction in the overall conductive thread length.

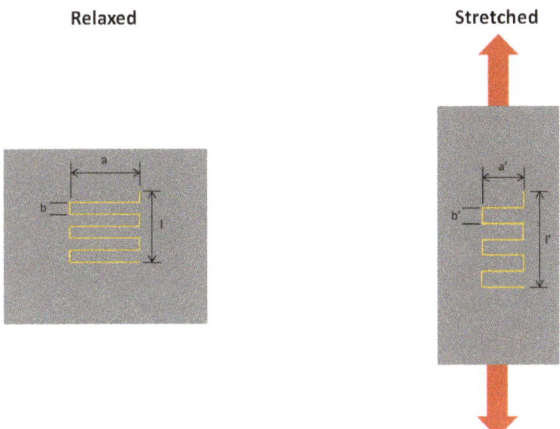

Figure 11. Deformation of the conductive thread.

Note that the main fabric is elongating as well as narrowing when it is stretched. This means that any portions of conductive thread that are in the direction of the overall stretch are also being elongated; however, those that are 90° to its length (i.e., lying across the fabric) will be shortening in length (compressing). In this stitch, there is more conductive thread length lying across the width than in the direction of stretch; therefore, overall the thread will be compressed, not stretched, hence reducing the sensor resistance.

Another explanation can be given if one examines the sewing thread tension performance when stretched on the tensile tester, shown in Figure 3, which leads to an increase in surface contact between conductive surfaces as the cross-section of the thread is expected to decrease under tension along its axis and improves in orientation along the direction of loading. Having said this, in this sensor, the sewing thread is not being stretched along its axis and hence any observation to that effect is difficult.

Another trend is that 4-ply stitches compared to their 2-ply counterparts have lower resistances. This would be expected due to the increased cross-sectional area of conducting available for the current to flow through. Stitches 506 2-ply and 605 2-ply had particularly high baseline resistances of 649 and 240 Ω, respectively. A higher baseline resistance means that, for a given gauge factor, changes in strain would produce larger changes in resistance. However, the high resistance sensors also tend to have a much lower gauge factor (percent change of resistance to strain), so offsetting this advantage.

Large differences in hysteresis were seen across the full sensor range, but when taken as a proportion of the output of the sensors, it is easier to see visually on the graphs. Within their working range, stitch type 605 showed low hysteresis, again taken as a proportion of its output, while stitch type 506 showed the highest hysteresis. Type 605 sensors also had good R^2 values within its working range, reflecting low hysteresis and high linearity.

Figure 12 shows graphs of percentage resistance change vs. strain of stitches between the 2nd and 99th cycles. Data derived from these graphs are tabulated in Tables 7 and 8.

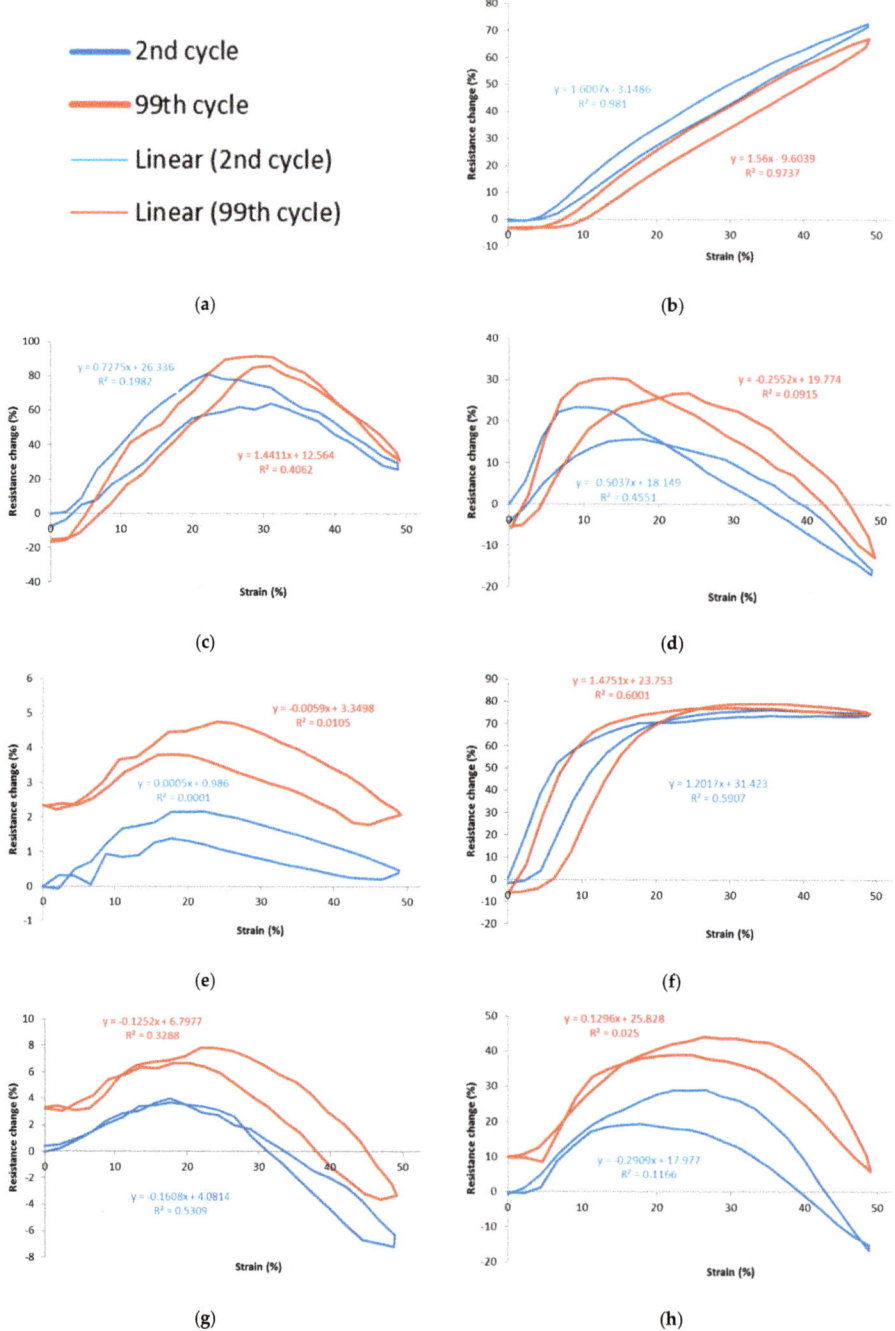

Figure 12. Graphs of percentage resistance change vs. strain, for the 2nd and 99th cycles. (**a**) Legend for all graphs; (**b**) 304: 2-ply; (**c**) 406: 2-ply; (**d**) 406: 4-ply; (**e**) 506: 2-ply; (**f**) 506: 4-ply; (**g**) 605: 2-ply; and (**h**) 605: 4-ply. Note the different vertical axis scales. 2-ply is thread 117/17, 33 tex and 4- ply is thread 235/34, 92 tex.

Table 7. Change in gauge factor from 2nd to 99th cycles. Calculated across the entire extension. 2-ply is thread 117/17, 33 tex and 4- ply is thread 235/34, 92 tex.

Stitch Type Thread Type	G.F. 2nd Cycle	G.F. 99th Cycle	G.F. Drift	Percentage G.F. Drift (%)
304 2-ply	1.60	1.56	−0.0497	−2.54
406 2-ply	0.728	1.44	0.714	98.1
406 4-ply	−0.504	−0.255	0.249	−49.3
506 2-ply	0.0005	−0.0059	−0.0064	−1280
506 4-ply	1.20	1.48	0.273	22.8
605 2-ply	−0.161	−0.125	0.0356	−22.1
605 4-ply	−0.291	0.130	0.421	−145

Table 8. Change in relaxed percentage resistance change and R^2 values from 2nd to 99th cycles calculated across entire extension. 2-ply is thread 117/17, 33 tex and 4- ply is thread 235/34, 92 tex.

Stitch Type Thread Type	Relaxed ΔR 2nd Cycle (Baseline) (%)	Relaxed ΔR 99th Cycle (%)	Relaxed ΔR Drift	R^2 2nd Cycle	R^2 99th Cycle	Drift in R^2
304 2-ply	0	−3.25	−3.25	0.981	0.974	−0.0073
406 2-ply	0	−15.2	−15.2	0.198	0.406	0.208
406 4-ply	0	−5.70	−5.70	0.451	0.0915	−0.360
506 2-ply	0	2.33	2.33	0.0001	0.0105	0.0104
506 4-ply	0	−5.68	−5.68	0.591	0.600	0.0094
605 2-ply	0	3.23	3.23	0.531	0.329	−0.202
605 4-ply	0	9.88	9.88	0.117	0.025	−0.0916

Five out of seven stitches have shown an overall increase in resistance between cycles 2 to 99. It was noted that, after stretching, the fabric does not return completely to its original length. This is because any contact between the loops of the stitch structure are reduced, hence lengthening the conductive path and increasing resistance. This means that the resistance in the relaxed state increases after each stretch cycle. Each stitch type is affected by this phenomenon to a different degree. This is because the conductive path is increased by differing amounts according to how much contact was made between loop before and after stretching.

It should be pointed out that other textile sensors may have similar gauge factor but their working range and other performance properties are very different [20].

The most uniform stitch result in the relaxed state was stitch 304 which is due to its loop contact being broken gradually. The least uniform result was stitch 605, which is due to its loop contact being broken suddenly. This phenomenon may be reduced in all sensor types by using a greater stitch density, so improving contact and making the subsequent contact breakage more gradual.

4. Conclusion

In this research, different types of textile stitch-based strain sensors have been investigated. The effects on their sensing properties related to their resistance change have been examined for a number of different stitch geometries. All sensors have shown significant levels of change in resistance depending on their stitch structure. It has been shown that stitch type variations have a significant effect on cyclic conductivity and resistance, revealing that each stitch design is more suitable for different sensor applications. Hence their suitability depends on the specific end-use requirements, i.e., limb articulation, heart monitoring, or respiration.

It should be pointed out that other textile sensors may have similar gauge factor but their working range and base line resistance are very different. Therefore, overall, this sensor is superior to other similar sensors that may have the same gauge factor but lack in range, linearity, and repeatability.

The sensor made with stitch type 304 and 2-ply 33 tex sewing threads was found to be the best performing in strain sensing for wearable garment end-uses, having investigated its performance in four different nylon/spandex single jersey fabrics It features an exceptionally wide working range (potentially well past 50%), good linearity (R^2 is 0.984), low hysteresis (6.25% ΔR), good gauge factor (1.61), and baseline resistance (125 Ω), as well as good repeatability (drift in R^2 is −0.0073). Therefore, overall, this sensor is superior to other similar sensors that may have the same gauge factor but lack in range, linearity, and repeatability.

Textile-based sensors have the ability to replace solid-state sensors and the next generation of garments will be capable of providing physiological measurement to users.

Author Contributions: G.S. conceptualized and supervised this project. O.T. carried out this research.

Funding: This research received no external funding.

Conflicts of Interest: The authors declare no conflict of interest

References

1. Kang, D.; Pikhitsa, P.V.; Choi, Y.W.; Lee, C.; Shin, S.S.; Piao, L.; Park, B.; Suh, K.-Y.; Kim, T.; Choi, M. Ultrasensitive mechanical crack-based sensor inspired by the spider sensory system. *Nature* **2014**, *516*, 222–226. [CrossRef] [PubMed]
2. Li, M.; Li, H.; Zhong, W.; Zhao, Q.; Wang, D. Stretchable conductive polypyrrole/polyurethane (PPy/PU) strain sensor with netlike microcracks for human breath detection. *ACS Appl. Mater. Interfaces* **2014**, *6*, 1313–1319. [CrossRef] [PubMed]
3. Yao, S.; Zhu, Y. Wearable multifunctional sensors using printed stretchable conductors made of silver nanowires. *Nanoscale* **2014**, *6*, 2345–2352. [CrossRef] [PubMed]
4. Gregory, R.; Kimbrell, W.; Kuhn, H. Electrically conductive non-metallic textile coatings. *J. Coat. Fabr.* **1991**, *20*, 167–175. [CrossRef]
5. Harms, H.; Amft, O.; Roggen, D.; Tröster, G. Rapid prototyping of smart garments for activity-aware applications. *J. Ambient Intell. Smart Environ.* **2009**, *1*, 87–101.
6. Gioberto, G. Measuring joint movement through garment-integrated wearable sensing. In Proceedings of the 2013 ACM Conference on Pervasive and Ubiquitous Computing Adjunct Publication, Zurich, Switzerland, 8–12 September 2013; pp. 331–336.
7. Metcalf, C.D.; Collie, S.; Cranny, A.W.; Hallett, G.; James, C.; Adams, J.; Chappell, P.H.; White, N.M.; Burridge, J.H. Fabric-based strain sensors for measuring movement in wearable telemonitoring applications. In *Proceedings of the IET Conference on Assisted Living 2009*; Curran Associates Inc.: London, UK, 2009; pp. 40–50. ISBN 9781615675890.
8. Gregory, R.; Kimbrell, W.; Kuhn, H. Conductive textiles. *Synth. Met.* **1989**, *28*, 823–835. [CrossRef]
9. Kuhn, H.; Kimbrell, W.; Worrell, G.; Chen, C. Properties of polypyrrole treated textile for advanced applications. In Proceedings of the ANTEC, Montréal, ON, Canada, 5–9 May 1991.
10. Oh, K.W.; Park, H.J.; Kim, S.H. Stretchable conductive fabric for electrotherapy. *J. Appl. Polym. Sci.* **2003**, *88*, 1225–1229. [CrossRef]
11. Thomson, W. On the electro-dynamic qualities of metals: Effects of magnetization on the electric conductivity of nickel and of iron. *Proc. R. Soc. Lond.* **1856**, *8*, 546–550.
12. Zhang, H.; Tao, X.; Wang, S.; Yu, T. Electro-mechanical properties of knitted fabric made from conductive multi-filament yarn under unidirectional extension. *Text. Res. J.* **2005**, *75*, 598–606. [CrossRef]
13. Zhang, H.; Tao, X.; Yu, T.; Wang, S. Conductive knitted fabric as large-strain gauge under high temperature. *Sens. Actuators A Phys.* **2006**, *126*, 129–140. [CrossRef]
14. Li, L.; Au, W.; Li, Y.; Wan, K.; Wan, S.; Wong, K. Electromechanical analysis of conductive yarn knitted in plain knitting stitch under unidirectional extension. In Proceedings of the 1st International Symposium on Textile Bioengineering and Informatics, Hong Kong, China, 14–16 August 2008; pp. 1188–1194.
15. Ehrmann, A.; Heimlich, F.; Brücken, A.; Weber, M.; Haug, R. Suitability of knitted fabrics as elongation sensors subject to structure, stitch dimension and elongation direction. *Text. Res. J.* **2014**, *84*, 2006–2012. [CrossRef]

16. Gioberto, G.; Dunne, L.E. Overlock-stitched stretch sensors: Characterization and effect of fabric property. *J. Text. Appar. Technol. Manag.* **2013**, *8*. Available online: https://ojs.cnr.ncsu.edu/index.php/JTATM/article/view/4417 (accessed on 11 February 2019).
17. Gioberto, G.; Compton, C.; Dunne, L. Machine-Stitched E-Textile Stretch Sensors. *Sens. Transducers J.* **2016**, *202*, 25–37.
18. Wang, J.; Long, H.; Soltanian, S.; Servati, P.; Ko, F. Electro-mechanical properties of knitted wearable sensors: Part 2—Parametric study and experimental verification. *Text. Res. J.* **2014**, *84*, 200–213. [CrossRef]
19. Gioberto, G.; Dunne, L. Theory and characterization of a top-thread coverstitched stretch sensor. In Proceedings of the 2012 IEEE International Conference on Systems, Man, and Cybernetics (SMC), Seoul, Korea, 14–17 October 2012.
20. Atalay, O.; Kennon, W.R. Knitted strain sensors: Impact of design parameters on sensing properties. *Sensors* **2014**, *14*, 4712–4730. [CrossRef] [PubMed]
21. Mattmann, C.; Clemens, F.; Tröster, G. Sensor for measuring strain in textile. *Sensors* **2008**, *8*, 3719–3732. [CrossRef] [PubMed]
22. Marozas, V.; Petrenas, A.; Daukantas, S.; Lukosevicius, A. A comparison of conductive textile-based and silver/silver chloride gel electrodes in exercise electrocardiogram recordings. *J. Electrocardiol.* **2011**, *44*, 189–194. [CrossRef] [PubMed]
23. Gioberto, G.; Coughlin, J.; Bibeau, K.; Dunne, L.E. Detecting bends and fabric folds using stitched sensors. In Proceedings of the 2013 International Symposium on Wearable Computers, Zurich, Switzerland, 8–12 September 2013.
24. Atalay, O.; Kennon, W.R.; Husain, M.D. Textile-based weft knitted strain sensors: Effect of fabric parameters on sensor properties. *Sensors* **2013**, *13*, 11114–11127. [CrossRef] [PubMed]

 © 2019 by the authors. Licensee MDPI, Basel, Switzerland. This article is an open access article distributed under the terms and conditions of the Creative Commons Attribution (CC BY) license (http://creativecommons.org/licenses/by/4.0/).

Article

Vehiculation of Active Principles as a Way to Create Smart and Biofunctional Textiles

Manuel J. Lis Arias [1],*, Luisa Coderch [2], Meritxell Martí [2], Cristina Alonso [2], Oscar García Carmona [1], Carlos García Carmona [1] and Fabricio Maesta [3]

1. Textile Research Institute of Terrassa (INTEXTER-UPC), 08222 Terrassa, Spain; oscargarciacarmona@gmail.com (O.G.C.); carlos.garcia.carmona@gmail.com (C.G.C.)
2. Catalonia Advanced Chemistry Institute (IQAC-CSIC), 08034 Barcelona, Spain; luisa.coderch@iqac.csic.es (L.C.); mmgesl@iiqab.csic.es (M.M.); cristina.alonso@iqac.csic.es (C.A.)
3. Textile Engineering Dept., Federal Technological University of Paraná, Apucarana 86812-460, Brazil; fabriciom@utfpr.edu.br
* Correspondence: manuel-jose.lis@upc.edu; Tel.: +34-937-398-045

Received: 20 September 2018; Accepted: 29 October 2018; Published: 1 November 2018

Abstract: In some specific fields of application (e.g., cosmetics, pharmacy), textile substrates need to incorporate sensible molecules (active principles) that can be affected if they are sprayed freely on the surface of fabrics. The effect is not controlled and sometimes this application is consequently neglected. Microencapsulation and functionalization using biocompatible vehicles and polymers has recently been demonstrated as an interesting way to avoid these problems. The use of defined structures (polymers) that protect the active principle allows controlled drug delivery and regulation of the dosing in every specific case. Many authors have studied the use of three different methodologies to incorporate active principles into textile substrates, and assessed their quantitative behavior. Citronella oil, as a natural insect repellent, has been vehicularized with two different protective substances; cyclodextrine (CD), which forms complexes with it, and microcapsules of gelatin-arabic gum. The retention capability of the complexes and microcapsules has been assessed using an in vitro experiment. Structural characteristics have been evaluated using thermogravimetric methods and microscopy. The results show very interesting long-term capability of dosing and promising applications for home use and on clothes in environmental conditions with the need to fight against insects. Ethyl hexyl methoxycinnamate (EHMC) and gallic acid (GA) have both been vehicularized using two liposomic-based structures: Internal wool lipids (IWL) and phosphatidylcholine (PC). They were applied on polyamide and cotton substrates and the delivery assessed. The amount of active principle in the different layers of skin was determined in vitro using a Franz-cell diffusion chamber. The results show many new possibilities for application in skin therapeutics. Biofunctional devices with controlled functionality can be built using textile substrates and vehicles. As has been demonstrated, their behavior can be assessed using in vitro methods that make extrapolation to their final applications possible.

Keywords: microencapsulation; biofunctional; drug-delivery

---

1. Introduction

In November 2017, the title of the International Symposium on Materials from Renewables (ISMR) was "Advanced, Smart, and Sustainable Polymers, Fibers and Textiles". Three specific sessions occurred under the denomination of "Smart Fibers and Textiles". That simple fact gives an idea of the importance of this work. However, what really are smart textiles? In the foreword of the book edited by Tao, X. [1], Lewis states clearly that these type of textiles are not only special finished fabrics. The main defining idea of smart textiles is related to the "active character" of them. Smart textiles "react to environmental *stimuli*, from mechanical, thermal, chemical, magnetic or others", including biotechnology, information technology, microelectronics, wearable computers, nanotechnology, and micromechanical machines.

Biofunctional textiles are fibrous substrates that have been modified to attain new properties and added value. The main idea is to modify their parameters, especially related to comfort, adapting the tissues' reaction to external or internal stimuli. Such textiles constitute appropriate substrates to be used for the delivery of active principles in cosmetic or pharmaceutical applications. Due to their specific response, biofunctional textiles are especially useful when the textile comes into close contact with the skin. As most of the human body is covered with some sort of textile, the potential of this type of textile is considerable. Textiles with functional properties used for delivery to skin have been studied and patented [2,3].

Three cases will be explored in this work as examples of biofunctional systems obtained using vehicles to transport different active principles to a textile substrate: Microcapsules, cyclodextrins, and liposomes.

1.1. Microcapsules

Microcapsules may be obtained by a series of techniques that involve liquids, gases, or solids in natural or synthetic polymeric membranes [4–7]. This process is known as microencapsulation, and requires a layer of an encapsulating agent, generally a polymeric material, that acts as a protective film insulating the active substance [8–10].

According to Souza and collaborators [11], this creates a physical barrier between the core (active principle) and the encapsulating material (shell).

This membrane is removed by a specific stimulus, releasing the active substance in the ideal place or moment.

The encapsulation technique can be used to fulfill diverse objectives, and has the following advantages, as listed by different authors [12–14]: Protection of the encapsulated materials against oxidation or deactivation due to reaction with the environment (light, oxygen, humidity); masking odors, tastes, and other active principles; insulation of the active principles of undesirable materials; retardation of alterations that might occur in loss of aroma, color, and flavor; separation of reactive or incompatible components; and reduction of the migration rate of the core to the external environment.

For this reason, the core and shell should have compatible physical-chemical properties [15].

The materials commonly used as encapsulating agents (shells) are polymers [16,17] and biologic-based materials [18–20].

Bosnea and collaborators [21] highlighted that shell materials based on natural polymers are promising for the formation of microcapsules due to their biodegradability, compatibility with other products, and low toxicity, as well as their wide availability from natural resources.

Furthermore, Al Shannaq and collaborators [22] accentuate in their work that the choice of encapsulation material is important to obtain better degradation of the microcapsule. Table 1 shows the main polymers of natural sources used in microencapsulation.

Table 1. Polymers used in microencapsulation [21–27].

Source	Polymer
Natural Polysaccharides	Starch, cellulose, chitosan, gum arabic, and alginate
Modified Polysaccharides	Dextrins, carboxymethylcellulose, ethylcellulose, methylcellulose, acetylcellulose, and nitrocellulose
Proteins	Gluten, casein, gelatin, and albumin
Waxes and lipids	Paraffin, tristearine, stearic acid, monoacyl, and diacyl

Besides the inherent parameters of the shell material choice, Santos, Ferreira, and Grosso [28] pointed out that the retention of the active principle is a fundamental factor for the realization of the process. Wang et al. [29] and Yang et al. [30] demonstrated in their works that the efficiency of the encapsulation should be a parameter that is taken into account in the choice of the core.

Sharipova et al. [31] showed that the value of the effectiveness of the encapsulation depends on the encapsulating method, the shell material, and the relationship with the core. Thus, the casing material should relate to the chemical nature (molar mass, polarity, functionality, volatility, and so forth) of the active principle so that a high yield of the process is possible.

There are several techniques for obtaining microcapsules, which can be divided into physical methods (spray drying [32], solvent evaporation [33], pan coating [34]); chemical methods (interfacial polymerization [35], suspension polymerization [36], in situ polymerization [37]); and physical-chemical methods (coacervation [38], ionic gelation [39], sol-gel [40]), among others.

The microencapsulation technique selection depends on the properties of the active principle, the morphology of the desired particle, the nature and capacity of releasing the components, reproducibility, ease of execution of the technique, and the cost/effectiveness ratio [41]. The chosen technique is a determining factor of the characteristics of the formed microcapsule, and will influence the release of the encapsulated agent via one of the following actions: Mechanical, temperature, pH, dissolution, or biodegradation [42].

Microcapsules in the textile field have been applied in various ways, giving very interesting results and showing very promising applications in several fields, including the use of flame retardant agents [43], protection of atmospheric agents [44–46], and functional finishing [47–50], along with the development of functional fabrics that might have a useful effect on the user and solve problems that conventional processes are not able to [51,52].

Nowadays, microcapsules are applied in textiles to transmit different embedded values, such as the liberation of oils with medicinal effects, protection against disease carriers, and antimicrobials, among others [53–65].

Microencapsulation of citronella oil, to be used as insect repellent and obtained by the coacervation method with the gelatin-arabic gum system, will be detailed as an example [38].

To assess the results obtained by the methodology applied, instrumental techniques will be used. Thermogravimetric analysis will be undertaken (TGA/DTGA), as well as application to textile fabrics by padding and analysis of the delivery kinetics. The results will show their potential for use as vehicles.

1.2. Cyclodextrines (CDs)

Cyclodextrines (CDs) are used in diverse industrial fields, such as food, drugs, cosmetics, domestic products, agrochemicals, the textile industry, the paper industry, chemical technology, and others [66–68]. This wide range of applications can be attributed to the fact that cyclodextrins have the capacity to form inclusion complexes with a broad range of substances, allowing the alteration of important properties in the complexed substances [69].

According to Matioli and collaborators [70], CDs are regularly produced from starch by the cyclation of linear chains of glucopyranoses using the enzyme cyclomaltodextrin-glucanotransferase (CGTase).

The three widest known natural cyclodextrins are alpha CD (α-CD), beta CD (β-CD), and gamma CD (γ-CD), composed of six, seven, and eight units of D-(+)-glucopyranose, respectively, and united by α-1,4 bonds.

Since the first publication regarding cyclodextrins in 1891, and the first patent in 1953, these molecules have been a source of great interest to researchers [71]. Nowadays, the widest use of cyclodextrins is their complexation with many classes of drugs [72–76], flavors [77–79], and aromas [80–82].

In order to compare them with the microencapsulation system, complexes of CDs will also be used to vehicularize citronella oil. The results will help us to understand how both systems may show possibilities for use on textile substrates to deliver this active compound, to be used as an insect repellent vector.

1.3. Liposomes

Liposomes are vesicles prepared with lipids that can encapsulate different ingredients; one of their applications could be onto textiles. Due to the liposome bilayer structure, liposomes have been applied as models for biological membranes in medical research. Another important use of this type of vesicle is as microcapsules for drug delivery in the cosmetic field [83–86]. The textile industry has used liposomes in the wool dyeing process as an auxiliary [87,88]. Liposomes could have different properties depending on the lipidic base used during their formulation. In this study, phosphatidylcholine (PC) and internal wool lipids (IWL) were used.

Internal wool lipids are a mix of cholesterol, free fatty acids, cholesterol sulphate, and ceramides, similar to those found in membranes of other keratinized tissues such as human hair or stratum corneum (SC) from skin. Wool is a fiber which is mainly keratinic but with a small amount of internal lipids [89,90]. Due to the IWL liposome bilayer structure's similarity to stratum corneum lipids, their application onto human skin has been assessed in previous studies. The results obtained have demonstrated the beneficial effect of this type of liposome when used with ceramides, topically applied onto intact skin in aging populations or in individuals with dry skin [91–93]. Therefore, we could consider IWL liposomes an optimal encapsulation route for cosmetic or dermopharmacy applications [94].

Using a solar filter as a tracer, ethyl hexyl methoxycinnamate (EHMC), PC-based, and IWL-based liposomes were prepared. The influence of the type of lipid in the vesicle on skin penetration has been demonstrated in previous studies. In particular, the crystalline liquid state of PC liposomes seems to play an important role in this characteristic. On the other hand, when using IWL liposomes, penetration into the skin is delayed—a fact that suggests some reinforcement of the barrier function of the skin's stratum corneum [95]. These two types of liposomes, with IWL and PC, were chosen to be applied to cotton and polyamide fabrics to design biofunctional textiles.

To evaluate the effectiveness of the textile in contact with the skin, a series of methodologies and an in vitro process were optimized to determine the penetration of the encapsulated active principle.

For the evaluation of the biofunctional textiles' beneficial capacity on skin, the transepidermal water loss change (TEWL) was used as an indicator of the barrier function state. TEWL measures water-holding capacity as changes in skin capacitance [95]. An in vitro methodology based on percutaneous absorption [96] was used to determine the amount of encapsulated principle that passed into the different skin layers (stratum corneum, epidermis, or dermis) from the textile.

An in vivo stripping was used as a minimally invasive methodology, where a series of strippings allowed quantification of the amount of active principal in the outermost layers of the SC [97].

These methodologies have shown that liposomes, especially IWL liposomes, are suitable for applying active principles onto biofunctional textiles.

2. Materials and Methods

2.1. Microcapsules

2.1.1. Materials

The materials used to produce the microcapsules were type A gelatin (GE, Sigma Chemical, Darmstadt, Germany) and arabic gum (GA, Sigma Chemical, Darmstadt, Germany), which were used as the microcapsule walls. Citronella essential oil (WNFt, Sao Jose dos Campos, Brazil) was used as the core material. Glutaraldehyde (50%), sodium lauryl sulfate (SLS), citric acid, sodium hydroxide, and all other chemicals used were of analytical grade. The textile substrate consisted of standard cotton fabric (COT, bleached and desized, style 400, 100 g cm^{-2}) and spun polyester type 54 fabric (PES, Style 777, 126 g m^{-2}). Both were obtained from Test Fabrics Inc. (West Pittston, PA, USA).

Optical microscopy was used to verify shape and homogeneity. A BX43 (Olympus, Tokyo, Japan) equipment and a JEOL-JSM 5610 (JEOL, Ltd., Tokyo, Japan) scanning electron microscope were used to analyze the distribution of microcapsules on the textile. The thermal characteristics of the microcapsules were studied by TGA (TGA, SDTA851, Mettler Toledo, Columbus, OH, USA) with the STARe software (version SW 9.01, Mettler-Toledo, Columbus, OH, USA). The samples were weighed and examined in the temperature range of 30–800 °C, using a heating rate of 10 °C min^{-1}.

2.1.2. Methodology

The procedure began with the formation of three emulsions. The first contained 3 g gelatin in 50 mL water, the second emulsion contained 5 mL citronella essential oil and 0.3 g sodium lauryl sulfate (SLS), and the last emulsion was prepared with 100 mL of water and 3 g of gum arabic. The three emulsions were prepared separately in aqueous solution at 50 °C and under stirring at 300 rpm, and the ratio GA:GE was 1:1, as has been used in several other works [38].

In the next stage, the first and second emulsions were blended. Then, the third emulsion was added to the previous mixture followed by adjustment of pH to 4 using citric acid. The mixture was left at rest for 90 min until complete stabilization. The resulting suspension was cooled to a temperature of <8 °C and remained unperturbed for 1 h. Then, the pH of this preparation was adjusted to 8–9 using NaOH (1 M), and 1 g glutaraldehyde (50%) was added dropwise. The system was left for 12 h under stirring at room temperature, resulting in microcapsules for application into fabric substrates.

2.2. Cyclodextrin Complexes

2.2.1. Materials

β-cyclodextrin (CD) was supplied by Wacker-Chemie GmbH, Germany, and citronella essential oil by WNFt, Brazil. Butane 1,2,3,4 tetracarboxylic acid (BTCA) and sodium hypophosphite monohydrate (SHP) were both supplied by Sigma-Aldrich, São Paulo, Brazil, and were of analytical grade.

Standard woven fabrics, namely cotton fabric (COT, bleached and desized cotton print cloth, style 400, 100 g m^{-2}, ISO 105-F02) and spun polyester type 54 fabric (PES, style 777, 126 g m^{-2}, ISO 105-F04) were used, both supplied by Test Fabrics Inc. (West Pittston, PA, USA).

2.2.2. Methodology

A solution of 50 mL of ethanol and water (1:3) and 3 g of CD was prepared and emulsified at 18,000 rpm for 5 min at 60 °C. High temperature and strong agitation allows cyclodextrin to become entirely soluble. Subsequently, 5 g of citronella oil was added. After 2 h agitation, the complexes are formed. This is a standard process proposed by several authors, including Wang and Cheng [61], Oliveira et al. [98], and Partanen et al. [81].

Application of the complexes to the fabrics was achieved by the pad-dry process. Dehabadi, Buschman and Gutmann [99] proposed following the application with a curing step at 170 °C for 3 min, to complete the process of fixation of the complexes by cross-linking.

The pad-dry machine had a width of 30 cm (ERNST BENZ AG KLD-HT and KTF/m250), and a pick-up of 90 ± 10%. The textile substrates were left to dry in environmental conditions.

COT and PES textile articles were impregnated for 1 min in 6 g L^{-1} cyclodextrin solution. The crosslinking agent used was butane 1,2,3,4 tetracarboxylic acid (BTCA) 6 g L^{-1} and the catalyst used was sodium hypophosphite at 6 g L^{-1}.

2.3. Liposomes

2.3.1. Materials

The standard fabrics used were plain cotton fabric, (bleached and desized cotton print cloth, style 400 ISO 105-F02), spun nylon 6 Du Pont type 200 woven fabric (PA) (style 361, 124 g/m^2,) spun polyester type 54 fabric (PES) (style 777, 126 g/m^2) (2,3,8), spun polyacrylic fabric (PAC) (style 864, 135 g/m^2), and knitted 100% wool fabric (WO) (style 537). The textile bandages used for skin penetration evaluation were knitted fabrics (plain stitch) of polyamide 78/68/1 (DeFiber, S.A., Tarragona, Spain).

Ethylhexyl methoxycinnamate (EHMC) (Escalol 557, ISP Global Headquarters, Wayne, NJ, USA) was used as the active principle, as a sun filter and as a tracer in the liposome preparation.

An antioxidant active principle, gallic acid (GA) (Sigma-Aldrich, São Paulo, Brazil), was employed, and isopropanol (Carlo Erba, Cornaredo, Italy) was used in its detection.

2.3.2. Methodology

The liposomes were prepared using two kinds of lipids: Internal wool lipids (IWL) and phosphatidylcholine (PC). The internal wool lipids are a mixture of cholesterol-esters (4%), free fatty acids (27%), cholesterol (19%), glycosylceramides (12%), cholesteryl sulfate (10%), and ceramides (29%). For this study, they were obtained in an extraction pilot plant [93,100].

The method used to prepare liposomes was film hydration, in which the first step was lipid solubilization in an organic phase. Secondly, the lipid solution was dried under a vacuum to form a lipid film on the flask wall, and third, a dispersed water solution of the active principle was added and with movement (energy) the vesicles were formed.

Liposomes of 4% Emulmetik 900 (PC) and 2% of the active principle (gallic acid (GA) or sun filter (EHMC)), were prepared using the well-known method of film hydration. PC (4 g) solubilized in 30 mL of chloroform was dried under a vacuum. The lipidic film was dispersed in 100 mL of 2% GA aqueous solution and then multilamellar vesicles (MLV) were obtained.

Liposomes prepared with phosphatydilcholine or internal wool lipids were also prepared to apply onto textiles.

In order to quantify the exact amount of GA encapsulated, 1 mL of liposome formulation was precipitated and separated from the supernatant by centrifugation using a 5415-Eppendorf centrifuge at 14,000 rpm for 15 min. After separation, the supernatant was retained. The initial and the supernatant were diluted in methanol and read spectrophotometrically at 269 nm. The entrapment efficacy of GA in the microspheres was determined by the amount of the active ingredient present in the whole solution, as well as in the supernatant.

Liposomes with active principle were applied by the foulard process using a pad-dry machine with a measured width of 30 cm (ERNST BENZ AG KLD-HT and KTF/m250), with the corresponding rolling pressure to obtain a pick-up of 90 ± 10% ((mass of bath solution taken by the textile/mass of dry textile) × 100). This was followed by drying in a curing and heat-setting chamber.

2.3.3. Assessment of the Absorption of Skin

Due to the specific skin-delivery applications of the vehicles formed using liposomes, the assessment of the final product's properties was different from the techniques used in former cases. The technique used in this case was percutaneous absorption, using a Franz Diffusion cell, which can be considered as an in vitro assessment in the field of skin care cosmetics.

Pig skin was used from the unboiled backs of large white/Landrace pigs weighing 30–40 kg. The pig skin was provided by the Clínic Hospital of Barcelona, Spain. After excision, the skin was dermatomed to a thickness of approximately 500 ± 50 µm with a Dermatome GA630 (Aesculap, Tuttlingen, Germany). Skin discs with a 2.5 cm inner diameter were prepared and fitted into static Franz-type diffusion cells.

Skin absorption studies were initiated by applying 10 μL of GA in liposomes, or by applying the textile substrates treated with the same formulations onto the skin surface. Both types of samples were allocated in the Franz-cell system. Between the textile and the skin, 20 μL of distilled water was added to ensure close contact. According to the OECD methodology [20], the skin penetration studies were performed for 24 h of close contact between the textile and the skin. To increase the contact pressure between the textile fabric and skin, permeation experiments were also carried out by placing a steel cylinder on the textile-skin substrate at a constant pressure, in accordance with standard conditions (125 g/cm^2) (see Figure 1).

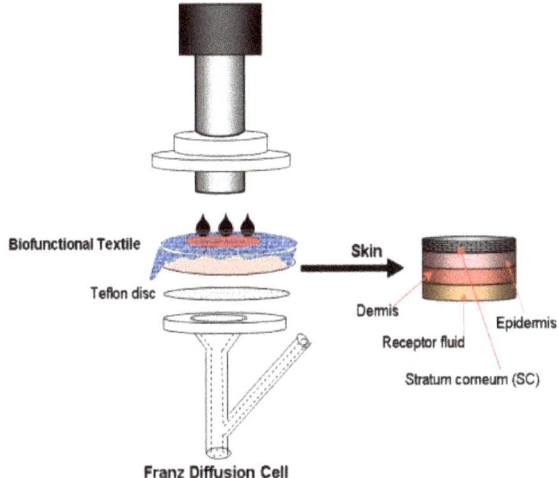

Figure 1. Diagram of in vitro percutaneous absorption experiments.

After the exposure time, the receptor fluid was collected and brought to a volume of 5 mL in a volumetric flask. In the case of the formulations, the skin surface was washed with a specific solution (500 μL SLES (sodium lauryl ether sulphate) (0.5%), and twice with 500 μL distilled water) and dried with cotton swabs. In the case of the textiles, the fabrics were removed from the skin surface and collected together with the top of the cell. In both cases, after eliminating the excess of actives from the skin surface, the stratum corneum of the skin was removed using adhesive tape (D-Squame, Cuderm Corporation, Dallas, TX, USA) applied under controlled pressure (80 g/cm^2 for 5 s). The epidermis was separated from the dermis after heating the skin to 80 °C for 5 s.

The actives were extracted from the different samples (surface excess, CO/PA, or skin layers) using a methanol:water (50:50) solution agitated in an ultrasonic bath for 30 min at room temperature. The receptor fluids were directly analyzed. After filtration on a Millex filter (0.22 μm, Millipore, Bedford, MA, USA), the solutions were assessed by HPLC-UV.

In Vivo Methodologies

In vivo methodologies were applied not only to evaluate the penetration of actives into the skin using the stripping technique, but also to determine the effectiveness of the actives, mainly using biophysical techniques.

The cosmeto-textiles (cotton or polyamide with EHMC in IWL/PC liposomes polyamide containing ME-GA, and control textiles (cotton or polyamide)) were applied on the subjects' arms, maintaining close contact with the skin over a period of four days as a bandage application. The volar forearm was previously evaluated by non-invasive biophysical techniques: Tewameter TM300 (Courage + Khazaka, Köln, Germany) and Corneometer CM825 (Courage + Kazaka, Köln, Germany).

Both parameters (TEWL and skin hydration) were recorded in accordance with established guidelines. After 24 h the textiles were removed and these skin properties were again determined.

The tape stripping of the SC of each forearm and arm was carried out on the fourth day in a conditioned room at 25 ± 1 °C and 50 ± 2% relative humidity using D-Squame tape (φ = 22 mm, CuDerm, Dallas, TX, USA) previously pressed onto the skin with a roller and stripped in one quick move. The weight of the SC removed was determined. The SC lipids from a group of three strips were extracted with methanol (Merck, Darmstadt, Germany) and sonicated in a Labsonic 1510 device (B. Braun, Melsungen, Germany) for 15 min, and the solutions were assessed by HPLC-UV.

3. Results and Discussion

3.1. Microcapsules

3.1.1. Microscopy: Optical and SEM

The first approach used to detect the existence of microcapsules was microscopy. As can be seen in Figure 2, the profile of microcapsules is easy to distinguish in the picture that was obtained.

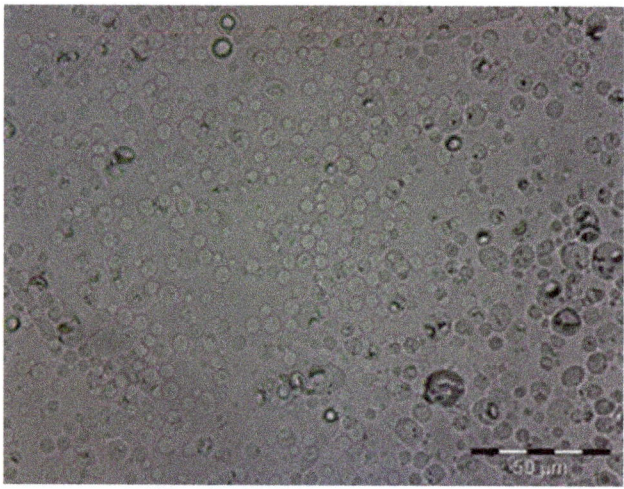

Figure 2. Optical microscopy image of microcapsules formed by complex coacervation (magnification 1000×).

From the image in Figure 2, it is possible to detect the uniform distribution of the microcapsules. There is a clear interface between them and the environment.

3.1.2. TGA and DTGA

The thermogravimetric curves shown in Figure 3 reveal that each component of the system has a different loss of weight when submitted to temperature changes in the oven of the instrument. Citronella oil shows a single stage loss of weight, while gelatin and arabic gum show two stages. Formation of the microcapsule gives the new organized system that presents three different stages of thermal rupture. This behavior is equivalent to that observed by Otálora et al. [101]. From these thermograms it can be seen that the microcapsule was formed, as was already reported in other papers by some of the same authors [38].

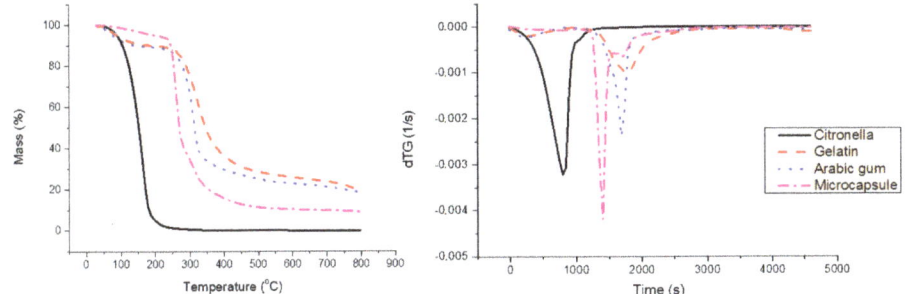

Figure 3. TGA/DTGA thermogram for different components of the system.

3.1.3. Application to PES and COT Fabrics Using Pad-Dry

Figure 4 shows the scanning electron microscopy of the microcapsules and applied cotton (a) and polyester (b). In these micrographs it is possible to verify the distribution of the dispersed microcapsules and their reduced size. These results were also observed in the work of Vahabzadeh, Zivdar, and Najafi [102].

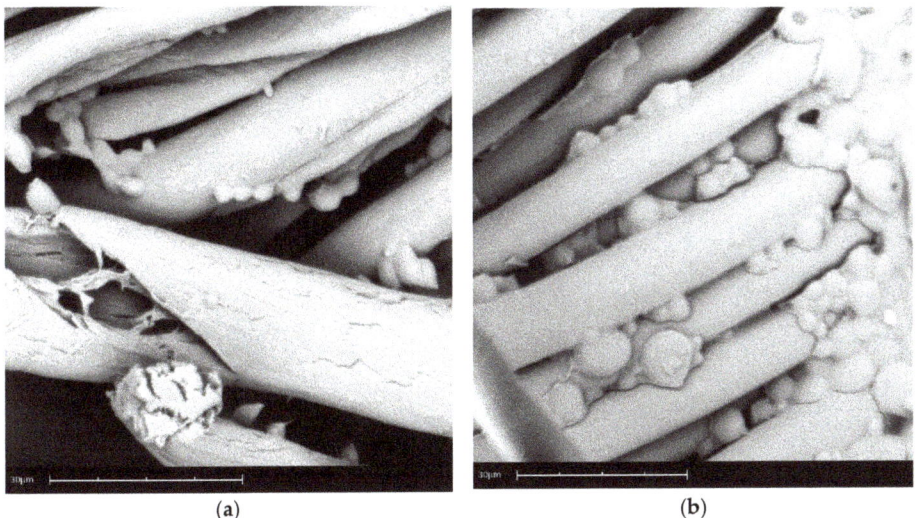

Figure 4. Scanning electron of microcapsules: (a) cotton, and (b) polyester.

Another factor that can be pointed out is the small size of the microcapsules, which facilitates the absorption and penetration of the fabric surface due to the occupation of the interstices of the textile article. Li et al. [23] related the advantages of controlled dosing and increased durability of the textile finish to this small size.

It is also evident that the microcapsules ensure effective protection of the encapsulated material, as already shown by the results of TGA. In short, it can be noted that the encapsulated material is not exposed to the elements, which has been the most serious problem of the application of oils in textiles, as pointed out by Chinta and Pooja [51] and Nelson [42].

3.1.4. Drug Delivery Kinetics

To assess the complexation process yield, spectroscopy at the UV region of the citronella oil in acetone solution was undertaken in order to detect the wavelength of maximum absorption (wavelength from 250 to 550 nm).

These many possibilities of fabric modification allow the textile to be combined with the controlled release system, thereby enabling the absorption of therapeutic or cosmetic compounds and the release of them to the skin [48,103–106].

The behavior observed in Figure 5, in the rate of release in water, allows us to affirm that until 200 min the retention of citronella oil is good enough to state that the delivery is controlled by the structure of the microcapsule. That means, as stated previously, that the structure established clear interactions between the microencapsulated oil (core) and the polymers used to build the microcapsule wall (shell). The nature of these interactions is related to chemical affinity. The majority of the components of the oil have a very organic character that makes them compatible with the organic part of polymers. Although these interactions are weak (see the high slope in Figure 5, compared with that obtained in the case of CD's), the compacted structure of the shell allows the different components of the oil to be retained.

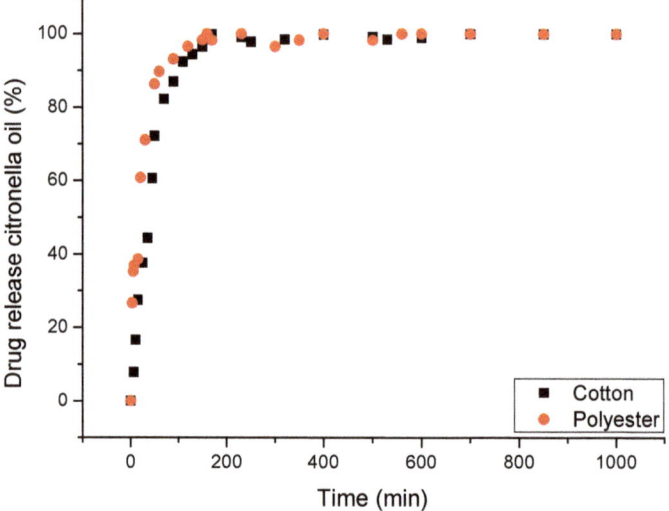

Figure 5. In vitro controlled release profiles in water at 37 °C for microencapsulated citronella oil applied on textiles.

In this case, the mechanism is diffusion-controlled, as shown in Figure 6. The rate of absorption of water is the determining step of the delivery to start transporting citronella oil to the bath, as was established previously in several other works [107].

Figure 6. Schematic representation of the diffusion of an active system with the polymeric membrane, adapted from Bezerra [108].

The adsorption of water molecules in the external (and less ordered) part of the shell promotes a swelling effect on the polymeric network. Due to this fact the molecules of oil in these less ordered zones are displaced to the external part and, afterwards, delivered to the fabric, where the chemical affinity between the fibers and active principle will govern the release mechanism. Meanwhile, water is diffusing into the inner part of the capsule due to the internal concentration gradient created (that depends, basically, on the hydrophilicity of the fiber), which extends the swelling effect to more ordered zones and, consequently, enables the delivery of the active principle.

In the diffusion process the matter is transported to the core of the system, resulting in molecular random movements that occur across small distances. Adolf Fick, in 1855 [109], quantified the diffusion process by adopting the mathematic equation proposed by Fourier (Heat Exchange Transfer Phenomenon) [109]. The proposed Equation (1) is as follows:

$$\frac{dQ}{dt} = -D\frac{dC}{dx} \qquad (1)$$

where:
$\frac{dQ}{dt}$ = diffusion speed of the asset transported in a given time;
D = diffusion coefficient;
$\frac{dC}{dx}$ = concentration of the substance diffusing in the spatial coordinate.

3.2. Cyclodextrin Complexes

3.2.1. Microscopy: Optical and SEM

The optical microscopy of the complexes (CD:Oil), shows their shape and distribution (Figure 7). Spherical and small cylinders are seen. The former are free citronella oil, and the latter are cyclodextrins like elongated cones. CDs can form inclusion complexes with volatile essential oils, protecting them against oxidation, enhancing their chemical stability, and reducing or eliminating eventual losses by evaporation. These results match those obtained in several other works [59,80].

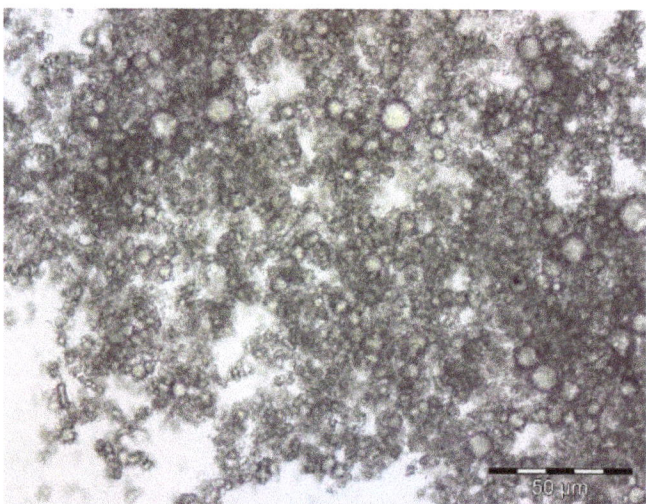

Figure 7. Optical microscopy image of complex. Magnification 500×.

In Figure 8, the application of the complexes on the cotton (a) and polyester (b) surface can be seen. The CDs in textiles can create interactions according to two different mechanisms: Physical bonding without strong chemical interactions, or covalent bonding [71,82]. CDs are permanently fixed in the substrate via covalent bonds [81]. In the case of the cotton fabric, an esterification reaction promoted by the carboxyl group of the BTCA with the hydroxyl groups of the cellulose and the cyclodextrin occurs. In the polyester application, the cyclodextrin covers the fiber, forming a net over the article. In this way, cyclodextrin can adsorb bioactive molecules, making the fabric a biofunctional article.

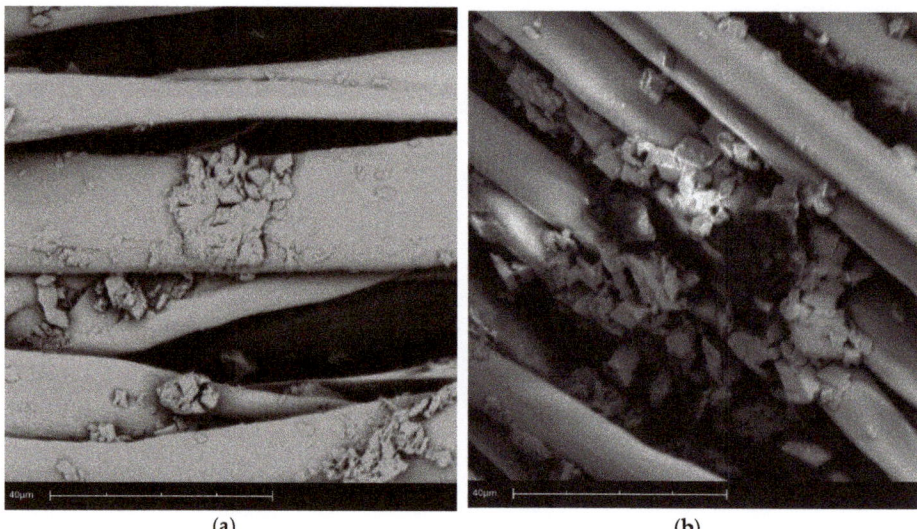

Figure 8. Scanning electron microscope image of complex: (a) cotton, and (b) polyester.

3.2.2. Drug Delivery Kinetics

The use of textile articles as supports for controlled release requires properties to emphasize a high contact area with the skin, drug-carrier capacity, ease of application, low cost, release via stimulus, biocompatibility, non-allergenicity, and non-toxicity, among others [110–114]. Figure 9 shows the controlled release profile of the cotton and polyester fabrics with the complexes (CD:Oil).

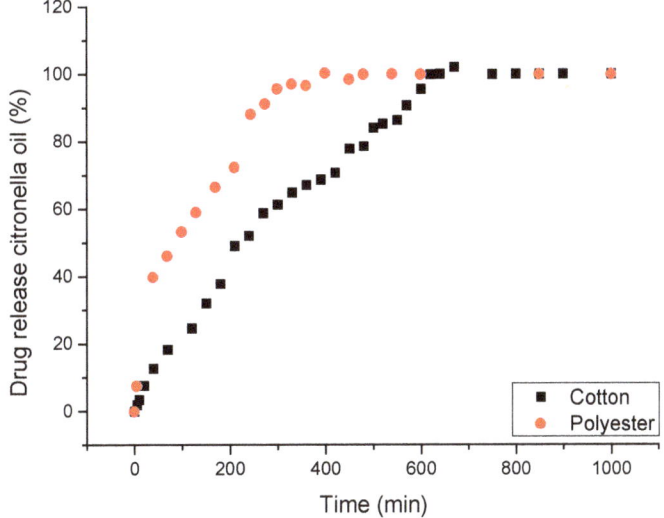

Figure 9. In vitro controlled release profiles in water at 37 °C for complexes (citronella: βCD) applied on textiles.

The analysis shows that the release is influenced by the textile matrix and the type of polymer that constitutes it. The rate of retention of citronella oil is higher in cotton than in polyester. Although the diffusion-controlled mechanism is the same as in the case of microcapsules, here the influence of the affinity of the core substance with respect to the support substrate is high enough to modify the delivery from the complexes. The influence of the textile substrate when used in combination with vehiculizers has been studied in previous cases with microencapsulated ibuprofen [38,48,107]. It is possible to modify the diffusion of the active agent and, furthermore, to indicate that textile supports show potential for use in systems of release [115,116].

3.3. Liposomes

Liposomes, alone or as mixed micelles, form a very stable microstructure that allows the vehiculization of active principles for application into different textiles. The chemical and physico-chemical interactions between them and the textile substrate are translated to an adequate substantivity for most of the studied cases. However, most synthetic substrates, such as acrylic and polyester fibers, showed high desorption. This fact confirmed the preferential application of cotton and polyamide as cosmetic biofunctional textiles due to their best affinity. This study shows that polyamide always presents with high affinity for the two phospholipid structures, and also for the antioxidant [117].

The in vitro percutaneous absorption tests of the different cosmeto-textile structures that were performed (CO and PA with GA vehicularized with liposomes and mixed micelles) have been developed to show, quantitatively, the GA penetration within the layers of the skin [118].

When GA was embedded into the biofunctional textile, it always promoted a reservoir effect (concentration gradient) that was much more marked for PA. This concentration gradient is, really, responsible for managing the mass transport process of the active principle molecules to the skin.

Similar penetration behavior was observed in the textiles treated with GA in MM or Lip in the different layers of skin: The stratum corneum, epidermis, and even the dermis. The GA was only detected in receptor fluid when CO was treated with the mixed micelles system (see Figure 10).

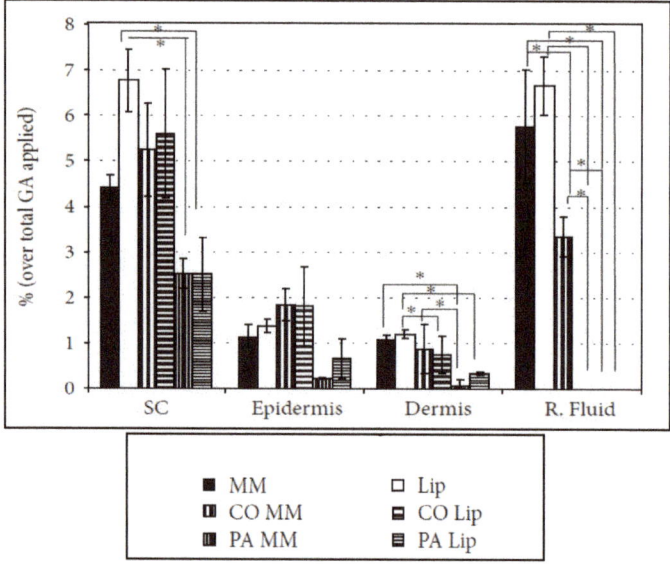

Figure 10. In vitro percutaneous absorption of gallic acid (GA) in liposomes (Lip) and mixed micelle (MM) formulations, and the polyamide (PA) and cotton (CO) cosmeto-textiles (SC: stratum corneum, R. Fluid: receptor fluid) (significant level accepted * $p < 0.01$).

The results show that this methodology may be useful in verifying the real amount of the encapsulated substances incorporated into textile materials that can reach the inner layers of human skin. Indeed, such materials can be considered as strategic delivery systems that can release a given active compound, controlling the final quantity delivered into the skin. If the reservoir effect is controlled, the textile substrate can act as a dosing device.

Similarly, a sun filter (EHMC) was encapsulated in different kinds of liposomic systems and applied to fabrics to create biofunctional textiles. The structured device remains stable due to the mutual affinity and the physico-chemical interactions, which can break as the fabric rubs the skin, releasing the active agents [117]. Moreover, ethyl hexyl methoxycinnamate (EHMC), was vehicularized using two liposome-based structures: Internal wool lipids (IWL) and phosphatidylcholine (PC) [118]. As in the case of GA, they were applied onto cotton and polyamide fabrics by exhaustion treatments. After topical applications on human volunteers, skin properties were evaluated by non-invasive biophysical techniques. The two methodologies, described previously, and based on percutaneous absorption, were applied to quantify the effectiveness of the penetration of the sun filter into the skin.

In both lipidic systems, the absorption over the fabrics was between 10 and 15%. Nevertheless, PC-based liposomes showed a higher affinity than IWL liposomes and cotton substrates showed slightly less absorption than polyamide. The TEWL and skin capacity results are shown in Figure 11.

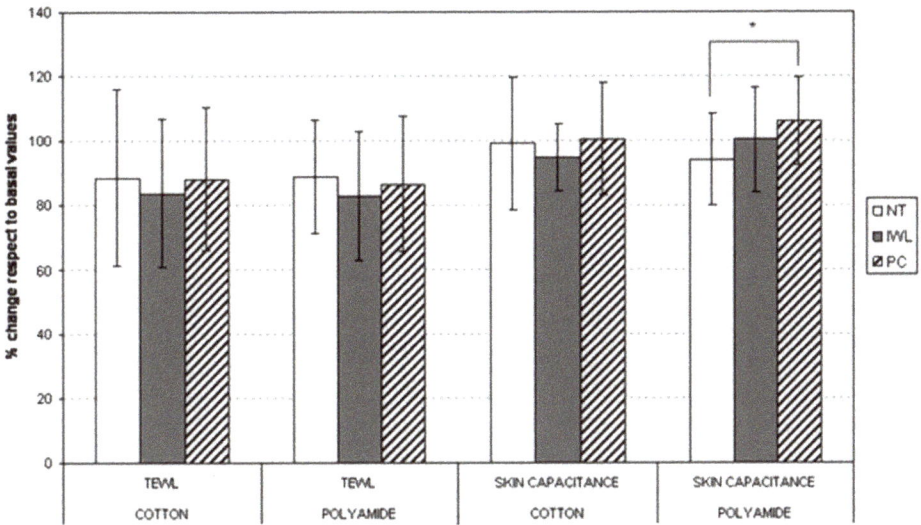

Figure 11. Variation of transepidermal water loss (TEWL) and skin capacitance (hydration) between initial and 24 h of skin application of cotton and polyamide fabrics. NT: non-treated fabric. Internal wool lipids (IWL): fabric treated with IWL liposomes. PC: fabric treated with PC liposomes (* $p < 0.05$ corresponds to significant difference between the marked columns).

TEWL showed a marked decrease for IWL liposome-treated fabrics. This effect is even more marked for liposome-treated polyamide. Non-treated and PC-treated polyamide fabrics showed significant differences which will influence the mechanism of drug permeation [118].

PA-treated substrates showed good permeation into the outermost skin layers in the in vivo experiments. This is consistent with the larger effect found in the decrease in TEWL for IWL liposomes, and in the increase of hydration for PC liposomes when absorbed into polyamide fabrics. Therefore, PA substrates showed the most promising properties for a delivering device, when EHMC is vehicularized using liposomic structures. Fiber behavior in the adsorption process is very good, and enhances the permeation of EHMC through the skin. There is a greater effect of skin barrier reinforcement with IWL, and better hydration when PC liposomes are used [118].

In summary, antioxidants vehicularized through liposomes can be better applied to cotton and polyamide due to their lower desorption compared to the other fibers assayed, such as acrylic or polyester. The two in vitro and in vivo methodologies used to determine the content of active principle penetration into the skin when in contact with the smart textiles indicates the influence of the physicochemical properties of the drugs. When GA is vehicularized in liposomes into the textile, there is a clear reservoir effect that is much more marked with PA [117]. However, when a clear lipophilic compound such as EHMC is also vehicularized in liposomes in the textile, a significantly greater release of the active to the skin was found. Polyamide fibers show the higher desorption capability [118]. This was corroborated by the in vivo results of percutaneous penetration, and the greater skin barrier reinforcement and hydration of the polyamide smart textiles. Therefore, it can be concluded that there are notably different release behaviors of hydrophilic drugs, which may be retained in the hydrophilic core of the liposome in front of lipophilic drugs which are embedded in the surface lipidic bilayer of the liposomes, favoring their release [119].

4. Conclusions

As can be seen from the experimental results obtained, many possibilities exist to make "active" textiles substrates in different environments, using well-designed vehicularizing systems to incorporate these into the fabrics. These complex structures can be, among others, microcapsules, cyclodextrins, or liposomes.

The textile substrate played a very important role in every system tested. The chemical affinity shown in every case clearly controlled the whole release behavior. The in vitro tests are useful to define every step of the delivery mechanisms.

The same citronella oil, vehicularized with different chemical structures, and applied to the same textile substrates, presented different release behaviors. This opens many possibilities for the design of biofunctional textiles.

The use of thermogravimetric methods allows us to define and establish the level of interactions between the active principle and the shell material.

In general, the use of vehicles and textile substrates improves the absorption of complex molecules and formulations by the skin.

The response of the smart fabric depends on the existing interactions between the active principle and the molecular covered structure, and on the interactions between these components and the textile substrate. The combination of these effects results in the development of biofunctional textiles capable of combining specific characteristics of bioactive molecules that cannot be inserted directly into the fabric, as is the case in essential oils that lose their effect due to their volatility.

Therefore, the use of biofunctional textiles allows the treatment of many skin diseases via skin–textile contact, displaying advantages in relation to the administration of the active substance.

Author Contributions: The global line of work, was conceived and designed by M.J.L.A., F.M. and L.C. Experiments of each part and data collection were performed by M.M., C.A., O.G.C. and C.G.C. The original draft was prepared by M.J.L.A., the review and editing by M.J.L.A. and F.M. Founding was acquired by M.J.L.A.

Funding: This work has been supported by each of the institutions, separately from their own resources.

Acknowledgments: M.J.L.A. gratefully acknowledges the financial support from The Forest Next.

Conflicts of Interest: The authors declare no conflict of interest.

References

1. Lewis, R.W. Foreword. In *Smart Fibres, Fabrics and Clothing*; Tao, X., Ed.; Woodhead Publishing Limited: Cambridge, UK; CRC Press: Boca Raton, FL, USA, 2001.
2. Wachter, R.; Weuthen, M.; Panzer, C.; Paff, E. Liposomes Are Used as Textile Finishes Which Not Only Improve Elasticity and Hand but Can Also Be Transferred to Skin Contact. Patent No. EP1510619-A2; DE10339358-A1; US2005058700-A1, 2 March 2005.
3. Guarducci, M. Product Having Particular Functional Properties for the Skin and Process for the Preparation Thereof. Patent No. WO/2006/106546, 13 March 2006.
4. Lei, M.; Jiang, F.; Cai, F.; Hu, S.; Zhou, R.; Li, G.; Wang, Y.; Wang, H.; He, J.; Xiong, X. Facile microencapsulation of olive oil in porous starch granules: Fabrication, characterization, and oxidative stability. *Int. J. Biol. Macromol.* **2018**, *111*, 755–761. [CrossRef] [PubMed]
5. Paulo, F.; Santos, L. Design of experiments for microencapsulation applications: A review. *Mater. Sci. Eng. C* **2017**, *77*, 1327–1340. [CrossRef] [PubMed]
6. Ramakrishnan, A.; Pandit, N.; Badgujar, M.; Bhaskar, C.; Rao, M. Encapsulation of endoglucanase using a biopolymer Gum Arabic for its controlled release. *Bioresour. Technol.* **2007**, *98*, 368–372. [CrossRef] [PubMed]
7. Reshetnikov, I.S.; Zubkova, N.S.; Antonov, V.A.; Potapova, E.V.; Svistunov, V.S.; Tuganova, M.A.; Khalturinskij, N.K. Microencapsulated fire retardants for polyolefins. *Mater. Chem. Phys.* **1998**, *52*, 78–82. [CrossRef]
8. Fadinia, A.L.; Alvim, I.D.; Ribeiro, I.P.; Ruzene, L.G.; Silva, L.B.; Queiroz, M.B.; Miguel, A.M.R.O.; Chaves, F.C.M.; Rodrigues, R.A.F. Innovative strategy based on combined microencapsulation technologies for food application and the influence of wall material composition. *LWT* **2018**, *91*, 345–352. [CrossRef]
9. Lamoudi, L.; Chaumeil, J.C.; Daoud, K. Effet des paramètres du procédé demicroencapsulation du piroxicam par coacervation complexe. *Ann. Pharm. Fr.* **2014**, *1*, 1–6.

10. Skeie, S. Developments in microencapsulation science applicable to cheese research and development. A review. *Int. Dairy J.* **1994**, *4*, 573–595. [CrossRef]
11. Souza, F.N.; Gebara, C.; Ribeiro, M.C.E.; Chaves, K.S.; Gigante, M.L.; Grosso, C.R.F. Production and characterization of microparticles containing pectin and whey proteins. *Food Res. Int.* **2012**, *49*, 560–566. [CrossRef]
12. Ren, P.W.; Ju, X.J.; Xie, R.; Chu, L.Y. Monodisperse alginate microcapsules with oil core generated from microfluidic device. *J. Colloid Interface Sci.* **2010**, *343*, 392–395. [CrossRef] [PubMed]
13. Ré, M.I. Microencapsulação—Em busca de produtos 'Inteligente'. *Ciência Hoje* **2000**, *27*, 24–29.
14. Yoshizawa, H. Trends in microencapsulation research. *KONA Powder Part.* **2004**, *22*, 23–31. [CrossRef]
15. Donbrow, M. *Microcapsules and Nanoparticles in Medicine and Pharmacy*; CRC Press: Boca Raton, FL, USA, 1992; 360p.
16. Song, X.; Zhao, Y.; Hou, S.; Xu, F.; Zhao, R.; He, J.; Cai, Z.; Li, Y.; Chen, Q. Dual agents loaded PLGA nanoparticles: Systematic study of particle size and drug entrapment efficiency. *Eur. J. Pharm. Biopharm.* **2008**, *69*, 445–453. [CrossRef] [PubMed]
17. Freitas, S.; Merkle, H.P.; Gander, B. Microencapsulation by solvent extraction/evaporation: Reviewing the state of the art of microsphere preparation process technology. *J. Control. Release* **2005**, *102*, 313–332. [CrossRef] [PubMed]
18. Rocha-Selmi, G.A.; Bozza, F.T.; Thomazini, M.; Bolini, H.M.A.; Fávaro-Trindade, C.S. Microencapsulation of aspartame by double emulsion followed by complex coacervation to provide protection and prolong sweetness. *Food Chem.* **2013**, *139*, 72–78. [CrossRef] [PubMed]
19. Nakagawa, K.; Nagao, H. Microencapsulation of oil droplets using freezing-induced gelatin-acacia complex coacervation. *Colloids Surf. A Physicochem. Eng. Asp.* **2012**, *411*, 129–139. [CrossRef]
20. Jun-Xia, X.; Hai-Yan, Y.; Jian, Y. Microencapsulation of sweet orange oil by complex coacervation with soybean protein isolate/gum Arabic. *Food Chem.* **2011**, *125*, 1267–1272. [CrossRef]
21. Bosnea, L.A.; Moschakis, T.; Biliaderis, C.G. Complex coacervation as a novel microencapsulation technique to improve viability of probiotics under different stress. *Food Bioprocess Technol.* **2014**, *7*, 2767–2781. [CrossRef]
22. Al-Shannaq, R.; Farid, M.; Al-Muhtaseb, S.; Kurdi, J. Emulsion stability and cross-linking of PMMA microcapsules containing phase change materials. *Sol. Energy Mater. Sol. Cells* **2015**, *132*, 311–318. [CrossRef]
23. Li, L.; Song, L.; Hua, T.; Au, W.M.; Wong, K.S. Characteristics of weaving parameters in microcapsule fabrics and their influence on loading capability. *Text. Res. J.* **2013**, *83*, 113–121. [CrossRef]
24. Semyonov, D.; Ramon, O.; Kaplun, Z.; Levin-Brener, L.; Gurevich, N.; Shimoni, E. Microencapsulation of *Lactobacillus paracasei* by spray freeze drying. *Food Res. Int.* **2010**, *43*, 193–202. [CrossRef]
25. Shalaka, D.S.R.; Amruta, N.A.; Parimal, K. Vitamin and loaded pectin alginate microspheres for cosmetic application. *J. Pharm. Res.* **2009**, *6*, 1098–1102.
26. Suave, J.; Dall'agnol, E.C.; Pezzini, A.P.T.; Silva, D.A.K.; Meier, M.M.; Soldi, V. Microencapsulação: Inovação em diferentes áreas. *Rev. Saúde Ambiente* **2006**, *7*, 12–20.
27. Santos, A.B.; Ferreira, V.P.; Grosso, C.R.F. Microesferas—Uma alternativa viável. *Biotecnol. Ciênc. Desenvolv.* **2000**, *16*, 26–30.
28. Wang, J.M.; Zheng, W.; Song, Q.W.; Zhu, H.; Zhou, Y. Preparation and characterization of natural fragrant microcapsules. *J. Fiber Bioeng. Inform.* **2009**, *4*, 293–300. [CrossRef]
29. Yang, Z.; Peng, Z.; Li, J.; Li, S.; Kong, L.; Li, P.; Wang, Q. Development and evaluation of novel flavor microcapsules containing vanilla oil using complex coacervation approach. *Food Chem.* **2014**, *145*, 272–277. [CrossRef] [PubMed]
30. Sharipova, A.A.; Aidarova, S.B.; Grigoriev, D.; Mutalieva, B.; Madibekova, G.; Tleuova, A.; Miller, R. Polymer-surfactant complexes for microencapsulation of vitamin E and its release. *Colloids Surf. B Biointerfaces* **2016**, *137*, 152–157. [CrossRef] [PubMed]
31. França, D.; Medina, A.F.; Messa, L.L.; Souza, C.F.; Faez, R. Chitosan spray-dried microcapsule and microsphere as fertilizer host for swellable—Controlled release materials. *Carbohydr. Polym.* **2018**, *196*, 46–55. [CrossRef] [PubMed]
32. Poncelet, D. Microencapsulation: Fundamentals, methods and applications. In *Surface Chemistry in Biomedical and Environmental Science*; Blitz, J., Gunoko, V., Eds.; Springer: Dordrecht, The Netherlands, 2006; pp. 23–34.

33. Jamekhorshida, A.; Sadramelia, S.M.; Farid, M. A review of microencapsulation methods of phase change materials (PCMs) as a thermal energy storage (TES) medium. *Renew. Sustain. Energy Rev.* **2014**, *31*, 531–542. [CrossRef]
34. Tsuda, N.; Ohtsubo, T.; Fuji, M. Preparation of self-bursting microcapsules by interfacial polymerization. *Adv. Powder Technol.* **2012**, *23*, 724–730. [CrossRef]
35. Alcázar, A.; Borreguero, A.M.; Lucas, A.; Rodríguez, J.F.; Carmona, M. Microencapsulation of TOMAC by suspension polymerisation: Process optimisation. *Chem. Eng. Res. Des.* **2017**, *117*, 1–10. [CrossRef]
36. Zuo, M.; Liu, T.; Han, J.; Tang, Y.; Yao, F.; Yuan, Y.; Qian, Z. Preparation and characterization of microcapsules containing ammonium persulfate as core by in situ polymerization. *Chem. Eng. J.* **2014**, *249*, 27–33. [CrossRef]
37. Bezerra, F.M.; Carmona, O.G.; Carmona, C.G.; Lis, M.J.; Moraes, F.F. Controlled release of microencapsulated citronella essential oil on cotton and polyester matrices. *Cellulose* **2016**, *23*, 1459–1470. [CrossRef]
38. Li, Y.; Wu, C.; Wu, T.; Wang, L.; Chen, S.; Ding, T.; Hu, Y. Preparation and characterization of citrus essential oils loaded in chitosan microcapsules by using different emulsifiers. *J. Food Eng.* **2018**, *217*, 108–114. [CrossRef]
39. Bae, J. Fabrication of carbon microcapsules containing silicon nanoparticles–carbon nanotubes nanocomposite by sol-gel method for anode in lithium ion battery. *J. Solid State Chem.* **2011**, *184*, 1749–1755. [CrossRef]
40. Barbosa-Cánovas, G.V.; Ortega-Rivas, E.; Juliano, P.; Yan, H. *Food Powders—Physical Properties, Processing, and Functionality*; Springer: New York, NY, USA, 2005; 372p.
41. Ulrich, K.; Eppinger, S. *Product Design and Development*; McGraw Hill: New York, NY, USA, 2003; 432p.
42. Nelson, G. Application of microencapsulation in textiles. *Int. J. Pharm.* **2002**, *242*, 55–62. [CrossRef]
43. Tekin, R.; Bac, N.; Erdogmus, H. Microencapsulation of fragrance and natural volatile oils application in cosmetics, and household cleaning products. *Macromol. Symp.* **2013**, *333*, 35–40. [CrossRef]
44. Zimet, P.; Livney, Y.D. Beta-lactoglobulin and its nanocomplexes with pectin as vehicles for ω-3 polyunsaturated fatty acids. *Food Hydrocoll.* **2009**, *23*, 1120–1126. [CrossRef]
45. Annan, N.T.; Borza, A.D.; Hansen, L.T. Encapsulation in alginate-coated gelatin microspheres improves survival of the probiotic Bifidobacterium adolescentis 15703T during exposure to simulated gastro-intestinal conditions. *Food Res. Int.* **2008**, *41*, 184–193. [CrossRef]
46. Prata, A.S.; Grosso, C.R.F. Production of microparticles with gelatina and chitosan. *Carbohydr. Polym.* **2015**, *116*, 292–299. [CrossRef] [PubMed]
47. Abdelkader, M.B.; Azizi, N.; Baffoun, A.; Chevalier, Y.; Majdoub, M. New microcapsules based on isosorbide for cosmetotextile: Preparation and characterization. *Ind. Crops Prod.* **2018**, *123*, 591–599. [CrossRef]
48. Rubio, L.; Alonso, C.; Coderch, L.; Parra, J.L.; Martí, M.; Cebrián, J.; Navarro, J.A.; Lis, M.; Valldeperas, J. Skin delivery of caffeine contained in biofunctional textiles. *Text. Res. J.* **2010**, *80*, 1214–1221. [CrossRef]
49. Cheng, S.Y.; Yuen, M.C.W.; Kan, C.W.; Cheuk, K.K.L.; Chui, C.H.; Lam, K.H. Cosmetic textiles with biological benefits: Gelatin microcapsules containing Vitamin C. *Int. J. Mol. Med.* **2009**, *24*, 411–419. [PubMed]
50. Meyer, A. Perfume microencapsulation by complex coacervation. *CHIMIA* **1992**, *46*, 101–102.
51. Chinta, S.K.; Pooja, P.W. Use of microencapsulation in textiles. *Indian J. Eng.* **2013**, *3*, 37–40.
52. Sanchez, P.; Sanchez-Fernandez, M.V.; Romero, A. Development of thermoregulating textiles using paraffin wax microcapsules. *Thermochim. Acta* **2010**, *498*, 16–21. [CrossRef]
53. Ma, Z.H.; Yu, D.G.; Branford-White, C.J. Microencapsulation of tamoxifen: Application to cotton fabric. *Colloids Surf. B* **2009**, *69*, 85–90. [CrossRef] [PubMed]
54. El Asbahani, A.; Miladi, K.; Badri, W.; Sala, M.; Aït Addi, E.H.; Casabianca, H.; El Mousadik, A.; Hartmann, D.; Jilale, A.; Renaud, F.N.R.; et al. Essential oils: From extraction to encapsulation. *Int. J. Pharm.* **2015**, *483*, 220–243. [CrossRef] [PubMed]
55. Lv, Y.; Yang, F.; Li, X.; Zhang, X.; Abbas, S. Formation of heat-resistant nanocapsules of jasmine essential oil via gelatin/gum Arabic based complex coacervation. *Food Hydrocoll.* **2014**, *35*, 305–314. [CrossRef]
56. Wang, B.; Adhikari, B.; Barrow, C.J. Optimisation of the microencapsulation of tuna oil in gelatin-sodium hexametaphosphate using complex coacervation. *Food Chem.* **2014**, *158*, 358–365. [CrossRef] [PubMed]
57. Patrick, K.E.; Abbas, S.; Lv, Y.; Ntsama, I.S.B.; Zhang, X. Microencapsulation by complex coacervation of fish oil using gelatin/SDS/NaCMC. *Pak. J. Food Sci.* **2013**, *23*, 17–25.
58. Piacentini, E.; Giorno, L.; Dragosavac, M.M.; Vladisavljevic, G.T.; Holdich, R.G. Microencapsulation of oil droplets using cold water fish gelatine/gum arabic complex coacervation by membrane emulsification. *Food Res. Int.* **2013**, *53*, 362–372. [CrossRef]

59. Thilagavathi, G.; Kannaian, T. Combined antimicrobial and aroma finishing treatment for cotton, using micro encapsulated geranium (*Pelargonium graveolens* L'Herit. Ex Ait.) leaves extract. *Indian J. Nat. Prod. Resour.* **2010**, *3*, 348–352.
60. Wang, C.X.; Chen, S.L. Aromachology and its application in the textile field. *Fibres Text. East. Eur.* **2005**, *13*, 41–44.
61. Solomon, B.; Sahle, F.F.; Gebre-Mariam, T.; Asres, K.; Neubert, R.H.H.H. Microencapsulation of citronela oil for mosquito-repellent application: Formulation and in vitro permeation studies. *Eur. J. Pharm. Biopharm.* **2012**, *80*, 61–66. [CrossRef] [PubMed]
62. Specos, M.M.M.; Garcia, J.J.; Tornesello, J.; Marino, P.; Della Vecchia, M.; Defain Tesoriero, M.V.; Hermida, L.G. Microencapsulated citronella oil for mosquito repelente finishing of cotton textiles. *Trans. R. Soc. Trop. Med. Hyg.* **2010**, *104*, 653–658. [CrossRef] [PubMed]
63. Gonsalves, J.K.M.C.; Costa, A.M.B.; Sousa, D.P.; Cavalcanti, S.C.H.; Nunes, R.S. Microencapsulação do óleo essencial de *Citrus sinensis* (L.) Osbeck pelo método da coacervação simples. *Sci. Plena* **2009**, *5*, 1–8.
64. Tawatsin, A.; Wratten, S.D.; Scott, R.R.; Thavara, U.; Techadamrongsin, Y. Reppelency of volatile oils from plants against three mosquito vectors. *J. Vector Ecol.* **2001**, *26*, 76–82. [PubMed]
65. Chatterjee, S.; Salaün, F.; Campagne, C. Development of multilayer microcapsules by a phase coacervation method based on ionic interations for textile applications. *Pharmaceutics* **2014**, *6*, 281–297. [CrossRef] [PubMed]
66. Szejtli, J. Introduction and general overview of cyclodextrin chemistry. *Chem. Rev.* **1998**, *98*, 1743–1753. [CrossRef] [PubMed]
67. Duchêne, D. *Cyclodextrins and Their Industrial Uses*; Edition Santé: Paris, France, 1987; 300p.
68. Bender, H. Production, characterization and application of CDs. *Adv. Biotechnol. Processs.* **1986**, *6*, 31–71.
69. Del Valle, E.M.M. Cyclodextrins and their uses: A review. *Process Biochem.* **2004**, *39*, 1033–1046. [CrossRef]
70. Matioli, G.; Moraes, F.F.; Zanin, G.M. *Ciclodextrinas e suas Aplicações em: Alimentos, Fármacos, Cosméticos, Agricultura, Biotecnologia, Química Analítica e Produtos Gerais*; Eduem: Maringá, Brazil, 2000; p. 124.
71. Bhaskara-Amrit, U.R.; Pramod, B.A.; Warmoeskerken, M.C.G. Applications of β-cyclodextrins in textiles. *AUTEX Res. J.* **2011**, *11*, 94–101.
72. Shlar, I.; Droby, S.; Rodov, V. Antimicrobial Coatings on Polyethylene Terephthalate Based on Curcumin/Cyclodextrin Complex Embedded in a Multilayer Polyelectrolyte Architecture. *Colloids Surf. B Biointerfaces* **2018**, *164*, 379–384. [CrossRef] [PubMed]
73. Lucio, D.; Irache, J.M.; Font, M.; Martíinez-Oharriz, M.C. Supramolecular structure of glibenclamide and cyclodextrins complexes. *Int. J. Pharm.* **2017**, *530*, 377–386. [CrossRef] [PubMed]
74. Suvarna, V.; Gujar, P.; Murahari, M. Complexation of phytochemicals with cyclodextrin derivatives—An insight. *Biomed. Pharm.* **2017**, *88*, 1122–1144. [CrossRef] [PubMed]
75. Zhang, L.; Man, S.; Qiu, H.; Liu, Z.; Zhang, M.; Ma, L.; Gao, W. Curcumin-cyclodextrin complexes enhanced the anti-cancer effects of curcumin. *Environ. Toxicol. Pharmacol.* **2016**, *48*, 31–38. [CrossRef] [PubMed]
76. Irie, T.; Uekama, K. Pharmaceutical applications of cyclodextrins. III. Toxicological issues and safety evaluation. *Pharm. Sci.* **1997**, *86*, 147–162. [CrossRef] [PubMed]
77. Sali, N.; Csepregi, R.; Kőszegi, T.; Kunsági-Máté, S.; Szente, L.; Poór, M. Complex formation of flavonoids fisetin and geraldol with β-cyclodextrins. *J. Lumin.* **2018**, *194*, 82–90. [CrossRef]
78. Lo Nostro, P.; Fratoni, L.; Ridi, F.; Baglioni, P. Surface treatments on Tencel fabric: Grafting with β-cyclodextrin. *J. Appl. Polym. Sci.* **2003**, *88*, 706–715. [CrossRef]
79. Partanen, R.; Ahro, M.; Hakala, M.; Kallio, H.; Forssell, P. Microencapsulation of caraway extract in β-cyclodextrin and modified starches. *Eur. Food Res. Technol.* **2002**, *214*, 242–247. [CrossRef]
80. Yildiz, Z.A.; Celebioglu, A.; Kilic, M.E.; Durgun, E.; Uyar, T. Menthol/cyclodextrin inclusion complex nanofibers: Enhanced water-solubility and high-temperature stability of menthol. *J. Food Eng.* **2018**, *224*, 27–36. [CrossRef]
81. Ciobanu, A.; Mallard, I.; Landy, D.; Brabie, G.; Nistor, S.; Fourmentin, S. Retention of aroma compounds from menthe piperita essential oil by cyclodextrins and crosslinked cyclodextrin polymers. *Food Chem.* **2013**, *138*, 291–297. [CrossRef] [PubMed]
82. Peila, R.; Migliavacca, G.; Aimone, F.; Ferri, A.; Sicardi, S. A comparison of analytical methods for the quantification of a reactive β-cyclodextrin fixed onto cotton yarns. *Cellulose* **2012**, *19*, 1097–1105. [CrossRef]

83. Teschke, O.; de Souza, E.F. Liposome structure imaging by atomic force microscopy: Verification of improved liposome stability during absorption of multiple aggregated vesicles. *Langmuir* **2002**, *18*, 6513–6520. [CrossRef]
84. Lian, T.; Ho, R.J.Y. Trends and developments in liposome drug delivery systems. *J. Pharm. Sci.* **2001**, *90*, 667–680. [CrossRef] [PubMed]
85. Betz, G.; Aeppli, A.; Menshutina, N.; Leuenberge, H. In vivo comparison of various liposome formulations for cosmetic application. *Int. J. Pharm.* **2005**, *296*, 44–54. [CrossRef] [PubMed]
86. Aggarwal, A.; Dayal, A.; Kumar, N. Microencapsulation processes and application in textile processing. *Colourage* **1998**, *45*, 15–24.
87. Martí, M.; de la Maza, A.; Parra, J.L.; Coderch, L. Dyeing wool at low temperatures: New method using liposomes. *Text. Res. J.* **2001**, *71*, 678–682. [CrossRef]
88. Montazer, M.; Validi, M.; Toliyat, T. Influence of temperature on stability of multilamellar liposomes in wool dyeing. *J. Liposome Res.* **2008**, *16*, 81–89. [CrossRef] [PubMed]
89. Ramirez, R.; Martí, M.; Manich, A.M.; Parra, J.L.; Coderch, L. Ceramides Extracted from Wool: Pilot Plant Solvent. *Text. Res. J.* **2008**, *78*, 73–80. [CrossRef]
90. Coderch, L.; Fonollosa, J.; Martí, M.; Garde, F.; de la Maza, A.; Parra, J.L. Extraction and Análisis of Ceramides from Internal Wool Lipids. *J. Am. Oil Chem. Soc.* **2002**, *79*, 1215–1220. [CrossRef]
91. de Pera, M.; Coderch, L.; Fonollosa, J.; de la Maza, A.; Parra, J.L. Effect of Internal Wool Lipid liposomes on Skin Repair. *Skin Pharmacol. Appl. Physiol.* **2000**, *13*, 188–195. [CrossRef] [PubMed]
92. Coderch, L.; de Pera, M.; Fonollosa, J.; de la Maza, A.; Parra, J.L. Efficacy of stratum corneum lipid supplementation on human skin. *Contact Dermat.* **2002**, *47*, 139–146. [CrossRef]
93. Ramírez, R.; Martí, M.; Barba, C.; Méndez, S.; Parra, J.L.; Coderch, L. Skin Efficacy of liposomes composed of internal wool lipids rich in ceramides. *J. Cosmet. Sci.* **2010**, *61*, 235–245. [CrossRef] [PubMed]
94. Coderch, L.; Fonollosa, J.; de Pera, M.; de la Maza, A.; Parra, J.L.; Martí, M. Compositions of Internal Lipid Extract of Wool and Use Thereof in the Preparation of Products for Skin Care and Treatment. Patent No. WO/2001/004244, 30 January 2001.
95. Ramón, E.; Alonso, C.; Coderch, L.; De la Maza, A.; López, O.; Parra, J.L.; Notario, J. Liposomes as Alternative Vesicles for Sun Filter Formulations. *Drug Deliv.* **2005**, *12*, 83–88. [CrossRef] [PubMed]
96. Williams, A.C. *Transdermal and Topical Drug Delivery: From Theory to Clinical Practice*; Pharmaceutical Press: London, UK, 2003.
97. Pinkus, H. Examination of the epidermis by the strip method of removing horny layers. I. Observation on thickness of the horny layer, and on mitotic activity after stripping. *J. Investig. Dermatol.* **1951**, *16*, 383–386. [CrossRef] [PubMed]
98. Oliveira, T.; Botelho, G.; Alves, N.; Mano, J.F. Inclusion complexes of alfa-cyclodextrines with poly(D,L-Lactic acid): Structural characterization, and glass transition dynamics. *Colloid Polym. Sci.* **2014**, *293*, 863–871. [CrossRef]
99. Dehabadi, V.A.; Buschmann, H.; Gutmann, J.S. A novel approach for fixation of β-cyclodextrin on cotton fabrics. *J. Incl. Phenom. Macrocycl. Chem.* **2013**, *79*, 459–464. [CrossRef]
100. Martí, M.; Martínez, V.; Rubio, L.; Coderch, L.; Parra, J.L. Biofunctional Textiles prepared with Liposomes: In Vivo and in Vitro assessment. *J. Microencapsul.* **2011**, *28*, 799–806. [CrossRef] [PubMed]
101. Otálora, M.C.; Carriazo, J.G.; Iturriaga, L.; Nazareno, M.A.; Osorio, C. Microencapsulation of betalains obtained from catos fruit (*Opuntia ficus-indica*) by spray-drying using cactus cladode mucilage and maltodextrin as encapsulating agents. *Food Chem.* **2015**, *187*, 174–181.
102. Vahabzadeh, F.; Zivdar, M.; Najafi, A. Microencapsulation of orange oil by complex coacervation and its release behavior. *IJE Trans. B Appl.* **2004**, *17*, 333–342.
103. Mihailiasa, M.; Caldera, F.; Li, J.; Peila, R.; Ferri, A.; Trotta, F. Preparation of functionalized cotton fabrics by means of melatonin loaded cyclodextrin nanosponges. *Carbohydr. Polym.* **2016**, *142*, 24–30. [CrossRef] [PubMed]
104. Soares-Latour, E.M.; Bernard, J.; Chambert, S.; Fleury, E.; Sintes-Zydowicz, N. Environmentally benign 100% bio-based oligoamide microcapsules. *Colloids Surf. A* **2017**, *524*, 193–203. [CrossRef]
105. Ghaheh, F.S.; Khoddami, A.; Alihosseini, F.; Jing, S.; Ribeiro, A.; Cavaco-Paulo, A.; Silva, C. Antioxidant cosmetotextiles: Cotton coating with nanoparticles containing vitamin E. *Process Biochem.* **2017**, *59*, 46–51. [CrossRef]

106. Ibrahim, N.A.; El-Ghany, N.A.A.; Eid, B.M.; Mabrouk, E.M. Green options for imparting antibacterial functionality to cotton fabrics. *Int. J. Biol. Macromol.* **2018**, *111*, 526–533. [CrossRef] [PubMed]
107. Carreras, N.; Acuña, V.; Martí, M.; Lis, M.J. Drug reléase system of ibuprofen in PCL-microespheres. *Colloid Polym. Sci.* **2013**, *291*, 157–165. [CrossRef]
108. Bezerra, F.M.; Croscato, G.S.; Valldeperas, J.; Lis, M.J.; Carreras, C.; Acuna, V. Aplicação de microesferas para desenvolvimento de novos acabamentos têxteis. *Química Têxtil* **2011**, *105*, 49–58.
109. Bird, R.B.; Stewart, W.E.; Lightfood, E.N. *Fenômenos de Transporte*; LTC Editora: Sao Paulo, Brasil, 2004; p. 838.
110. Manadas, R.; Pina, M.E.; Veiga, F. A dissolução in vitro na previsão da absorção oral de fármacos em formas farmacêuticas de liberação modificada. *Braz. J. Pharm. Sci.* **2002**, *38*, 375–399. [CrossRef]
111. Sun, X.; Wang, X.; Wu, J.; Li, S. Development of thermosensitive microgel-loaded cotton fabric for controlled drug release. *Appl. Surf. Sci.* **2017**, *403*, 509–518. [CrossRef]
112. Guan, Y.; Zhang, L.; Wang, D.; West, J.L.; Fu, S. Preparation of thermochromic liquid crystal microcapsules for intelligent functional fiber. *Mater. Des.* **2018**, *147*, 28–34. [CrossRef]
113. Mocanu, G.; Nichifor, M.; Mihai, D.; Oproiu, L.C. Bioactive cotton fabrics containing chitosan and biologically active substances extracted from plants. *Mater. Sci. Eng. C* **2013**, *33*, 72–77. [CrossRef] [PubMed]
114. Todorova, S.B.; Silva, C.J.S.; Simeonov, M.P.; Cavaco-Paulo, A. Cotton fabric: A natural matrix suitable for controlled release systems. *Enzym. Microb. Technol.* **2007**, *40*, 1646–1650. [CrossRef]
115. Radu, C.D.; Parteni, O.; Popa, M.; Muresan, I.E.; Ochiuz, L.; Bulgariu, L.; Munteanu, C.; Istrate, B.; Ule, E. Comparative Study of a Drug Release from a Textile to Skin. *J. Pharm. Drug Deliv. Res.* **2015**, *4*, 2–8. [CrossRef]
116. Martin, A.; Tabary, N.; Leclercq, L.; Junthip, J.; Degoutin, S.; Aubert-Viard, F.; Cazaux, F.; Lyskawa, J.; Janus, L.; Bria, M.; et al. Multilayered textile coating based on a cyclodextrin polyelectrolyte for the controlled release of drugs. *Carbohydr. Polym.* **2013**, *93*, 718–730. [CrossRef] [PubMed]
117. Martí, M.; Martínez, V.; Lis, M.J.; Valldeperas, J.; de la Maza, A.; Parra, J.L.; Coderch, L. Gallic Acid vehiculized through liposomes or mixed micelles in biofunctional textiles. *J. Text. Inst.* **2014**, *105*, 175–186. [CrossRef]
118. Martí, M.; Alonso, C.; Lis, V.M.J.; de la Maza, A.; Parra, J.L.; Coderch, L. Cosmetotextiles with Gallic Acid: Skin reservoir effect. *J. Drug Release* **2013**, *2013*, 456248. [CrossRef] [PubMed]
119. Martí, M.; Rodríguez, R.; Carreras, N.; Lis, M.J.; Valldeperas, J.; de la Maza, A.; Coderch, L.; Parra, J.L. Monitoring of the microcapsule/liposome application on textile fabrics. *J. Text. Inst.* **2012**, *103*, 19–27. [CrossRef]

© 2018 by the authors. Licensee MDPI, Basel, Switzerland. This article is an open access article distributed under the terms and conditions of the Creative Commons Attribution (CC BY) license (http://creativecommons.org/licenses/by/4.0/).

Article

Surface Area Evaluation of Electrically Conductive Polymer-Based Textiles

Lukas Vojtech [1], Marek Neruda [1], Tomas Reichl [2,*], Karel Dusek [2] and Cristina de la Torre Megías [1]

[1] Department of Telecommunication Engineering, Faculty of Electrical Engineering, Czech Technical University in Prague, 166 27 Prague, Czech Republic; vojtecl@fel.cvut.cz (L.V.); nerudmar@fel.cvut.cz (M.N.); cristinatm07@gmail.com (C.d.l.T.M.)

[2] Department of Electrotechnology, Faculty of Electrical Engineering, Czech Technical University in Prague, 166 27 Prague, Czech Republic; dusekk1@fel.cvut.cz

* Correspondence: reichto3@fel.cvut.cz; Tel.: +420-224-352-167

Received: 13 August 2018; Accepted: 9 October 2018; Published: 10 October 2018

Abstract: In this paper, the surface area of coated polymer-based textiles, i.e., copper and nickel plated woven polyester fabric, copper and acrylic coated woven polyester fabric, and copper and acrylic coated non-woven polyamide fabric, is investigated. In order to evaluate the surface area of the woven fabrics, Peirce's geometrical model of the interlacing point and measurement using an electron microscope are used. Non-woven fabrics are evaluated using an optical method, handmade method, and MATLAB functions. An electrochemical method, based on the measurement of the resistance between two electrodes, is used for relative comparison of the effective surface area of the coated woven and non-woven fabrics. The experimental results show that the measured and calculated warp lengths do not differ within the standard deviation. The model for the surface area evaluation of the Pierce's geometrical model for monofilament (non-fibrous) yarns is extended to multifilament yarns and to a uniform sample size. The experimental results show the increasing trend of surface area evaluation using both modeling and electrochemical methods, i.e., the surface area of the copper and acrylic coated woven Polyester fabric (PES) is the smallest surface area of investigated samples, followed by the surface area of the copper and acrylic coated non-woven fabric, and by copper and nickel plated woven PES fabric. These methods can be used for surface area evaluation of coated polymer-based textiles in the development of supercapacitors, electrochemical cells, or electrochemical catalysts.

Keywords: electrically conductive textiles; polymers; smart textiles; surface area evaluation

1. Introduction

Electrically conductive textile materials have attracted considerable attention from the scientific community and have found their use in many applications. These materials are used in shielding against electromagnetic radiation, e.g., protective clothing, cable shielding, shielding of high frequency sources, etc. [1–5], in development of textile antennas [6–8] and sensors, e.g., textile sensor for Electrocardiography monitoring (ECG), and pressure and humidity sensors [9–11], as well as in the development of other electrical components and devices, e.g., supercapacitors, electrochemical cells, electrochemical catalysts, etc. [12–14].

There are many ways to improve the parameters of the last mentioned components [15]. The reduction of the internal resistance decreases the thermal loss, increases the capacity of electrochemical cells, and causes a better current distribution. One of the possible ways to reduce the internal resistance is to increase the relative effective surface area in 3-D of the sub-parts of these components (electrodes), i.e., the enlargement of the reaction area upon which the chemical reactions occur. It will result in greater electrochemical interconnection between the electrode and the electrolyte. In electrochemical catalysts, the increase of surface area, i.e., reaction area, causes a larger reaction area to be available during the chemical reaction, e.g., in fuel cells. In supercapacitors, the increase of surface area increases the capacity [15].

The surface area, i.e., reaction area, can be increased by using electrically conductive polymer-based textiles. The electrically conductive polymer-based textiles are an interesting class of materials that combine some mechanical properties of polymers with electrical properties typical of metals [16]. This is especially true for conductivity and permanence in applications where the chemical process takes place [15,16]. Additionally, these polymers have become popular because they are lightweight and economical, and have relatively high adjustable electrical conductivity, flexibility, chemical stability, and biocompatibility. The biggest advantage of conductive polymers is their processability [16]. For the future construction and engineering of the mentioned components, it is necessary to know the value of the surface area of their sub-parts. This is typically in the form of a grid, expanded metal, or plate-shaped. However, the shape of the surface area of the electrically conductive polymer-based textile, e.g., coated polymer textile, is not a typical one. The method most commonly used to determine the specific area of a solid porous substance is the Brunaer, Emmett, and Teller method (BET) [17,18]. This method relies on the BET equation for specific isothermal gas adsorption [17]. Although the BET method remains a popular choice for assessing the specific surface areas of nanostructured materials [18], it is speculated that BET does not provide a true representation of the geometric area. It is known that the applicability of the BET method to the porosity characterization is problematic because it is unable to capture complex adsorption mechanisms due to microporous effects [18]. BET regions are highly susceptible to the size of pores, heterogeneity of structures, and adsorbent–adsorbent interactions. There is no simple correlation between BET and geometric surface areas of nanoporous materials [18].

In this paper, three different coated polymer-based textiles, i.e., copper and nickel plated woven polyester (PES) fabric, copper and acrylic coated woven PES fabric, and copper and acrylic coated non-woven polyamide fabric, are tested for the surface area evaluation. The arrangement of the fiber in the woven fabric is investigated using the Peirce's geometrical model of the interlacing point. This model can be applicable for calculation of the surface area of monofilament (non-fibrous) yarns. The model for surface area evaluation of the multifilament yarns of woven fabric is calculated from the theoretical lengths of yarns, which are obtained by means of an electron microscope. Three methods are used for evaluation of the surface area of non-woven fabrics, i.e., handmade selection method, automatic image conversion to binary method, and threshold selection method. Results are compared with an electrochemical method, which is based on measurement of resistance between two immersed electrodes inside an electrolyte. These methods evaluate the 3-D surface area of the coated polymer-based textiles and can be used in the development of electrical and electrochemical components and devices such as supercapacitors, electrochemical cells, or electrochemical catalysts. The knowledge of surface area evaluation of a conductive 3-D layer of coated textiles can help in adjusting the weight, energy density, and capacity of these future components and equipment.

2. Materials and Methods

Three different types of polymer fabrics were tested, i.e., two types of woven fabrics and one type of nonwoven fabric, Table 1. CerexCuIv4 is a copper- and acrylic-coated non-woven fabric with a raw material Polyamide Cerex fabric 36 g/m^2 and surface resistivity 0.02 ohm/square [19]. RSKCu+Ni is a copper- and nickel-plated woven PES fabric with a plain weave (parachute silk) and a surface resistivity 0.02 ohm/square [20]. RSKCuIv4 is a copper- and acrylic-coated woven PES fabric with a plain weave (parachute silk) and a surface resistivity 0.05 ohm/square [20]. Samples were examined using a scanning electron microscope (Phenom pro X ThermoFisher SCIENTIFIC, Eindhoven, The Netherlands).

Table 1. Specifications of the used commercial polymer coated fabrics [20].

Material as Named by the Producer/Producer	Description	Surface Resistivity	Note
CerexCuIv4/LORIX Ltd., Budapest, Hungary	copper + acrylic coated non-woven fabric	max avg. 0.02 ohm/square	Raw material: Polyamide Cerex fabric 36 g/m^2
RSKCu+Ni/LORIX Ltd., Budapest, Hungary	copper + nickel plated woven PES fabric	max avg. 0.02 ohm/square	Weave: plain weave (parachute silk)
RSKCuIv4/LORIX Ltd., Budapest, Hungary	copper + acrylic coated woven PES fabric	max avg. 0.05 ohm/square	Weave: plain weave (parachute silk)

2.1. Theory of Peirce's Geometrical Model of the Interlacing Point

The arrangement of the fiber in the fabric can be mathematically described in several ways. The simplest, but fully satisfactory, is the so-called the Peirce's geometrical model of the interlacing point [21,22] (Figure 1). This model describes the fiber as a system of a regular repeating part of a torus (a body formed by rotating a circle around a straight line) and a cylinder of height d. The disadvantage of the Peirce's geometrical model of the interlacing point is that it neglects the deformation of the fiber caused by forces that act on the fibers, such that shape of the fiber is always circular [21,22]. The main reason for using this model is a satisfactory balance between the efficiency of using it and its accuracy [22].

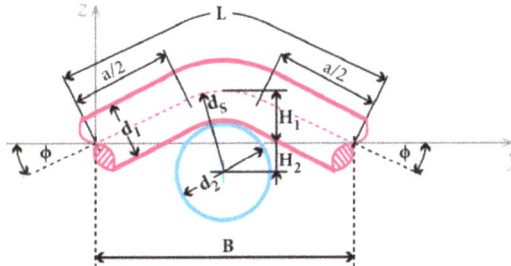

Figure 1. Peirce's geometrical model of the interlacing point [22]. Legend: a is the length of the line connecting the arcs, d_1 is the diameter of the warp, d_2 is the diameter of the weft, d_s is the sum of radii of the warp and weft yarns, H_1 is the height of warp interlacing wave, H_2 is the height of weft interlacing wave, B is the spacing of weft yarns, Φ is the warp interlacing angle, and L is the length of warp yarn.

The relationship between the warp interlacing angle Φ and other geometric parameters (spacing of weft yarns B, height of the warp interlacing wave H_1, and the sum of radii of the warp and weft yarns d_S) can be expressed using Equations (1) and (2) resulting from the section in the y-axis Equation (1) and in the z-axis Equation (2) [22]:

$$2 \cdot d_S \cdot sin(\Phi) + a \cdot cos(\Phi) = B \quad (1)$$

$$2 \cdot d_S \cdot [1 - cos(\Phi)] + a \cdot sin(\Phi) = 2 \cdot H_1 \quad (2)$$

The sum of the radii of the warp and weft yarns d_s is calculated using Equation (3), Figure 1:

$$d_S = (d_1 + d_2)/2 \quad (3)$$

Combining Equations (1) and (2), a quadratic equation for the warp interlacing angle Φ is obtained, as shown in Equation (4):

$$\left[B^2 + 4\cdot(H_1 - d_S)^2\right]\cdot\cos^2(\Phi) + 8\cdot d_S\cdot(H_1 - d_S)\cdot\cos(\Phi) + \left(4\cdot d_S^2 - B^2\right) = 0 \quad (4)$$

The warp interlacing angle Φ can be expressed from Equation (4) as in Reference [23], given here as Equation (5):

$$\Phi = \arccos\frac{2Md \pm 16r^2}{2(M^2 + 4r^2)} \quad (5)$$

where M is the spacing of yarns, r is the radius of yarns, and d is the height of cylinder of yarns.

The height of the cylinder of yarns d is then calculated using Equation (6):

$$d = \sqrt{M^2 - 12r^2} \quad (6)$$

The theoretical length of the warp L_{WARP} and weft L_{WEFT} is calculated using Equations (7) and (8):

$$L_{WARP} = 2\cdot d_S\cdot\Phi + a_{WARP} \quad (7)$$

$$L_{WEFT} = 2\cdot d_S\cdot\Phi + a_{WEFT} \quad (8)$$

where a_{WARP} is the length of the line connecting the arcs in warp, a_{WEFT} is the length of the line connecting the arcs in weft.

2.2. Surface Area Evaluation Based on Peirce's Geometrical Model of the Interlacing Point

To calculate the effective surface area of the conductive fabrics, i.e., the surface area that is involved in chemical reactions, the warp and weft lengths of Pierce's geometrical model of the interlacing point are first determined. The theoretical length of weft and warp are calculated from Equations (7) and (8). The length of the line connecting the arcs, diameter of warp, and diameter of weft are measured by means of an electron microscope, Phenom pro X (ThermoFisher SCIENTIFIC). From Equations (5) and (6), we obtain the warp interlacing angle and weft angle. These angles are substituted into Equations (7) and (8), which gives the theoretical length of the warp and weft. The surface area of the warp and weft is then calculated using Equations (9) and (10):

$$S_{WARP} = 2\pi r_1 L_{WARP} \quad (9)$$

$$S_{WEFT} = 2\pi r_2 L_{WEFT} \quad (10)$$

The surface area of the interlacing points is calculated using Equation (11):

$$S_{Inter\_Point} = 2\pi r_1 r_2 \quad (11)$$

The surface area of the investigated Pierce's geometrical model of the interlacing point of the textile is a sum of the warp and weft surface area, from which the surface area of interlacing points is subtracted (Equation (12)). The obtained value is then divided by two due to the 50% of the depth efficiency of the electrochemical process, which is related to the object used in the fabric structure. This ratio will be different for the electrochemical reaction and electrolyzer construction in practice. For comparison of examined methods, the ratio 50% is chosen, i.e., "visible part of the surface area" during the observation of the structure in the direction of the z-axis is considered.

$$S_{Effect} = (S_{WARP} + S_{WEFT})/2 - S_{Inter\_Point} \quad (12)$$

Equation (12) can be applicable for the calculation of the surface area of monofilament (non-fibrous) yarns. The surface area of multifilament yarns in warp S_1 is obtained by multiplying the theoretical length of the warp L_{WARP} and the width in weft b and coefficient K, which takes into account the amorphousness of the fibers, as seen in Equation (13):

$$S_1 = L_{WARP} \cdot b \cdot K \tag{13}$$

where S_1 is the surface area of the multifilament yarns in warp, b is the width in weft, and K is the coefficient amorphousness of the fibers.

The same procedure as for S_1 is used to calculate the surface area of the multifilament yarns in weft S_2, as seen in Equation (14):

$$S_2 = L_{WEFT} \cdot c \cdot K \tag{14}$$

where c is the width in warp.

The K is calculated as the ratio of the width of the multifilament yarns $K_{straight}$ and the length of coating of multifilament yarns in cross-section K_{curve}, which is shown in Figure 2 and given in Equation (15). The difference in the shape of the yarns, i.e., difference between elliptical shape of yarns as assumed by the Pierce's geometrical model of the interlacing point and the real shape of yarns as shown in Figure 2, is assumed to be negligible. This assumption is verified in Section 3.2 (Table 7) and Section 3.3 (Table 12).

$$K = \frac{K_{curve}}{K_{straight}}. \tag{15}$$

where $K_{straight}$ is the width of the multifilament yarns and K_{curve} is the length of the coating of multifilament yarns in cross-section.

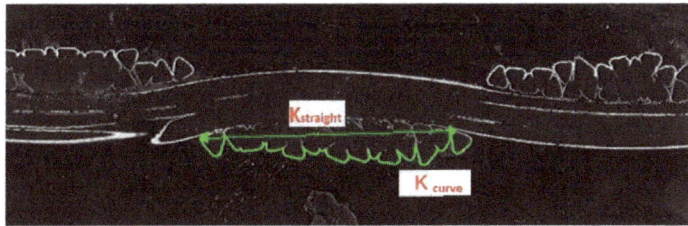

Figure 2. Evaluation of the width of the multifilament yarns $K_{straight}$ and the length of coating of multifilament yarns in cross-section K_{curve}.

2.3. Optical Method for the Non-Woven Fabric Evaluation

2.3.1. Handmade Selection Method

One of the methods used for surface area evaluation is a time-consuming handmade selection method. It is based on choosing all fibers individually by hand in the obtained picture from the electron microscope image (Figure 3). It illustrates individual fibers and the background, represented by the light and dark areas, respectively.

The white bar that is given at the bottom of the picture is used as a reference value in µm to match the measured length of each fiber. For this reason, the length of this scale bar has to be saved as a variable. The measurement of the scale bar is accomplished by selecting the length of the bar each time the method is executed. It is carried out with the function of MATLAB "imline" [24].

Once the bar length is selected, the diameter of one fiber is selected by hand [24]. We assume the diameter is the same in all fibers. Therefore, fiber selection for measuring the diameter does not have the relevance. To increase accuracy, it is recommended to follow these two recommendations:

- Select the diameter in the upper fibers, which are sharper than the rest.

- Choose fibers that are directly in contact with the background since the boundaries of these fibers can be differentiated more easily.

After this, a line that covers the whole length of each fiber is traced to obtain the fiber length. It is important to take into account that the lines are determined by hand, using the function "imfreehand" of MATLAB [25]. Lines are drawn in the center of every fiber with the highest precision the user can perform.

Next, the length of the fiber and the diameter are compared with the scale bar length and the area is calculated using the known mathematical equation for the volume of a cylinder, i.e., half of 3-D surface area is calculated [26]. Finally, the total area is the sum of the area of all fibers.

As it can be appreciated in the described method, accuracy of the length measurements depend on the user. At least lengths referring to the bar and diameter are straight lines and are taken only once, but every fiber length is completely determined by hand, which increases the error. This error can be mitigated by creating a function that calculates the exact points of the center of the yarn. It must also be taken into account that the mistake that may appear in the selection of fibers that are missing in each step, i.e., the user can forget some fibers or parts of them that seem part of background.

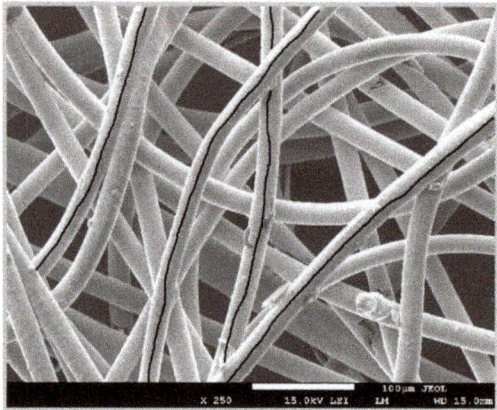

Figure 3. Selection of the fiber lengths by hand.

2.3.2. Automatic Image Conversion to Binary

Another method for the surface area evaluation of non-woven fabrics is the conversion of the obtained microscope picture of the sample to a binary image. It consists of changing all fibers to white and the background to black to differentiate clearly which parts of the image are conductive.

The same sample is used for the surface area evaluation as described in the previous section. As per the previous method, the length of scale bar is measured by hand. However, the used standard MATLAB function "imtool" provides the possibility to check the value of the pixels, 0 (black) or 255 (white) [27]. When the user selects the length manually, they check the pixel value at the same time. If it is white, the user starts to measure the length and stops the measurement when they see the change from 255 to 0.

The procedure of the diameter measurement is the same as the bar measurement. The user can appreciate how the intensity of gray varies, since white values of pixels are higher than black values, therefore the user knows when to start and stop the measurement. As explained in the previous section, the diameter is selected according to the two described recommendations.

After that, the original image is converted to a binary image (Figure 4). The input image is converted to a grayscale format (if it is not already an intensity image), and then to binary. This procedure is directly performed using the function "im2bw (I, 0.5)".

Next, the user selects the part of image that contains fibers, and pixels in white are counted [28]. In case two consecutive pixels are white, the distance between them is calculated and saved into a variable. This variable, and also the diameter, are compared with the scale bar, and then the area is calculated using the same equation as in the previous method.

As shown in Figure 4, when the image is converted to binary, not all the fibers are completely white, which is the main cause of the error in this method. In this case the improved function consists of delimiting the boundaries of all fibers and fills the pixels inside these limits. Therefore, pixels inside fibers that are not in white using function "im2bw" will be converted using this improved function. The other cause comes from the user when the user takes measurements of the scale bar and diameter. Another source of error appears when the image is cropped. The user can make a mistake if they select a smaller size of the required image [28].

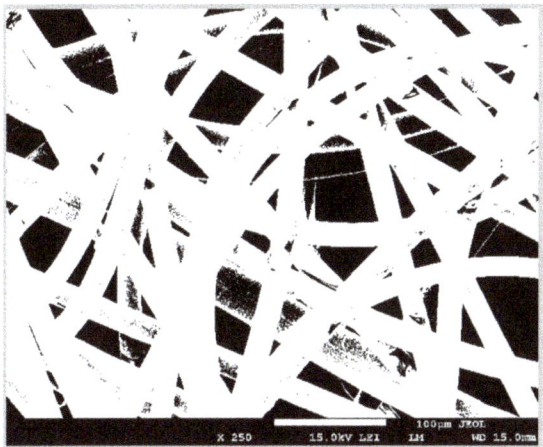

Figure 4. Binary image.

2.3.3. Threshold Selection Method

The third method used for evaluating the surface area of non-woven fabrics is the transformation of the color of the background using thresholding. This method is based on finding the appropriate threshold in the gray scale to change the background pixels to black. Therefore, the fibers can be seen more clearly in the image. The same image sample is used as in the previous cases (Figure 3). Measurement of the scale bar and diameter is performed according to the same procedure as in the previous method, i.e., the function "imtool" is used.

When the measured values of the length of the scale bar and diameter are saved, the image is cropped to work only with the part that contains the fibers. The difference between this method and previous one is that the original image is not converted to binary. In this case, a threshold is used whose value can be changed to achieve the best results. The background color is changed to black to check the amount of pixels that are considered background (Figure 5).

Figure 5. Filled image using a threshold to change the background color.

As can be seen from Figure 5, some parts inside the fibers are painted in black because they are darker, such that the threshold condition is met due to deformation of the fiber. On the contrary, not all parts of the background are painted in black because all backgrounds do not have the same color or the same intensity of black color. This is a major error for this method and should also be considered a user-side error as described in the previous method. In conclusion, the preferred method is the threshold selection method, due to its higher accuracy and lower assistance requirement from the user.

2.4. Experimental Electrochemical Method for Surface Area Evaluation

The surface area of the coated woven and non-woven fabrics can be also evaluated using an electrochemical method, based on the measurement of the resistance between two electrodes. This method is effective, i.e., easy to use, for the relative comparison of surface area values, and is not suitable for a total surface area evaluation for several reasons. The contact resistance error, i.e., systematic error, is eliminated by comparative measurement. The electrode polarization errors are eliminated by using an alternating current and a small current (i < 100 µA). Chemical changes of the electrolyte are eliminated by the low conductivity of the electrolyte, the small current used, and by the high volume of electrolyte. The conductivity of the metal coating of textiles is much higher than the conductivity of the electrolyte (H_2O), therefore the voltage drop in the electrolyte is much higher than the voltage drop on the coated fabric, and thus different coatings of the textiles can be neglected. The electrodes are placed in a plastic holder that is immersed into the electrolyte (H_2O), with temperature 25 °C). The first (base) electrode is fixed in a stable position in a plastic holder. The material of this electrode is a Cu plate, i.e., fully copper-plated printed circuit board with a surface area of copper 3.06×10^{-3} m^2. It is chosen because of a good surface flatness and negligible thickness of the conductive copper layer. The following four types of the second electrode are used:

- Cu plate (same sample properties as the first (base) electrode)
- CerexCuIv4
- RSKCu+Ni
- RSKCuIv4

In case of fabric samples, the samples were stuck on non-conductive plate using double-sided adhesive tape. The size of the surface area of all electrodes was the same, i.e., 3.06×10^{-3} m^2. An immersed plastic holder with electrodes inside the electrolyte is shown in Figure 6. The distance between the electrodes is changed during the measurement of resistance between electrodes. The resistance is measured by Battery Analyzer, 60 V Battery Analyzer BA6010 (BK Precision, Yorba Linda, CA, USA) at 1×10^3 Hz for three different distances between electrodes, i.e., 0.03, 0.04, and 0.05 m.

Comparison of the surface area of samples was based on the electrolyte resistivity constant and the distance between the electrodes:

$$R \cdot S = \rho \cdot l = const. \qquad (16)$$

where R is the measured resistance, S is the surface area, ρ is the resistivity of electrolyte, and l is the distance between electrodes.

Equation (16) shows the comparison of the surface area of samples was relative, i.e., it identifies the larger or smaller surface area in comparison with other samples. This comparative measurement method eliminates systematic errors.

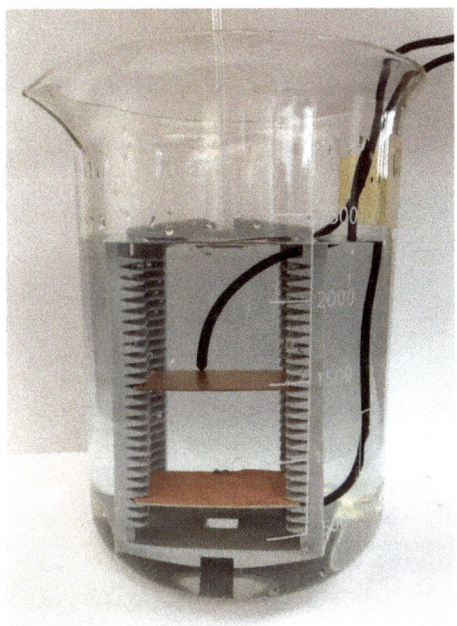

Figure 6. Immersed plastic holder with electrodes inside the electrolyte to give a measurement of the resistance between two electrodes.

3. Results and Discussion

3.1. Surface Area Evaluation of CerexCuIv4

The surface area of CerexCuIv4, i.e., coated non-woven fabric, was evaluated by the optical method described in Section 2.3. Figure 7 shows a detail of a sample, which was used for the calculation of the effective surface area. The results are shown for a handmade selection method in Table 2, for the automatic image conversion to binary method in Table 3, and for the threshold selection method in Table 4. The surface area evaluated by the handmade selection method was $2.67 \times 10^{-7} \pm 1.89 \times 10^{-8}$ m^2, by automatic image conversion to binary method was $2.30 \times 10^{-7} \pm 1.49 \times 10^{-8}$ m^2, and using the threshold selection method was $2.70 \times 10^{-7} \pm 1.60 \times 10^{-8}$ m^2.

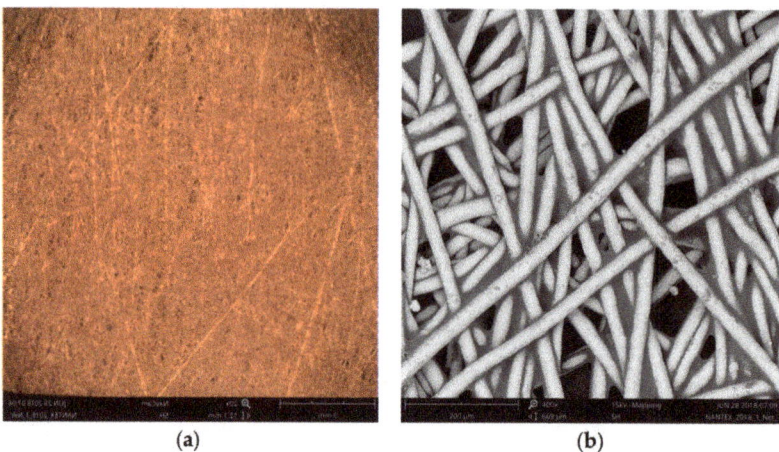

Figure 7. A sample of CerexCuIv4 (coated non-woven fabric) (**a**), and detail of the sample structure (**b**).

Table 2. Results for the handmade selection method.

Parameter	Average [1]
Scale bar (pxs)	2.66×10^2
Diameter (pxs)	5.82×10^1
Surface area (m^2)	2.67×10^{-7}
Surface area standard deviation σ (m^2)	1.89×10^{-8}

[1] Number of measurements: 10.

Table 3. Results for automatic image conversion to binary method.

Parameters	Average [1]
Scale bar (pxs)	2.66×10^2
Diameter (pxs)	5.61×10^1
Surface area (m^2)	2.30×10^{-7}
Surface area standard deviation σ (m^2)	1.49×10^{-8}

[1] Number of measurements: 10.

Table 4. Results for threshold selection method.

Parameters	Average [1]
Scale bar (pxs)	2.66×10^2
Diameter (pxs)	5.65×10^1
Surface area (m^2)	2.70×10^{-7}
Surface area standard deviation σ (m^2)	1.60×10^{-8}

[1] Number of measurements: 10.

3.2. Surface Area Evaluation of RSKCuIv4

The surface area of RSKCuIv4 was evaluated using the method based on Peirce's geometrical model of the interlacing point described in Sections 2.1 and 2.2 (Figure 8).

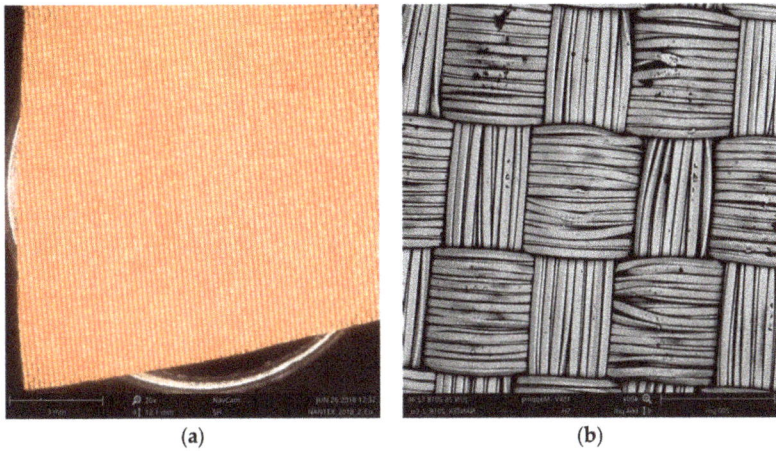

Figure 8. A sample of RSKCuIv4 (**a**), and detail of the sample structure (**b**).

Figure 9 shows the measurement of input parameters for the theoretical length of the warp and weft calculation using Equations (7) and (8). The length of the line connecting the arcs in warp and weft are marked *a* and *b*, respectively. The sum of the radii of the warp and weft yarns d_s was obtained from measurement of diameters of the warp and weft yarns, marked *d-warp* and *d-weft*, respectively. The warp interlacing angle Φ was obtained from measurement of the spacing of yarns, marked *M*; from calculation of the height of cylinder of yarns, Equation (6), i.e., parameters *M*, *d-warp*, and *d-weft* were considered; and from calculation of the radius of yarns, i.e., parameters *d-warp* and *d-weft* were considered. The theoretical length of the warp L_{WARP_CALC} and L_{WEFT_CALC} was then calculated (Table 5). The length of the warp L_{WARP_MEAS} was also measured (Figure 10 and Table 6).

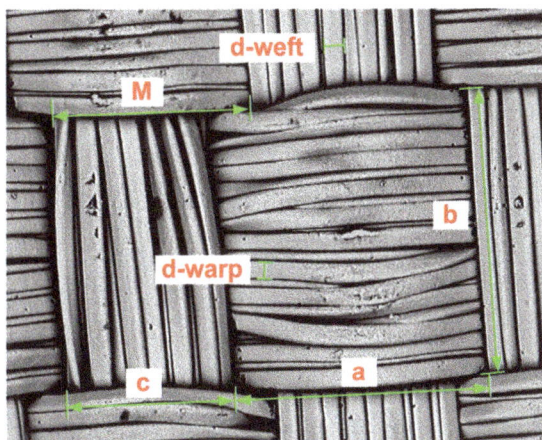

Figure 9. Measurement of input parameters for the theoretical length of the warp and weft calculation for the sample RSKCuIv4.

Table 5. The measurement results of input parameters a, b, c, M, d-warp, d-weft, L_{WARP_CALC}, and L_{WEFT_CALC}-RSKCuIv4.

Meas. No.	a (m)	b (m)	c (m)	M (m)	d-warp (m)	d-weft (m)	L_{WARP_CALC} (m)	L_{WEFT_CALC} (m)
1	2.01×10^{-4}	2.12×10^{-4}	1.45×10^{-4}	1.55×10^{-4}	1.08×10^{-5}	1.02×10^{-5}	2.03×10^{-4}	2.13×10^{-4}
2	2.03×10^{-4}	2.12×10^{-4}	1.44×10^{-4}	1.58×10^{-4}	1.04×10^{-5}	1.00×10^{-5}	2.04×10^{-4}	2.13×10^{-4}
3	2.04×10^{-4}	2.12×10^{-4}	1.45×10^{-4}	1.58×10^{-4}	1.08×10^{-5}	1.08×10^{-5}	2.06×10^{-4}	2.13×10^{-4}
4	2.02×10^{-4}	2.17×10^{-4}	1.47×10^{-4}	1.60×10^{-4}	1.08×10^{-5}	1.05×10^{-5}	2.03×10^{-4}	2.18×10^{-4}
5	2.00×10^{-4}	2.22×10^{-4}	1.46×10^{-4}	1.59×10^{-4}	1.02×10^{-5}	1.09×10^{-5}	2.01×10^{-4}	2.23×10^{-4}
6	2.00×10^{-4}	2.23×10^{-4}	1.47×10^{-4}	1.61×10^{-4}	1.03×10^{-5}	1.02×10^{-5}	2.01×10^{-4}	2.24×10^{-4}
7	1.97×10^{-4}	2.25×10^{-4}	1.46×10^{-4}	1.59×10^{-4}	1.09×10^{-5}	1.00×10^{-5}	1.98×10^{-4}	2.26×10^{-4}
8	1.96×10^{-4}	2.23×10^{-4}	1.44×10^{-4}	1.59×10^{-4}	1.08×10^{-5}	1.01×10^{-5}	1.97×10^{-4}	2.24×10^{-4}
9	1.95×10^{-4}	2.20×10^{-4}	1.49×10^{-4}	1.60×10^{-4}	1.03×10^{-5}	1.02×10^{-5}	1.96×10^{-4}	2.21×10^{-4}
10	1.97×10^{-4}	2.17×10^{-4}	1.45×10^{-4}	1.62×10^{-4}	1.08×10^{-5}	1.03×10^{-5}	1.98×10^{-4}	2.18×10^{-4}
Average	2.00×10^{-4}	2.18×10^{-4}	1.46×10^{-4}	1.59×10^{-4}	1.06×10^{-5}	1.03×10^{-5}	2.01×10^{-4}	2.19×10^{-4}
Stand. deviation σ	2.90×10^{-6}	4.80×10^{-6}	1.47×10^{-6}	1.80×10^{-6}	3.00×10^{-7}	3.00×10^{-7}	3.00×10^{-6}	4.70×10^{-6}

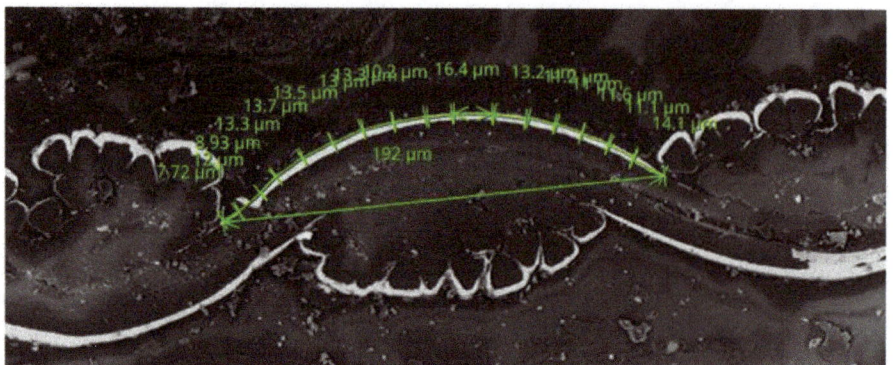

Figure 10. Measurement of the length of the warp for the sample RSKCuIv4.

Table 6. The measurement results of the length of the warp L_{WARP_MEAS} and the length of the line connecting the arcs in warp a-RSKCuIv4.

Meas. No.	Measured Parts of the arc (m)						L_{warp_MEAS} (m)	a (m)
1	7.70×10^{-6} 1.30×10^{-5} 1.10×10^{-5}	1.20×10^{-5} 1.33×10^{-5} 1.16×10^{-5}	8.90×10^{-6} 1.02×10^{-5} 1.16×10^{-5}	1.33×10^{-5} 1.44×10^{-5} 1.11×10^{-5}	1.37×10^{-5} 1.32×10^{-5} 1.41×10^{-5}	1.35×10^{-5} 1.12×10^{-5} -	2.04×10^{-4}	1.92×10^{-4}
2	1.00×10^{-5} 1.30×10^{-5} 1.05×10^{-5}	1.30×10^{-5} 1.05×10^{-5} 1.15×10^{-5}	7.00×10^{-6} 1.12×10^{-5} 1.30×10^{-5}	1.20×10^{-5} 1.70×10^{-5} 1.30×10^{-5}	1.10×10^{-5} 1.40×10^{-5} 1.25×10^{-5}	1.40×10^{-5} 1.30×10^{-5} -	2.06×10^{-4}	1.93×10^{-4}
3	1.20×10^{-5} 1.25×10^{-5} 1.02×10^{-5}	1.10×10^{-5} 1.05×10^{-5} 1.10×10^{-5}	1.05×10^{-5} 9.50×10^{-6} 1.12×10^{-5}	1.30×10^{-5} 1.45×10^{-5} 1.35×10^{-5}	1.40×10^{-5} 1.40×10^{-5} 1.30×10^{-5}	1.38×10^{-5} 1.09×10^{-5} -	2.05×10^{-4}	1.96×10^{-4}
4	1.40×10^{-5} 1.27×10^{-5} 1.05×10^{-5}	1.10×10^{-5} 1.01×10^{-5} 9.00×10^{-6}	1.35×10^{-5} 1.37×10^{-5} 1.22×10^{-5}	1.10×10^{-5} 1.27×10^{-5} 1.35×10^{-5}	1.25×10^{-5} 1.13×10^{-5} 1.22×10^{-5}	1.27×10^{-5} 1.15×10^{-5} -	2.06×10^{-4}	1.97×10^{-4}
5	1.30×10^{-5} 1.40×10^{-5} 1.12×10^{-5}	1.25×10^{-5} 1.36×10^{-5} 1.10×10^{-5}	1.17×10^{-5} 1.90×10^{-5} 1.28×10^{-5}	1.20×10^{-5} 1.28×10^{-5} 1.17×10^{-5}	1.27×10^{-5} 1.30×10^{-5} 9.70×10^{-6}	1.30×10^{-5} 9.00×10^{-6} -	2.13×10^{-4}	1.98×10^{-4}
Average	-	-	-	-	-	-	2.07×10^{-4}	1.95×10^{-4}
Stand. deviation σ	-	-	-	-	-	-	3.09×10^{-6}	2.30×10^{-6}

The theoretical warp length L_{WARP_CALC} was $2.01 \times 10^{-4} \pm 3.00 \times 10^{-6}$ m and the measured warp length L_{WARP_MEAS} was $2.07 \times 10^{-4} \pm 3.09 \times 10^{-6}$ m. The ratio of L_{WARP_CALC} and L_{WARP_MEAS} is presented in Table 7. It shows the difference in the units of percentage, i.e., the deformation of the yarns could be neglected, thus the circular shape of cross-section of warp and weft yarns of the Peirce's geometrical model of the interlacing point could be used. The theoretical warp lengths L_{WARP_CALC} and L_{WEFT_CALC} were inserted in the Equations (9) and (10) to find S_{WARP} and S_{WEFT}, respectively. The surface area of the investigated Peirce's geometrical model of the sample was then obtained using Equation (12). The surface area of the investigated multifilament model of the sample was then obtained using Equations (13) and (14) (Table 8).

Table 7. The comparison of measurement and calculated results of the length of the warp-RSKCuIv4.

Meas. No.	L_{warp_MEAS} (m)	L_{warp_CALC} (m)	$L_{warp\_MEAS}/L_{warp\_CALC}$ (%)
1	2.04×10^{-4}	2.02×10^{-4}	100.99
2	2.06×10^{-4}	2.04×10^{-4}	100.98
3	2.05×10^{-4}	2.05×10^{-4}	100.00
4	2.06×10^{-4}	2.03×10^{-4}	101.48
5	2.13×10^{-4}	2.01×10^{-4}	105.97
Average	2.07×10^{-4}	2.03×10^{-4}	101.88
Stand. deviation σ	3.09×10^{-6}	1.43×10^{-6}	2.10

Table 8. The modeling results for RSKCuIv4.

Parameter	Average	Standard Deviation
L_{WARP_CALC} (m)	2.01×10^{-4}	3.00×10^{-6}
L_{WEFT_CALC} (m)	2.19×10^{-4}	4.70×10^{-6}
S_{WARP} (m^2)	6.70×10^{-15}	1.90×10^{-16}
S_{WEFT} (m^2)	7.11×10^{-15}	2.40×10^{-16}
S_{Inter_point} (m^2)	1.7×10^{-16}	1.00×10^{-17}
S_{Effect} (m^2)	6.73×10^{-15}	1.30×10^{-16}
$S1_{WARP\_Multi}$ (m^2)	6.50×10^{-14}	1.03×10^{-15}
$S2_{WEFT\_Multi}$ (m^2)	4.74×10^{-14}	1.03×10^{-15}

3.3. Surface Area Evaluation of RSKCu+Ni

Surface area of RSKCu+Ni was evaluated using the method based on Peirce's geometrical model of the interlacing point described in Sections 2.1 and 2.2 (Figure 11).

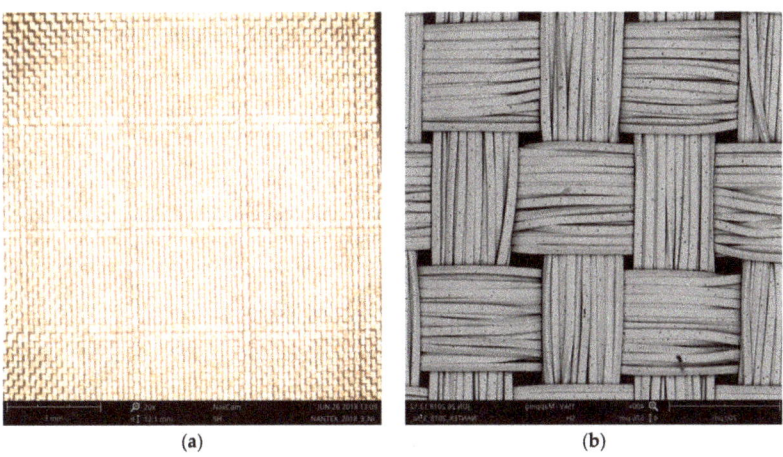

Figure 11. A sample of RSKCu+Ni (a), and detail of the sample structure (b).

Figure 12 shows the measurement of the input parameters for the theoretical length of the warp and weft calculation using Equations (7) and (8). The length of the line connecting the arcs in warp and weft are marked a and b, respectively. The sum of the radii of the warp and weft yarns d_s was obtained from the measurement of diameters of the warp and weft yarns, marked d-$warp$ and d-$weft$, respectively. The warp interlacing angle Φ was obtained from measurement of the spacing of yarns, marked M; from calculation of the height of cylinder of yarns, Equation (6), i.e., parameters M, d-$warp$ and d-$weft$ were considered; and from calculation of the radius of yarns, i.e., parameters d-$warp$ and d-$weft$ were considered (Table 9). The theoretical length of the warp L_{WARP_CALC} and the weft L_{WEFT_CALC} was then calculated (Table 10). The length of the warp L_{WARP_MEAS} was also measured (Figure 13 and Table 11).

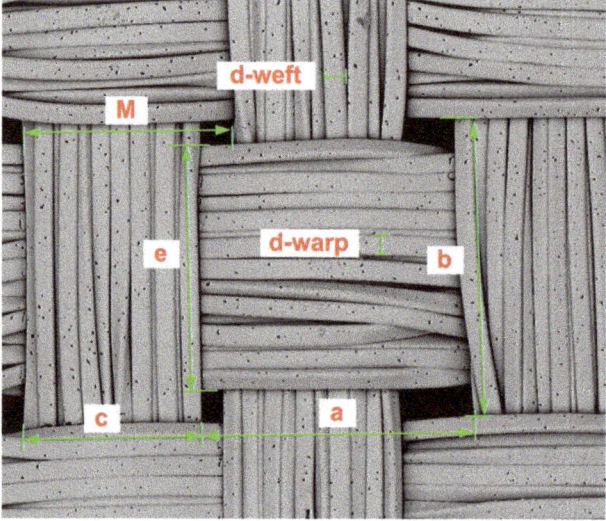

Figure 12. Measurement of input parameters for the theoretical length of the warp and weft calculation for the sample RSKCu+Ni.

Table 9. The measurement results of input parameters a, b, c, e, M, d-$warp$, and d-$weft$-RSKCu+Ni.

Meas. No.	a (m)	b (m)	c (m)	e (m)	M (m)	d-warp (m)	d-weft (m)
1	1.99×10^{-4}	2.38×10^{-4}	1.46×10^{-4}	1.76×10^{-4}	1.74×10^{-4}	1.05×10^{-5}	1.02×10^{-5}
2	2.07×10^{-4}	2.34×10^{-4}	1.41×10^{-4}	1.83×10^{-4}	1.72×10^{-4}	1.03×10^{-5}	1.00×10^{-5}
3	2.07×10^{-4}	2.39×10^{-4}	1.53×10^{-4}	1.80×10^{-4}	1.70×10^{-4}	1.01×10^{-5}	1.05×10^{-5}
4	2.06×10^{-4}	2.39×10^{-4}	1.44×10^{-4}	1.99×10^{-4}	1.69×10^{-4}	1.03×10^{-5}	1.01×10^{-5}
5	2.06×10^{-4}	2.39×10^{-4}	1.41×10^{-4}	1.93×10^{-4}	1.69×10^{-4}	1.03×10^{-5}	1.01×10^{-5}
6	2.08×10^{-4}	2.39×10^{-4}	1.48×10^{-4}	1.93×10^{-4}	1.67×10^{-4}	1.03×10^{-5}	1.02×10^{-5}
7	2.10×10^{-4}	2.38×10^{-4}	1.51×10^{-4}	1.91×10^{-4}	1.66×10^{-4}	1.05×10^{-5}	1.09×10^{-5}
8	2.12×10^{-4}	2.38×10^{-4}	1.47×10^{-4}	1.94×10^{-4}	1.64×10^{-4}	1.03×10^{-5}	1.01×10^{-5}
9	2.15×10^{-4}	2.37×10^{-4}	1.53×10^{-4}	1.91×10^{-4}	1.61×10^{-4}	1.04×10^{-5}	1.07×10^{-5}
10	2.15×10^{-4}	2.37×10^{-4}	1.42×10^{-4}	1.91×10^{-4}	1.61×10^{-4}	1.05×10^{-5}	1.03×10^{-5}
Average	2.09×10^{-4}	2.38×10^{-4}	1.47×10^{-4}	1.89×10^{-4}	1.67×10^{-4}	1.04×10^{-5}	1.03×10^{-5}
Stand. deviation σ	4.50×10^{-6}	1.50×10^{-6}	4.41×10^{-6}	6.74×10^{-6}	4.10×10^{-6}	1.00×10^{-7}	3.00×10^{-7}

Table 10. The modeling results for L_{WARP_CALC} and L_{WEFT_CALC} – RSKCu+Ni.

Meas. No.	L_{WARP_CALC} (m)	L_{WEFT_CALC} (m)
1	2.00×10^{-4}	2.13×10^{-4}
2	2.08×10^{-4}	2.13×10^{-4}
3	2.08×10^{-4}	2.13×10^{-4}
4	2.07×10^{-4}	2.18×10^{-4}
5	2.07×10^{-4}	2.23×10^{-4}
6	2.09×10^{-4}	2.24×10^{-4}
7	2.11×10^{-4}	2.26×10^{-4}
8	2.13×10^{-4}	2.24×10^{-4}
9	2.16×10^{-4}	2.21×10^{-4}
10	2.16×10^{-4}	2.18×10^{-4}
Average	2.10×10^{-4}	2.19×10^{-4}
Stand. deviation σ	4.60×10^{-6}	4.70×10^{-6}

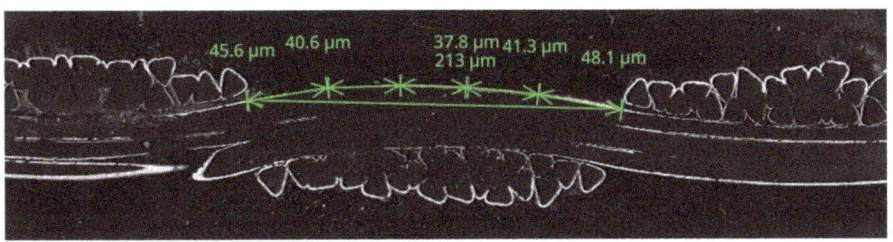

Figure 13. Measurement of the length of the warp for the sample RSKCu+Ni.

Table 11. The measurement results of the length of the warp L_{WARP_MEAS} and the length of the line connecting the arcs in warp a-RSKCu+Ni.

Meas. No.	Measured Parts of the arc (m)					L_{warp_MEAS} (m)	a (m)
1	4.56×10^{-5}	4.06×10^{-5}	3.78×10^{-5}	4.13×10^{-5}	4.81×10^{-5}	2.13×10^{-4}	2.13×10^{-4}
2	4.47×10^{-5}	3.99×10^{-5}	3.72×10^{-5}	4.29×10^{-5}	4.90×10^{-5}	2.14×10^{-4}	2.13×10^{-4}
3	4.39×10^{-5}	4.12×10^{-5}	3.76×10^{-5}	4.10×10^{-5}	5.00×10^{-5}	2.14×10^{-4}	2.14×10^{-4}
4	4.49×10^{-5}	4.15×10^{-5}	3.73×10^{-5}	4.11×10^{-5}	4.87×10^{-5}	2.14×10^{-4}	2.13×10^{-4}
5	4.60×10^{-5}	4.04×10^{-5}	3.80×10^{-5}	4.14×10^{-5}	4.86×10^{-5}	2.14×10^{-4}	2.13×10^{-4}
Average	4.50×10^{-5}	4.07×10^{-5}	3.76×10^{-5}	4.15×10^{-5}	4.89×10^{-5}	2.14×10^{-4}	2.13×10^{-4}
Stand. deviation σ	7.00×10^{-7}	6.00×10^{-7}	3.00×10^{-7}	7.00×10^{-7}	6.00×10^{-7}	3.00×10^{-7}	4.00×10^{-7}

The theoretical warp length L_{WARP_CALC} was $2.10 \times 10^{-4} \pm 4.58 \times 10^{-6}$ m and the measured warp length L_{WARP_MEAS} was $2.14 \times 10^{-4} \pm 3.00 \times 10^{-7}$ m. The ratio of L_{WARP_CALC} and L_{WARP_MEAS} is presented in Table 12. It also shows the difference in the units of percentage, i.e., the deformation of the yarns could be neglected, thus the circular shape of cross-section of warp and weft yarns of the Peirce's geometrical model of the interlacing point could be used. The theoretical warp lengths L_{WARP_CALC} and L_{WEFT_CALC} were inserted into Equations (9) and (10) to find S_{WARP} and S_{WEFT}, respectively. The surface area of the investigated Pierce's geometrical model of the sample S_{Effect} was then obtained using Equation (12). The surface area of the investigated multifilament model of the sample RSKCu+Ni was then obtained using Equations (13) and (14) (Table 13).

Table 12. The comparison of measurement and calculated results of the length of the warp-RSKCu+Ni.

Meas. No.	L_{warp_MEAS}	L_{warp_CALC}	$L_{warp\_MEAS}/L_{warp\_CALC}$
	(m)	(m)	(%)
1	2.13×10^{-4}	2.00×10^{-4}	106.50
2	2.14×10^{-4}	2.08×10^{-4}	102.88
3	2.14×10^{-4}	2.08×10^{-4}	102.88
4	2.14×10^{-4}	2.07×10^{-4}	103.38
5	2.14×10^{-4}	2.07×10^{-4}	103.38
Average	2.14×10^{-4}	2.06×10^{-4}	103.81
Stand. deviation σ	3.00×10^{-7}	3.02×10^{-6}	1.36

Table 13. The modeling results for RSKCu+Ni.

Parameter	Average	Standard Deviation
L_{WARP_CALC} (m)	2.10×10^{-4}	4.58×10^{-6}
L_{WEFT_CALC} (m)	2.39×10^{-4}	1.50×10^{-6}
S_{WARP} (m^2)	6.82×10^{-15}	1.80×10^{-16}
S_{WEFT} (m^2)	7.73×10^{-15}	2.20×10^{-16}
S_{Inter_point} (m^2)	1.70×10^{-16}	1.00×10^{-17}
S_{Effect} (m^2)	7.11×10^{-15}	1.60×10^{-16}
$S1_{WARP\_Multi}$ (m^2)	4.88×10^{-14}	1.07×10^{-15}
$S2_{WEFT\_Multi}$ (m^2)	4.30×10^{-14}	2.60×10^{-16}

3.4. Relative Comparison of Surface Areas of Samples Using the Electrochemical Method

The surface area of the Cu plate, RSKCuIv4, CerexCuIv4, and RSKCu+Ni was evaluated using the comparative measurement method described in Section 2.4. Table 14 and Figure 14 show measurement results of the resistance between the first (base) electrode, i.e., Cu plate, and the second electrode, i.e., Cu plate, RSKCuIv4, CerexCuIv4, and RSKCu+Ni. The resistance results were inserted into Equation (16) and surface areas were obtained (Table 14).

Table 14. The measurement results of the resistance between two electrodes: Cu plate + (Cu plate, CerexCuIv4, RSKCu+Ni, or RSKCuIv4).

Distance between Electrodes (m)	Resistance (Ω)			
	Cu plate	RSKCuIv4	CerexCuIv4	RSKCu+Ni
1.00×10^{-2}	6.5×10^{-1}	6.9×10^{-1}	6.4×10^{-1}	6.3×10^{-1}
2.00×10^{-2}	1.2×10^{-2}	1.2×10^{-2}	1.2×10^{-2}	1.1×10^{-2}
3.00×10^{-2}	1.7×10^{-2}	1.6×10^{-2}	1.6×10^{-2}	1.5×10^{-2}
4.00×10^{-2}	2.0×10^{-2}	1.9×10^{-2}	1.9×10^{-2}	1.8×10^{-2}
5.00×10^{-2}	2.2×10^{-2}	2.1×10^{-2}	2.1×10^{-2}	2.1×10^{-2}
10.00×10^{-2}	3.1×10^{-2}	3.1×10^{-2}	3.2×10^{-2}	3.0×10^{-2}
	Surface area (m^2)			
Average	3.06×10^{-3}	3.18×10^{-3}	3.22×10^{-3}	3.28×10^{-3}
Stand. deviation σ	0.00×10^{0}	4.50×10^{-5}	8.22×10^{-5}	7.48×10^{-5}

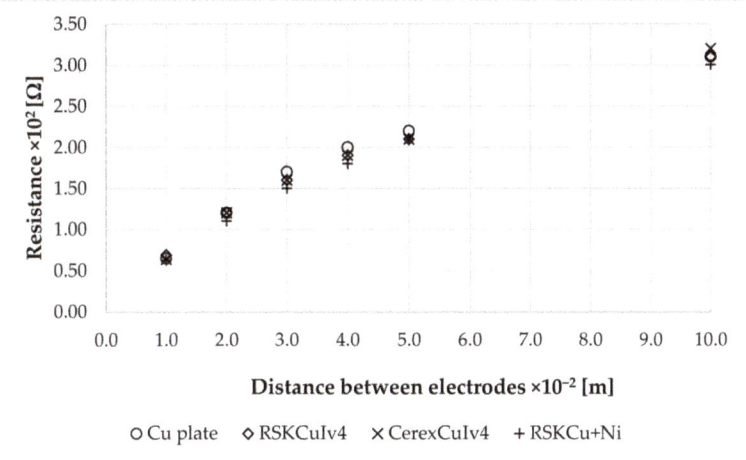

Figure 14. Measurement results of the resistance values in relation to the distance between electrodes.

Resistance values increased with the increased distance between electrodes. This corresponded with the theory as shown in Equation (16), i.e., the surface area was obtained for each sample. The largest surface area was obtained for RSKCu+Ni, followed by CerexCuIv4. The lowest surface area was obtained for RSKCuIv4.

3.5. Summary of the Surface Area Evaluation

The theoretical warp length L_{WARP_CALC} was $2.01 \times 10^{-4} \pm 3.00 \times 10^{-6}$ m and the measured warp length L_{WARP_MEAS} was $2.07 \times 10^{-4} \pm 3.09 \times 10^{-6}$ m for the RSKCuIv4. The theoretical warp length L_{WARP_CALC} was $2.10 \times 10^{-4} \pm 4.58 \times 10^{-6}$ m and the measured warp length L_{WARP_MEAS} was $2.14 \times 10^{-4} \pm 3.00 \times 10^{-7}$ m for the RSKCu+Ni. The measured and the calculated values did not differ from each other within the standard deviation, and the theoretical warp lengths L_{WARP_CALC} and L_{WEFT_CALC} were inserted into Equations (9) and (10) to find the surface area of the warp S_{WARP} and weft S_{WEFT}. The surface area of the investigated Pierce's geometrical model S_{effect} for monofilament (non-fibrous) yarns was then calculated for RSKCuIv4 to be $6.73 \times 10^{-15} \pm 1.30 \times 10^{-16}$ m^2 and $7.11 \times 10^{-15} \pm 1.60 \times 10^{-16}$ m^2 for RSKCu+Ni. The surface area of multifilament yarns in warp S_1 and weft S_2 was determined for RSKCuIv4 to be $S_1 = 6.50 \times 10^{-14} \pm 1.03 \times 10^{-15}$ m2 and $S_2 = 4.74 \times 10^{-14} \pm 1.03 \times 10^{-15}$ m^2, and $S_1 = 4.88 \times 10^{-14} \pm 1.07 \times 10^{-15}$ m^2 and $S_2 = 4.30 \times 10^{-14} \pm 2.60 \times 10^{-16}$ m^2 for RSKCu+Ni.

The surface area of Cu plate, RSKCuIv4, CerexCuIv4, and RSKCu+Ni was evaluated using the comparative calculation. The results from the individual surface area of the presented models of samples S_1 and S_2 were recalculated to a uniform surface area $S_{default}$ 3.06×10^{-3} m^2 using Equation (17) (Table 15).

$$S_{Surface\_area\_model} = (S_1 + S_2) * \frac{S_{default}}{S_{background}}. \quad (17)$$

where $S_{default}$ is the surface area about size 3.06×10^{-3} m^2 and $S_{background}$ is the surface area under of Peirce's geometrical model of the interlacing point of individual samples.

Modeling results of the surface area of samples shown in Table 15 were compared with the measurement results shown in Table 14 (Figure 15).

Table 15. The modeling results for surface area of samples.

Modeling No.	RSKCuIv4 (m^2)	CerexCuIv4 (m^2)	RSKCu+Ni (m^2)
1	4.65×10^{-3}	5.37×10^{-3}	5.78×10^{-3}
2	4.67×10^{-3}	5.36×10^{-3}	5.73×10^{-3}
3	4.69×10^{-3}	5.31×10^{-3}	5.66×10^{-3}
4	4.67×10^{-3}	5.43×10^{-3}	5.68×10^{-3}
5	4.64×10^{-3}	5.33×10^{-3}	5.68×10^{-3}
6	4.64×10^{-3}	5.36×10^{-3}	5.66×10^{-3}
7	4.60×10^{-3}	5.32×10^{-3}	5.65×10^{-3}
8	4.59×10^{-3}	5.36×10^{-3}	5.62×10^{-3}
9	4.58×10^{-3}	5.41×10^{-3}	5.60×10^{-3}
10	4.61×10^{-3}	5.36×10^{-3}	5.60×10^{-3}
Average	4.63×10^{-3}	5.36×10^{-3}	5.67×10^{-3}
Stand. deviation σ	3.62×10^{-5}	3.54×10^{-5}	5.23×10^{-5}

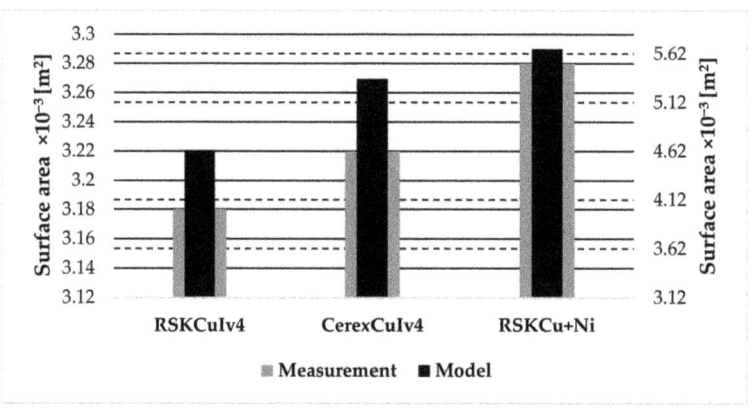

Figure 15. Comparison of modeling and measurement results of the surface area of samples.

The surface area of RSKCuIv4 was evaluated to be $S_{model\_RSKCuIv4} = 4.63 \times 10^{-3} \pm 3.62 \times 10^{-5}$ m^2 and $S_{measurement\_RSKCuIv4} = 3.18 \times 10^{-3} \pm 4.50 \times 10^{-5}$ m^2. The surface area of CerexCuIv4 was evaluated to be $S_{model\_CerexCuIv4} = 5.36 \times 10^{-3} \pm 3.54 \times 10^{-5}$ m^2 and emph$S_{measurement\_CerexCuIv4} = 3.22 \times 10^{-3} \pm 8.22 \times 10^{-5}$ m^2. The surface area of RSKCu+Ni was evaluated to be $S_{model\_RSKCu+Ni} = 5.67 \times 10^{-3} \pm 5.23 \times 10^{-5}$ m^2 and $S_{measurement\_RSKCu+Ni} = 3.28 \times 10^{-3} \pm 7.48 \times 10^{-5}$ m^2. The modeling results for the surface area evaluations of the samples showed an increasing trend, i.e., the surface area of the RSKCuIv4 was the smallest surface area of investigated samples, followed by the surface area of the CerexCuIv4, and by RSKCu+Ni. This increasing trend, i.e., the surface area of the RSKCuIv4 was the smallest one, followed by CerexCuIv4 and by RSKCu+Ni, was also shown using the experimental electrochemical method. Both the modeled and the measured surface areas were larger than the uniform surface area, i.e., the sample Cu plate, which corresponded to the theory that the coated polymer-based textiles have a larger surface area than the standard metal plate. The measurement results were of the same order as the modeling results, but showed higher values. This was caused by the used method, which is suitable only for the relative comparison of surface area values as described in Section 2.4.

4. Conclusions

Calculation of the surface area of coated woven and non-woven textiles finds its application in development of electrical and electrochemical components and devices such as supercapacitors, electrochemical cells, and electrochemical catalysts. In this paper, several methods for the real effective surface area evaluation of coated woven and non-woven polymer-based textiles were presented. The proposed model for woven textiles was based on Pierce's geometrical model and on the evaluation of samples by optical methods. The evaluation of non-woven textiles was performed using optical methods and image processing methods. All these proposed models were compared with an experimental electrochemical method for surface area evaluation measurement. The experimental results confirmed the results obtained from modeling. The results show that the largest effective surface area was obtained for copper- and nickel-plated woven polyester fabric and the lowest was obtained for copper- and acrylic-coated woven polyester fabric.

Author Contributions: All authors contributed to the writing of the article. All authors contributed to computations and models. L.V. and K.D. designed and made an experimental verification of the model.

Funding: This research was funded by the Ministry of Industry and Trade grant number FV30171.

Conflicts of Interest: The authors declare no conflict of interest.

References

1. Šafářová, V.; Militký, J. Multifunctional metal composite textile shields against electromagnetic radiation—Effect of various parameters on electromagnetic shielding effectiveness. *Polym. Compos.* **2017**, *38*, 309–323. [CrossRef]
2. Vojtech, L.; Neruda, M. Design of radiofrequency protective clothing containing silver nanoparticles. *Fibres Text. East. Eur.* **2013**, *5*, 141–147.
3. Lopez, A.; Vojtech, L.; Neruda, M. Comparison among models to estimate the shielding effectiveness applied to conductive textiles. *Adv. Electr. Electron. Eng.* **2013**, *11*, 387–391. [CrossRef]
4. Neruda, M.; Vojtech, L.; Rohlik, M.; Hajek, J.; Holý, R.; Kalika, M. Application of shielding textile materials in electric vehicles. In Proceedings of the 24th Wireless and Optical Communication Conference (WOCC), Taipei, Taiwan, 23–24 October 2015.
5. Mahltig, B.; Zhang, J.; Wu, L.; Darko, D.; Wendt, M.; Lempa, E.; Rabe, M.; Haase, H. Effect pigments for textile coating: a review of the broad range of advantageous functionalization. *J. Coat. Technol. Res.* **2017**, *14*, 35–55. [CrossRef]
6. Vojtech, L.; Neruda, M. Modelling of surface and bulk resistance for wearable textile antenna design. *Prz. Elektrotech.* **2013**, *89*, 217–222.
7. Ferreira, D.; Pires, P.; Rodrigues, R.; Caldeirinha, R.F. Wearable textile antennas: Examining the effect of bending on their performance. *IEEE Antennas Propag. Mag.* **2017**, *59*, 54–59. [CrossRef]
8. Singh, N.K.; Singh, V.K.; Naresh, B. Textile antenna for microwave wireless power transmission. *Procedia Comput. Sci.* **2016**, *85*, 856–861. [CrossRef]
9. Vojtech, L.; Bortel, R.; Neruda, M.; Kozak, M. Wearable textile electrodes for ECG measurement. *Adv. Electr. Electron. Eng.* **2013**, *11*, 410–414. [CrossRef]
10. Liu, M.; Pu, X.; Jiang, C.; Liu, T.; Huang, X.; Chen, L.; Du, C.; Sun, J.; Hu, W.; Wang, Z.L. Large-Area All-Textile Pressure Sensors for Monitoring Human Motion and Physiological Signals. *Adv. Mater.* **2017**, *29*, 1703700. [CrossRef] [PubMed]
11. Zhou, G.; Byun, J.H.; Oh, Y.; Jung, B.M.; Cha, H.J.; Seong, D.G.; Um, M.K.; Hyun, S.; Chou, T.W. Highly sensitive wearable textile-based humidity sensor made of high-strength, single-walled carbon nanotube/poly (vinyl alcohol) filaments. *ACS Appl. Mater. Interfaces* **2017**, *9*, 4788–4797. [CrossRef] [PubMed]
12. Laforgue, A. All-textile flexible supercapacitors using electrospun poly(3,4-ethylenedioxythiophene) nanofibers. *J. Power Sources* **2011**, *196*, 559–564. [CrossRef]
13. Zhang, Z.; Guo, K.; Li, Y.; Li, X.; Guan, G.; Li, H.; Luo, Y.; Zhao, F.; Zhang, Q.; Wei, B.; et al. A colour-tunable, weavable fibre-shaped polymer light-emitting electrochemical cell. *Nat. Photonics* **2015**, *9*, 233. [CrossRef]
14. Vojtech, L.; Hajek, J.; Neruda, M.; Zatloukal, M. Monometallic textile electrodes for "green" batteries. *Elektron. Elektrotech.* **2014**, *20*, 25–28. [CrossRef]
15. Reddy, T.B.; Linden, D. *Linden's Handbook of Batteries*, 4th ed.; McGraw-Hill Education: New York, NY, USA, 2010.

16. Grancarić, A.M.; Jerković, I.; Koncar, V.; Cochrane, C.; Kelly, F.M.; Soulat, D.; Legrand, X. Conductive polymers for smart textile applications. *J. Ind. Text.* **2018**, *48*, 612–642. [CrossRef]
17. Tian, Y.; Wu, J. A comprehensive analysis of the BET area for nanoporous materials. *AIChE J.* **2018**, *64*, 286–293. [CrossRef]
18. De Lange, M.F.; Thijs, J.H.V.; Gascon, J.; Kapteijn, F. Adsorptive characterization of porous solids: Error analysis guides the way. *Microporous Mesoporous Mater.* **2014**, *200*, 199–215. [CrossRef]
19. Non-Woven Metallized Textiles. Available online: https://lorix.hu/en/products/non-woven-metallized-textile (accessed on 30 July 2018).
20. RS Type Metallized Fabrics. Available online: https://lorix.hu/en/products/rs-type-metallized-fabrics (accessed on 30 July 2018).
21. Peirce, F.T. 5—The geometry of cloth structure. *J. Text. Inst. Trans.* **1937**, *28*, T45–T96. [CrossRef]
22. Dvořák, J.; Karel, P.; Žák, J. *Study of Interactions between Weaving Process and Weaving Machine Systems*; VÚTS, a.s.: Liberec, Czech Republic, 2018.
23. Bartoš, P.; Špatenka, P.; Volfová, L. Deposition of TiO$_2$-Based Layer on Textile Substrate: Theoretical and Experimental Study. *Plasma Process. Polym.* **2009**, *6*, 897–901. [CrossRef]
24. Create Draggable, Resizable Line. Available online: http://www.mathworks.com/help/images/ref/imline.html (accessed on 20 July 2018).
25. Create Draggable Freehand Region. Available online: http://www.mathworks.com/help/images/ref/imfreehand.html (accessed on 20 July 2018).
26. Volume Enclosed by a Cylinder. Available online: http://www.mathopenref.com/cylindervolume.html (accessed on 20 July 2018).
27. Open Image Viewer app. Available online: http://www.mathworks.com/help/images/ref/imtool.html (accessed on 20 July 2018).
28. Crop Image. Available online: http://www.mathworks.com/help/images/ref/imcrop.html (accessed on 20 July 2018).

© 2018 by the authors. Licensee MDPI, Basel, Switzerland. This article is an open access article distributed under the terms and conditions of the Creative Commons Attribution (CC BY) license (http://creativecommons.org/licenses/by/4.0/).

Article
A Hybrid Textile Electrode for Electrocardiogram (ECG) Measurement and Motion Tracking

Xiang An * and George K. Stylios

Research Institute for Flexible Materials, Heriot-Watt University, Edinburgh TD1 3HF, UK; G.Stylios@hw.ac.uk
* Correspondence: xa30@hw.ac.uk

Received: 27 July 2018; Accepted: 28 September 2018; Published: 2 October 2018

Abstract: Wearable sensors have great potential uses in personal health monitoring systems, in which textile-based electrodes are particularly useful because they are comfortable to wear and are skin and environmentally friendly. In this paper, a hybrid textile electrode for electrocardiogram (ECG) measurement and motion tracking was introduced. The hybrid textile electrode consists of two parts: A textile electrode for ECG monitoring, and a motion sensor for patient activity tracking. In designing the textile electrodes, their performance in ECG measurement was investigated. Two main influencing factors on the skin-electrode impedance of the electrodes were found: Textile material properties, and electrode sizes. The optimum textile electrode was silver plated, made of a high stitch density weft knitted conductive fabric and its size was 20 mm × 40 mm. A flexible motion sensor circuit was designed and integrated within the textile electrode. Systematic measurements were performed, and results have shown that the hybrid textile electrode is capable of recording ECG and motion signals synchronously, and is suitable for ambulatory ECG measurement and motion tracking applications.

Keywords: textile electrode; ECG; motion sensor; skin-electrode impedance

1. Introduction

With the miniaturization of electronics, improvements in performance of low-power microprocessors, and the development of artificial intelligence, personal health monitoring systems are becoming possible. Wearable electronics, wireless communications, textile sensors, mobile computing, and cloud computing are becoming increasingly important in personal health monitoring systems. Wearable sensors and textile electrodes are particularly suitable for some long-term health monitoring applications, such as electrocardiogram (ECG) measurement and motion tracking.

Textile electrodes are usually made of conductive yarns by weaving, knitting or embroidering processes; or by coating or printing conductive polymers on non-conductive fabrics. In the studies of textile electrodes, most textile electrodes are knitted structure [1–3]. Priniotakis et al. [4] compared the knitted and woven textile electrodes by using an electrochemical cell; the results show that the knitted structure has the lowest contact resistance. Woven and embroidered textile electrodes have also been researched with some success [5]. However, there is no consistent conclusion as to which type of textile structure (knit, woven, embroidered) performs best in ECG recording, because it involves many factors, such as the structure of the fibers and yarns, the fabric density, and the manufacturing process. The conductive material type for making the textile electrode is another important factor that affects the performance of the electrode. Many studies have used silver plated textile materials to make textile electrodes [2,5,6]. Other conductive materials have also been studied [7–10]. Rattfalt [11] made textile electrodes with 100% stainless steel and 20% stainless steel, which showed the acceptable stability of electrode potentials. However, stainless steel is highly direct current voltage (DC) polarizable and very alloy dependent [12]. Jang et al. [13] explored the possibilities of copper (Cu) sputtered fabric as

ECG electrode. Conductive polymers have also been used for making textile electrodes. Pani et al. [14] made textile electrodes with poly(3,4-ethylene dioxythiophene):poly(styrene sulfonate) (PEDOT:PSS) coated woven fabric to monitor ECG signals.

Compared with conventional silver/silver chloride (Ag/AgCl) rigid metal electrodes, textile electrodes have the advantage of being soft, flexible and breathable, allowing the wearer to feel more comfortable than conventional metal plate electrodes in long-term monitoring. In addition, as textile electrodes can be easily integrated into a garment by weaving, knitting or sewing, there is no need for any adhesive to attach on the body, so they are skin friendly (no skin irritation or discomfort) and environmentally friendly (electrodes are reusable). Based on these advantages, many researchers have used textile electrodes in the development of wearable ECG systems [7,15–18].

In this paper, a hybrid textile electrode is proposed. It consists of two parts: A textile electrode for ECG measurement, and a motion sensor for patient activity tracking. Although there are some studies that combine motion sensors and textile electrodes into a wearable system [17–20], this is the first time that the motion sensor is directly integrated into the textile electrode. There are good reasons for designing this hybrid textile electrode. First of all, motion signals that recorded in synchrony with ECG signals are beneficial in the diagnosis of heart disease. Some studies [21–24] have found that heavy physical exertion can be the trigger of the onset of arrhythmia and acute myocardial infarction. Furthermore, changes in posture (sitting up or standing up) may also be the cause of arrhythmia, known as postural tachycardia syndrome (PoTS) [25]. Therefore, the motion signals recorded in synchronization with the ECG signals can help the cardiologist find the cause of the heart disease by providing information about the patient's physical activity when the ECG shows an abnormality. Moreover, tracking daily physical activity and ECG can also help prevent the sudden death in patients with coronary heart disease, because some studies have shown that sudden death is related to physical exertion [26–28]. Secondly, since the hybrid textile electrode is placed on the patient's chest to measure the ECG, the motion sensor on the hybrid electrode can also obtain information about the patient's respiration by tracking the movement of the chest while the patient remains stationary (sitting or standing). The measured respiration along with the ECG can also be used to diagnosis a common heart disease—respiratory sinus arrhythmia.

Due to the absence of conductive gel/paste, textile electrodes usually have much higher and more unstable skin-electrode impedance than conventional Ag/AgCl wet electrodes. And the complexity and instability of fabric structure itself also make the characteristics of textile electrodes different from conventional metal plate electrodes. Therefore, in this paper, the electrical properties of the dry textile electrode at the skin-electrode interface were first studied. Based on that, the electrode material and size were investigated, and an optimum textile electrode was made. In order to integrate a motion sensor with a textile electrode, a small flexible printed circuit board (FPCB) was designed. A hybrid textile electrode was finally fabricated by integrating an optimum textile electrode with the small flexible motion sensor circuit board.

2. The Skin-Electrode Interface of Textile Electrode

Textile electrodes, like conventional metallic plate electrodes, are in contact with human skin as electrical conductors. The difference from metallic plate electrodes is that the conductive metal is electroplated onto the textile substance or blended into the yarn. Therefore, the electrochemical reactions occurring at the interface between the conventional metal electrode and the skin also occur at the interface between the textile electrode and the skin. The interface is called the skin-electrode interface. Neuman [29] proposed an equivalent circuit for modelling the electrical characteristics of the skin-electrode interface for conventional metal electrodes, as shown in Figure 1.

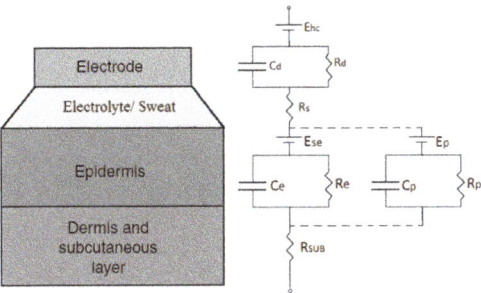

Figure 1. The electrical equivalent circuit of the skin-electrode interface [29].

In the case of dry textile electrodes, although they do not have a conductive gel/paste on the electrode surface, the skin moisture and perspiration can also be considered as a thin electrolyte layer between the textile electrode and the skin. "Dry" electrodes are really only dry when first applied, skin moisture and perspiration will quickly accumulate under the electrode [30]. Therefore, the equivalent circuit for conventional metal electrodes is also applicable to textile electrodes. According to the equivalent circuit, the skin-electrode impedance of the textile electrode $Z_{Textile}$ can be calculated as follows:

$$Z_{Textile} = R_s + \frac{R_d}{1 + j\omega R_d C_d} + R_{sub} + \frac{R_e}{1 + j\omega R_e C_e}, \quad (1)$$

where R_d represents the charge transfer resistance and C_d represents the capacitance across the electrode-electrolyte interface, R_e represents the resistance of epidermis layer, C_e represents the capacitance induced by the nonconductive stratum corneum layer, R_s represents the resistance of the sweat, R_{sub} represents the overall resistance of the tissue underneath the epidermis layer.

The skin-electrode impedance of the dry textile electrode is usually much higher than the conventional wet electrode. In most cases, the value of $Z_{Textile}$ is up to several hundred kΩ. Due to the fact that the human skin has a highly nonhomogeneous multi-layered structure, the electrical properties of the skin vary along different body parts, which also mean that the skin-electrode impedances of the two electrodes at different skin locations are generally different. Webster [29] has found that the impedance imbalance introduces noise into ECG signals. Olsen [31] has found that the impedance imbalance was typically 50 percent of the individual skin-electrode impedance. Therefore, the most effective way to reduce the impedance imbalance of dry textile electrodes is to reduce the skin-electrode impedance. Thus, the optimum textile electrode should be made of a material having low skin-electrode impedance characteristics.

3. Electrode Material

Various materials have been used to produce conductive textiles that are either embedded into fabrics as conductive yarns, or plated with electrically conductive components, such as carbon, copper, nickel, or silver. However, when choosing materials that will come into contact with the human skin, as in the case of ECG electrodes, their biocompatibility becomes very important as the electrode is directly applied onto the human body. Different from most other materials, silver is not only innocuous to human skin, but also antibacterial [32–35]. Therefore, conductive fabrics made from silver plated nylon yarns are favored for making textile electrodes by weaving or knitting. When compared with woven fabrics, knitted fabrics usually are more flexible, stretchable, and can take up easily the curvature of the body when attached. So, in this paper, four different knitted conductive fabrics made from silver plated nylon yarn were considered as electrode materials, shown in Table 1, and their electrical properties were investigated on a skin dummy. Electrode material TE1 is a silver plated knitted fabric purchased from Shieldex (MedTex P-130, Shieldex, Bremen, Germany), material TE2 is made of 4 ply silver plated nylon yarn (235/34 dtex 4-ply, Shieldex, Bremen, Germany), material TE3 is made of 2 ply silver plated nylon yarn (117/17 dtex 2-ply, Shieldex, Bremen, Germany) and material TE4 is a silver plated spacer fabric purchased from Shieldex (Spacer Fabric B, Shieldex, Bremen, Germany). Figure 2 shows the scanning electron microscope (SEM) micrograph of a silver plate nylon yarn. The average diameter of a silver plated nylon monofilament is about 0.028 mm.

Table 1. The properties of the four selected conductive knitted fabrics.

Electrode Materials		Components	Structure	Fabric Thickness (mm)	Yarn Diameter (mm)	Wales/cm	Courses/cm
TE1		78% silver plated nylon 66 + 22% Elastomer	Weft knitted	0.45 ± 10%	0.13 ± 20%	28/cm	30/cm
TE2		100% silver plated nylon 66	Weft knitted	1.25 ± 10%	0.60 ± 20%	5/cm	6/cm
TE3		100% silver plated nylon 66	Weft knitted	0.70 ± 10%	0.30 ± 20%	8/cm	12/cm
TE4		94% silver plated nylon 66 + 6% elastomer	Weft knitted 3D spacer	2.50 ± 10%	0.18 ± 20%	17/cm (surface)	28/cm (surface)

Figure 2. The scanning electron microscope (SEM) micrograph of a silver plated nylon yarn.

3.1. Experimental Method

The electrical properties of human skin have great variations and dependent upon time and location of the skin [36]. Thus, a skin dummy (Figure 3a) is used to measure the skin-electrode impedance to avoid the unwanted impedance variation induced by the human skin. The design of the skin dummy is based on Westbroek's electrochemical cell [4,37,38], which consists of a Polyvinyl chloride (PVC) tube filled with 0.9% of NaCl solution to simulate the body fluid. Two polyvinylidene fluoride (PVDF) membranes are installed on the two open ends of the PVC tube to simulate the skin barrier between the body fluid and the textile electrodes. The PVDF membranes were obtained from

Merck® (Darmstadt, Germany), and are the same membranes that were used in P.J. Xu's [39] dynamic evaluation system. The pore size of the PVDF membranes in our evaluation system is 100 nm, as this size is large enough to allow the electrolyte to flow freely through the perforated membrane [40].

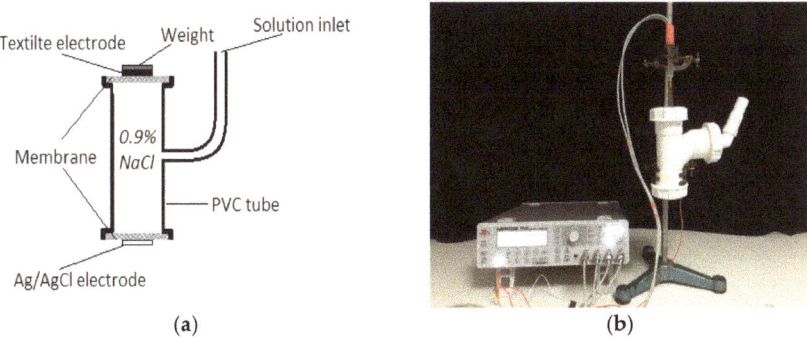

Figure 3. Skin-electrode impedance measurement on a skin dummy: (a) Skin dummy; (b) test setup.

Four textile electrodes made from materials TE1, TE2, TE3 and TE4 were tested on the skin dummy, as shown in Table 1. The structure of the textile electrodes used in the measurement, shown in Figure 4, it consists of two parts: A square area of size 15 mm × 15 mm, which is the electrode surface that is in contact with the skin dummy; a rectangular area of size 7 mm × 50 mm is the electrode wire that is connected to an impedance meter. In the measurement, the textile electrode is placed on the top surface of the skin dummy and is fixed with a weight of 100 g, which applied a force of 0.98 N to the electrode. A self-adhesive Ag/AgCl electrode is placed on the bottom surface to serve as a reference electrode. A high-precision LCR-Bridge meter (HM8118, HAMEG instruments, Mainhausen, Germany) is used in this system to measure the impedance. Two measurements were done in here. First, impedances were measured at the frequency of 100 Hz, and measurements were last for one hour. Second, impedances were measured within a frequency range of 20 Hz to 20 kHz when the skin-electrode impedance is stabilized. Measurements were performed in a conditioned laboratory where the room temperature was controlled at 20 ± 2 °C and the relative humidity at 65 ± 2%.

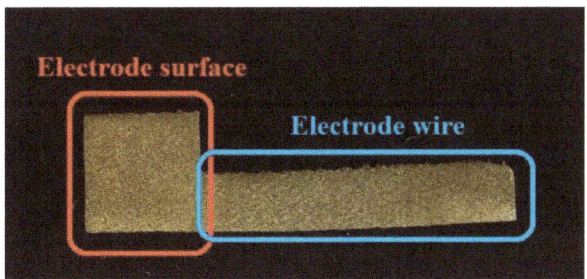

Figure 4. Structure of a textile electrode for the skin-electrode measurement.

3.2. Results and Discussion

As shown in Figure 5, the skin-electrode impedances of all four textile electrodes show a similar trend: Impedances are rapidly decreasing within the first few minutes and then gradually become stable. However, their differences are also noticeable. In the first few minutes of measurement, the electrode TE1 has the fastest impedance drop among the four electrodes, and its impedance tends to be stable in the shortest time. Moreover, after the stabilization period, the skin-electrode impedances of the four electrodes are different, from the highest to the lowest impedance TE2, TE3, TE4, TE1 respectively.

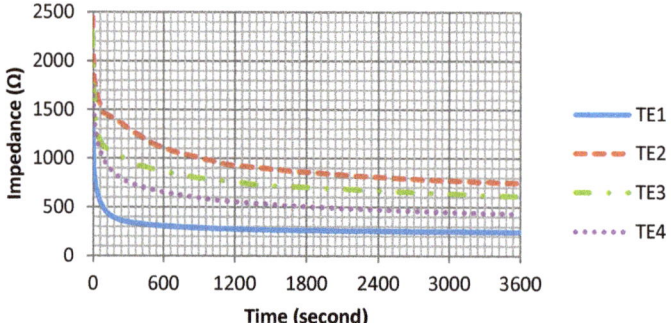

Figure 5. Skin-electrode impedance in 1 h.

Figure 6 shows the skin-electrode impedance over the frequency range. The impedance of the four electrodes is frequency-dependent: As the frequency increases, the impedance decreases, which is consistent with the capacitive behavior of the skin-electrode interface. However, the differences in these four electrode materials are also clearly shown in the impedance curves. Electrode TE1 has the smallest impedance, as well as the smoothest impedance-frequency curve.

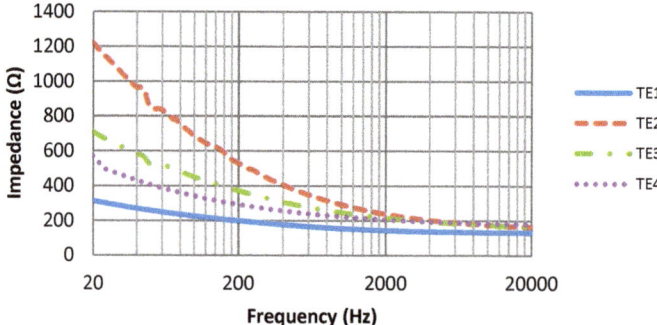

Figure 6. Skin-electrode impedance versus frequency.

The difference in skin-electrode impedance can be explained by the stitch density of these fabrics. As demonstrated in Table 1, the four electrode materials are all made of silver plated yarn, and they are all made by knitting. Textile Electrodes TE1, TE2, TE3 are all weft knitted structures, electrode TE4 is a knitted 3D spacer structure, but its surface layer is also weft knitted, as shown in Figure 7. The most significant difference between these four fabrics is their stitch density and the yarn diameter. Electrode TE1 has the highest stitch density and the smallest yarn diameter, whilst electrode TE2 has the lowest stitch density and the largest yarn diameter. The measured skin-electrode impedance is positively related to the yarn diameter and negatively related to stitch density.

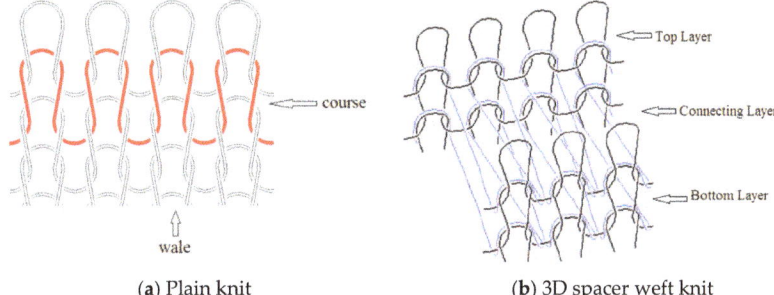

(a) Plain knit (b) 3D spacer weft knit

Figure 7. Knitted fabric structure.

According to the geometrical model of a plain knitted fabric, suggested by Munden [41], the basic structure of a knitted fabric is a loop that consists of parts of circles joined by straight lines, as shown in Figure 8. This model is based on Peirce's assumptions [42]: The bending resistance of the yarns was negligible and that the yarn was circular in cross-section.

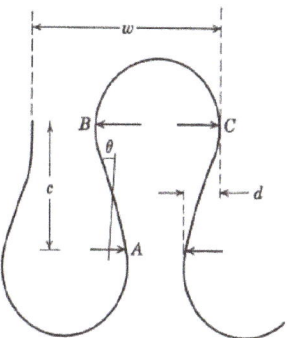

Figure 8. Loop model according to Munden [41].

The length of a single loop can be calculated by Equation (2):

$$l = 2 \cdot l_{AB} + 2 \cdot l_{BC} \tag{2}$$

where l_{AB} is the straight length between point A and point B on the loop, l_{BC} is the semicircle length between point B and point C, c is the course spacing, d is the yarn diameter. Equation (3) is the calculation of l_{AB} that based on the Peirce's model of plain weave [43]. Equation (4) is the calculation of l_{BC} that proposed by Munden [41].

$$l_{AB} = c\left[1 + \frac{9}{16}\left(\frac{d}{c}\right)^2\right] \tag{3}$$

$$l_{BC} = 0.544\frac{c^2}{d} \tag{4}$$

Noting that the loop is a 3D structure, and the section of the loop between B and C is actually covered by a higher loop and only the section between A and B and its mirror can directly get contact with the skin. So, the effective skin contact length in a single loop can be estimated by Equation (5):

$$l_1 = 2c\left[1 + \frac{9}{16}\left(\frac{d}{c}\right)^2\right]. \tag{5}$$

Therefore, the effective contact area per square centimeter can be estimated by Equation (6):

$$S = 2 \cdot C \cdot W \cdot c \left[1 + \frac{9}{16}\left(\frac{d}{c}\right)^2\right] \cdot D, \quad (6)$$

where W is the number of wales per cm, C is the number of courses per cm, c is the course spacing equals to $10/C$, D is the effective contact width of the yarn and skin. D is related to the yarn diameter, fiber diameter, and the deformation rate of the yarn and the skin.

According to Equation (6), the stitch density $C \cdot W$ is positively related to the effective skin contact area. When the stitch density increases, it actually increases the effective skin-electrode contact area. As we know from the electrical equivalent circuit of the skin-electrode interface in Figure 1, the skin-electrode interface has both resistive behavior and capacitive behavior. The resistive behavior can be expressed by Equation (7), and the capacitive behavior can be expressed by Equation (8):

$$R_e = \rho \frac{l}{A}, \quad (7)$$

$$C_e = \varepsilon \frac{A}{d}, \quad (8)$$

where ρ is the electrical resistivity of the material, l is the fabric thickness, A is the skin-electrode contact area, ε is the permittivity of the dielectric layer, d is the distance between the electrode and the skin. According to Equations (7) and (8), the increase in the effective skin-electrode contact area will results in a decrease in resistance and an increase in capacitance and according to Equation (1), reduction in resistance and increase in capacitance will eventually lead to a reduction in the skin-electrode impedance. Therefore, increasing the stitch density can effectively reduce the skin-electrode impedance. In addition, the increased stitch density also helps to accumulate sweat under the electrode, so that the skin-electrode impedance drops in a shorter time, meaning a shorter impedance stabilization period.

In ECG monitoring, small skin-electrode impedance means small noise interference; a smooth impedance-frequency curve means that low-frequency signals have less amplitude distortion. So according to our results, electrode material TE1 shows the best performance among all four electrode materials, not only because it has the smallest skin-electrode impedance, but also because it has the smoothest impedance-frequency curve and the shortest impedance stabilization period. Therefore, conductive fabric TE1 is an optimum material for making the hybrid textile electrode.

4. Electrode Size

The size of the electrode has also been reported as having a significant influence on skin-electrode impedance and on the ECG signal's quality [44]. Puurtinen et al. [45] studied different sizes of textile electrodes and found that the skin-electrode impedance increases with decreasing of the electrode size. Marozas et al. [9] also found that textile electrodes with a contact area smaller than 4 cm^2 might cause distortions to the signal's low-frequency spectrum. Therefore, in order to choose the optimum electrode size of the proposed hybrid electrode, the electrode size and its influence on the ECG signal have to be investigated.

4.1. Experimental Method

Three different electrode sizes were investigated, the electrodes were all made from conductive fabric TE1. Conventional wet ECG electrodes (2228, 3M, Minnesota, USA) were also used to measure the ECG for comparison purposes. Table 2 listed the areas and dimensions of the four different electrodes.

Table 2. Electrodes in different size.

	Large Textile Electrode	Medium Textile Electrode	Small Textile Electrode	Conventional Wet Electrode
Shape of Electrodes				
Electrode Area (cm^2)	8	4.5	2.25	2.27
Electrode Dimension	2 × 4 (w × L, cm)	1.5 × 3.0 (w × L, cm)	1.5 × 1.5 (w × L, cm)	1.7 (φ, cm)

ECG signals were measured with these four different electrodes on the chest of a female subject. All ECG signals were recorded using ADS1292ECG-FE (Texas Instruments, Dallas, TX, USA), with only the 50 Hz notch filter operating and all other filters switched off. The sample rate of the signal was 500 Hz, a reference electrode was used to reduce the common-mode noise. All electrodes were secured on the skin with a 30 mmHg pressure applied by an elastic chest band.

4.2. Results and Discussion

Figure 9 shows the original ECG signals recorded with three pairs of dry textile electrodes and one pair of wet electrodes, Figure 10 shows the power spectral density of all ECG signals. As can be seen from Figure 9, baseline drift exists in all ECG signals, this is mainly due to body respiration and its effect on body volume change causing skin-electrode impedance imbalance. As the size of the electrode increases, the baseline drift effect caused by respiration is significantly reduced. Small size textile electrodes have the largest baseline drift compared to other electrodes. Figure 10 shows that the energy of the drifted baseline is in the frequency range of 0–0.5 Hz, and that the smallest textile electrode has the largest baseline drift noise. Besides the baseline drift, high-frequency noise can also be observed in all graphs, because of the presence of electromagnetic fields in the vicinity of the patient. The smallest dry textile electrode introduces more high-frequency noise than all others. However, the high-frequency interference is also decreased with increased electrode size. Although the ECG signal has been filtered by a 50 Hz notch filter, the 50 Hz alternating current (AC) power line interference and its third harmonic (150 Hz) are clearly observed in the spectrum of the smallest size electrode. However, in the spectrum of the largest size electrode, only the 150 Hz harmonic frequency can be observed, and its energy level is similar with that of the conventional wet electrode.

In comparison with wet electrodes, dry textile electrodes usually introduce more noise (including the baseline drift noise and the AC power line interference) into the ECG signals. However, the result for the largest size textile electrode is comparable to that of the wet electrode, meaning that the dry textile electrode having a large electrode size can perform equally well for ECG monitoring. Therefore, the electrode size of 2 cm × 4 cm is an optimum size for making our hybrid textile sensor.

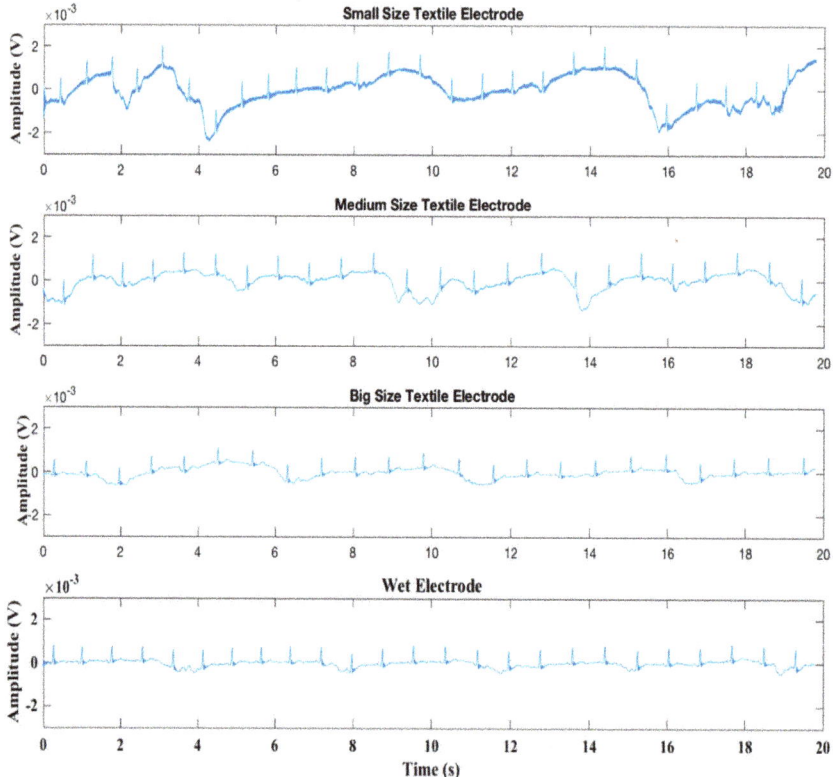

Figure 9. Resting electrocardiogram (ECG) signals recorded by different electrodes.

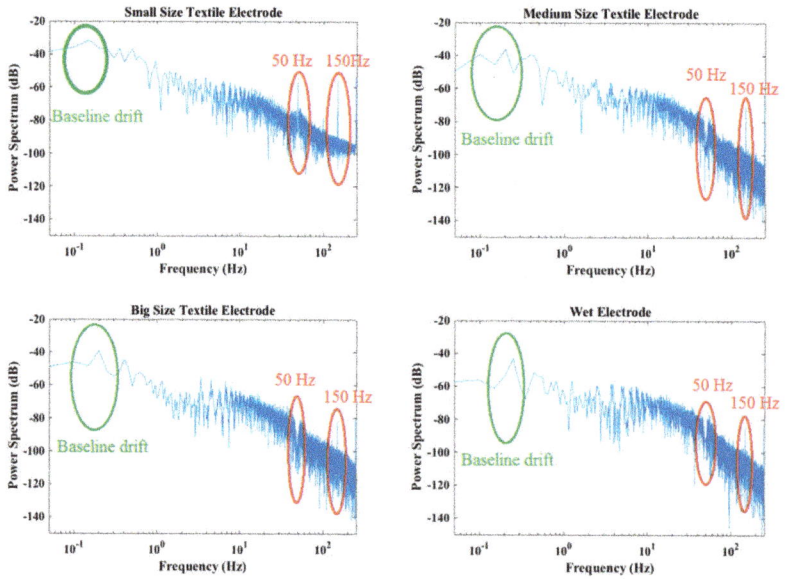

Figure 10. Power spectral density.

5. Motion Sensor FPCB

To track human activity, microelectromechanical (MEMS) motion sensor MPU-9250 (InvenSense, Calgary, AB, Canada) was used in the design of the hybrid textile electrode, because of its miniature size and powerful features. The MPU-9250 is a multi-chip module consisting of a 3-Axis gyroscope, 3-axis accelerometer, 3-axis magnetometer and a digital motion processor all in a small $3 \times 3 \times 1$ mm package. In order to integrate the motion sensor with the textile electrode, a flexible printed circuit board (FPCB) is specially designed. The FPCB offers power supplies to the MPU-9250 and transmits the detected motion data from the sensor to the microcontroller (MCU). The printed transmission lines in the FPCB also serve as an electrode lead between the textile electrode and the input of the amplifier. As shown in Figure 11, the FPCB consists of two parts: The first part is a mini-circuit board with electronic components on it; the second part is printed transmission lines for transmitting bio-potential signals and motion data to the MCU. The size of the first part is 10 mm $\times$ 7 mm. Figure 12: (a) shows the top view of the FPCB; (b) shows the rear view of the FPCB; and (c) shows the flexibility of the FPCB.

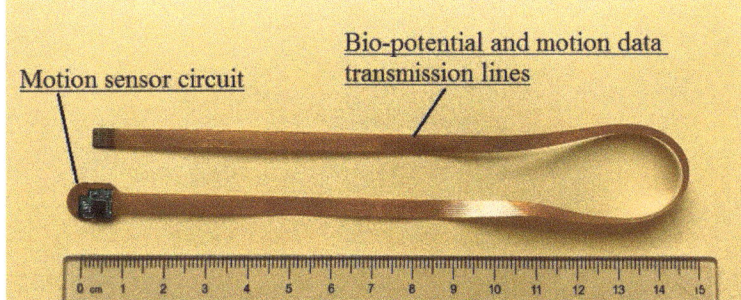

Figure 11. The flexible printed circuit.

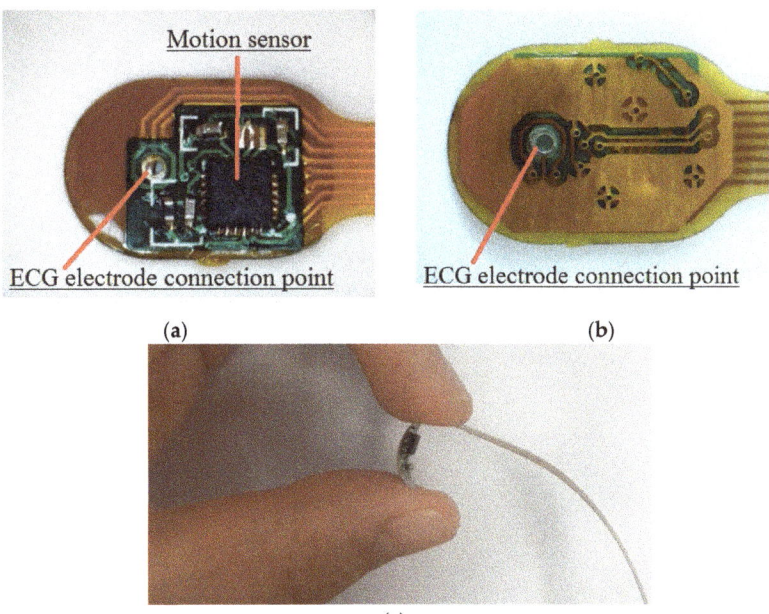

Figure 12. The motion sensor mini flexible printed circuit board (FPCB): (**a**) Top view; (**b**) rear view; (**c**) the flexibility of the FPCB.

6. Fabrication of the Hybrid Textile Electrode

In order to integrate the mini FPCB into the textile electrode, we used silver conductive adhesive (CW2400, Chemtronics, Kennesaw, GA, USA) to bond the rear side of the mini circuit board with the rear side of the conductive fabric together, to establish a coherent ECG signal transmission line, as demonstrated in Figure 13. The biopotential signal is transmitted from the textile electrode through the silver conductive adhesive layer to the ECG electrode connection point of the FPCB, and it then reaches the input of the biopotential amplifier through the printed transmission line. Electrical insulation is also important for making the hybrid textile electrode. A thick, transparent, flexible silicon coating (FSC, Electrolube, Leicestershire, UK) is brushed on the top surface of the mini circuit board to provide insulation of the electronic components with the outside world. The flexible silicon coating is a solvent based conformal coating designed to protect printed circuit boards. As seen in Figure 14, the FPCB is integrated with the conductive fabric.

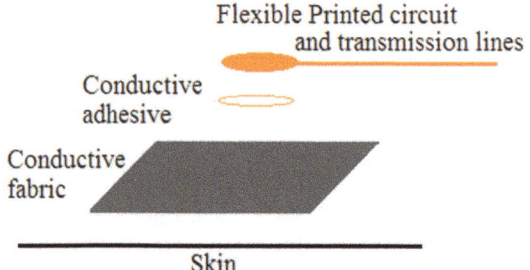

Figure 13. The layout of the hybrid textile electrode.

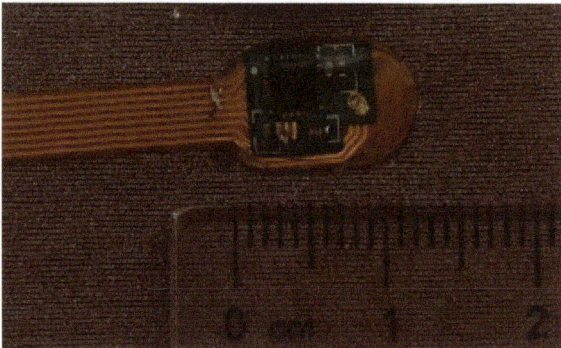

Figure 14. The integration of the circuit board and the conductive fabric.

Non-conductive fabric and non-conductive sponge filler fabric are also used for making the hybrid textile electrode. Non-conductive fabric serves as an insulating layer. Non-conductive sponge fabric provides support for the flexible circuit board and the conductive fabric. Figure 15 illustrates the top and the cross-sectional views of the hybrid textile electrode. The middle rectangular area is the conductive fabric surrounded by the non-conductive fabric. The skin contact area of the conductive fabric is 20 mm × 40 mm. Figure 16 shows the top view and the rear view of the hybrid textile electrode.

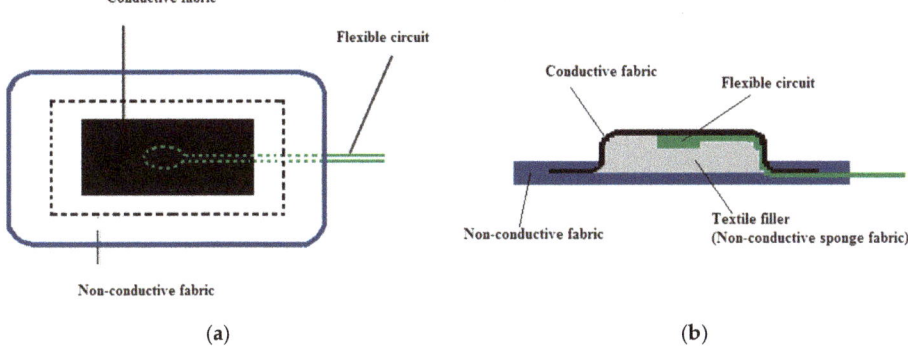

Figure 15. The illustration of the structure of textile electrode: (**a**) Top view; (**b**) side view.

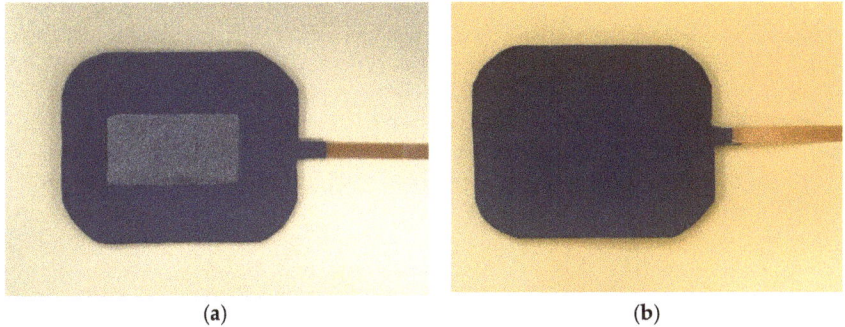

Figure 16. Hybrid textile electrode: (**a**) Top view; (**b**) rear view.

7. Hardware Setup

The measurement system configuration is shown in Figure 17. It is based on an MSP 430 microcontroller (Texas Instruments, Dallas, TX, USA) with build in high-performance 12-bit Analog-to-digital converter (ADC) for data acquisition and system control. A low power, 24-bit Analog front-end ADS1292R (Texas Instruments, Dallas, TX, USA) is used for the ECG measurements; and a motion tracking device MPU-9250 (InvenSense, Calgary, AB, Canada) is embedded in the textile electrode to track movement. The measured ECG and motion signals are transferred into a Bluetooth module RN41 (Microchip, Chandler, Arizona, USA) and send to a remote computer or mobile device wirelessly.

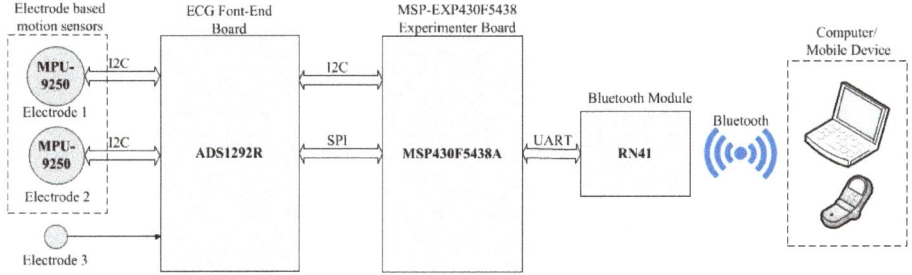

Figure 17. The system setup.

8. The Implementation of the Hybrid Textile Electrode for ECG Monitoring and Motion Tracking

In order to test the performance of the hybrid textile electrodes, we have undertaken measurement using a female subject. Two hybrid textile electrodes and one textile electrode with the same structure were sewn onto an elastic chest band, as seen in Figure 18. The Electrodes were placed on the subject's chest and were secured onto the skin with a 30 mmHg pressure, applied by an elastic chest band. Signals were recorded under two everyday activities: Sitting and walking.

Figure 18. The chest band.

Figure 19 shows the original, unfiltered signal that is recorded with the hybrid textile electrodes. Figure 19a shows the resting ECG and its corresponding motion data; Figure 19b shows the exercise ECG and its corresponding motion data. All ECG signals present baseline drift and high-frequency noise. The baseline drift is mainly caused by the respiration, as the hybrid textile electrodes were integrated into a chest band, the chest movement during respiration induces motion artefacts into the ECG signals. As shown in Figure 19a, chest movement during respiration was captured by the motion sensor in synchronization with the ECG. The accelerometer's data on the Z axis has the same trend as the baseline drift of the resting ECG. The gyroscope's data on the X axis also shows a similar trend. The high-frequency noise in ECG signals is mainly induced from power line interference, because textile electrodes have high and unbalanced skin-electrode impedance, thus introducing differential mode noise into the ECG signals. The waveform of exercise ECG shows more interference than the resting ECG, due to motion artefacts caused by walking motion. As shown in Figure 19b, walking motion was captured by the motion sensor in synchronization with the ECG. The accelerometer data and the gyroscope data illustrate the motion pattern.

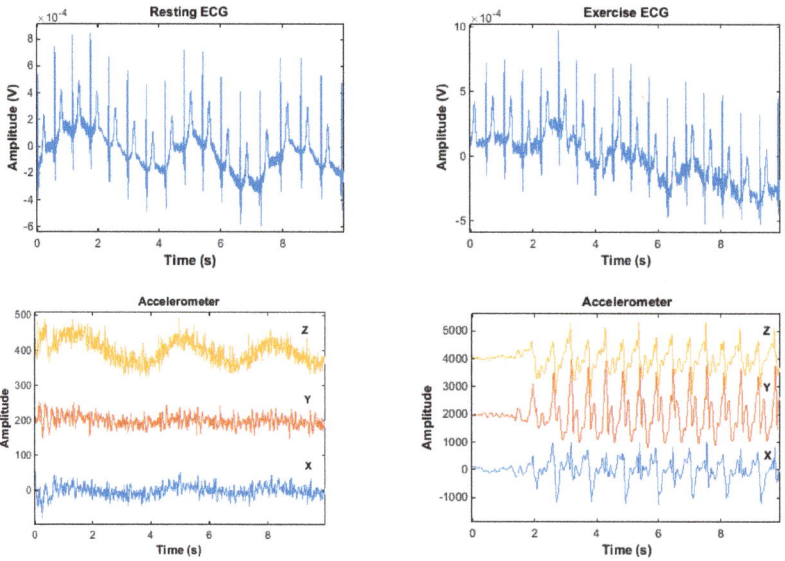

Figure 19. *Cont.*

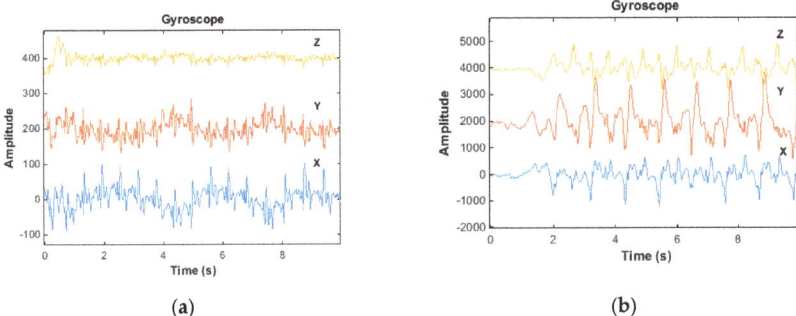

Figure 19. Signals recorded by the hybrid textile electrode: (**a**) Sit; (**b**) walk.

According to the results presented in Figure 19, the hybrid textile electrode is capable of recording ECG and tracking motion at the same time. Although the recorded ECG signals were contaminated by baseline drift and high-frequency noises, the magnitude of the noise did not corrupt the morphology of the recorded ECG. Therefore, the presence of these noises is tolerable and the performance of the hybrid textile electrode in ECG monitoring is reasonable, and the integrated motion sensor was very accurate in capturing movement by the hybrid textile electrode.

9. Conclusions

This paper presents a new hybrid textile electrode that integrates motion sensor MPU9250 with a textile-based electrode. This proposed hybrid textile electrode is not only suitable for long-term ECG monitoring, but also capable of tracking the patient activity simultaneously.

In the design of the hybrid textile, the performances of textile electrodes have been studied. Four electrode materials were investigated, and the conductive fabric TE1 was chosen to be the optimum electrode material. According to the skin-electrode impedance measurement, the conductive fabric TE1 not only has the smallest skin-electrode impedance, but also has the smoothest impedance-frequency curve and the shortest impedance stabilization period. The study on the electrode size has proven that dry textile electrodes with the size of 2 cm × 4 cm can perform equally well as commercial wet electrodes in ECG monitoring. Therefore, the size of the hybrid textile electrode was found optimum at 2 cm × 4 cm. In order to integrate the motion sensor MPU9250 with the textile base, a flexible printed circuit board (FPCB) has been specially designed for this purpose. The size of the FPCB is only 1 cm × 0.7 cm, which makes it easy to integrate into the textile electrode.

The combination of motion signals and ECG signals offers great potential for cardiology clinical trials cardiac rehabilitation patient care, and in general wellbeing and sports. The motion signals recorded in synchronization with the ECG signals can help the cardiologist find the cause of the heart disease by providing information about the patient's physical activity when the ECG shows an abnormality. By tracking daily physical activity and alerting the patient when the patient is over-exercised can also help prevent the sudden death in patients with coronary heart disease. Furthermore, sports and exercise can also be monitored with these new technologies that can provide better data than the commonplace heart beat monitors.

Author Contributions: Conceptualization, X.A. and George K.S.; Methodology, X.A.; Software, X.A.; Validation, X.A. and G.K.S.; Formal Analysis, X.A.; Investigation, X.A.; Resources, G.K.S.; Data Curation, X.A.; Writing—Original Draft Preparation, X.A.; Writing—Review & Editing, G.K.S.; Visualization, X.A.; Supervision, G.K.S.

Funding: This research received no external funding.

Conflicts of Interest: The authors declare no conflict of interest.

References

1. Scilingo, E.P.; Gemignani, A.; Paradiso, R.; Taccini, N.; Ghelarducci, B.; De Rossi, D. Performance evaluation of sensing fabrics for monitoring physiological and biomechanical variables. *IEEE Trans. Inf. Technol. Biomed.* **2005**, *9*, 345–352. [CrossRef] [PubMed]
2. Mestrovic, M.A.; Helmer, R.J.; Kyratzis, L.; Kumar, D. Preliminary study of dry knitted fabric electrodes for physiological monitoring. In Proceedings of the Intelligent Sensors, Sensor Networks and Information 3rd International Conference, Melbourne, Australia, 3–6 December 2007; pp. 601–606.
3. Paradiso, R.; Loriga, G.; Taccini, N. A wearable health care system based on knitted integrated sensors. *IEEE Trans. Inf. Technol. Biomed.* **2005**, *9*, 337–344. [CrossRef] [PubMed]
4. Priniotakis, G.; Westbroek, P.; Van Langenhove, L.; Hertleer, C. Electrochemical impedance spectroscopy as an objective method for characterization of textile electrodes. *Trans. Inst. Meas. Control* **2007**, *29*, 271–281. [CrossRef]
5. Pola, T.; Vanhala, J. Textile electrodes in ECG measurement. In Proceedings of the Intelligent Sensors, Sensor Networks and Information 3rd International Conference, Melbourne, Australia, 3–6 December 2007; pp. 635–639.
6. Beckmann, L.; Kim, S.; Dueckers, H.; Luckhardt, R.; Zimmermann, N.; Gries, T.; Leonhardt, S. Characterization of textile electrodes for bioimpedance spectroscopy. In Proceedings of the Ambience: International Scientific Conference Smart Textile—Technology and Design, Boras, Sweden, 19–20 September 2008; pp. 79–83.
7. Paradiso, R.; Loriga, G.; Taccini, N.; Gemignani, A.; Ghelarducci, B. WEALTHY-a wearable healthcare system: New frontier on e-textile. *J. Telecommun. Inf. Technol.* **2005**, 105–113.
8. Catrysse, M.; Puers, R.; Hertleer, C.; Van Langenhove, L.; Van Egmond, H.; Matthys, D. Towards the integration of textile sensors in a wireless monitoring suit. *Sensor Actuat. A-Phys.* **2004**, *114*, 302–311. [CrossRef]
9. Marozas, V.; Petrenas, A.; Daukantas, S.; Lukosevicius, A. A comparison of conductive textile-based and silver/silver chloride gel electrodes in exercise electrocardiogram recordings. *J. Electrocardiol.* **2011**, *44*, 189–194. [CrossRef] [PubMed]
10. Ankhili, A.; Tao, X.; Cochrane, C.; Coulon, D.; Koncar, V. Washable and Reliable Textile Electrodes Embedded into Underwear Fabric for Electrocardiography (ECG) Monitoring. *Materials* **2018**, *11*, 256. [CrossRef] [PubMed]
11. Rattfält, L.; Lindén, M.; Hult, P.; Berglin, L.; Ask, P. Electrical characteristics of conductive yarns and textile electrodes for medical applications. *Med. Biol. Eng. Comput.* **2007**, *45*, 1251–1257. [CrossRef] [PubMed]
12. Martinsen, O.G.; Grimnes, S. *Bioimpedance and Bioelectricity Basics*; Academic Press: Cambridge, MA, USA, 2011.
13. Jang, S.; Cho, J.; Jeong, K.; Cho, G. Exploring Possibilities of ECG Electrodes for Bio-monitoring Smartwear with Cu Sputtered Fabrics. In *Human-Computer Interaction. Interaction Platforms and Techniques*. HCI 2007; Lecture Notes in Computer Science; Jacko, J.A., Ed.; Springer: Berlin/Heidelberg, Germany, 2007; Volume 4551.
14. Pani, D.; Dessì, A.; Saenz-Cogollo, J.F.; Barabino, G.; Fraboni, B.; Bonfiglio, A. Fully textile, PEDOT: PSS based electrodes for wearable ECG monitoring systems. *IEEE Trans. Biomed. Eng.* **2016**, *63*, 540–549. [CrossRef] [PubMed]
15. Paradiso, R. Wearable health care system for vital signs monitoring. In Proceedings of the Information Technology Applications in Biomedicine, Birmingham, UK, 2003; pp. 283–286.
16. Carvalho, H.; Catarino, A.P.; Rocha, A.M.; Postolache, O. Health monitoring using textile sensors and electrodes: An overview and integration of technologies. In Proceedings of the International Symposium on Medical Measurements and Applications, Lisbon, Portugal, 2014; pp. 70–75.
17. Castiglioni, P.; Faini, A.; Parati, G.; Di Rienzo, M. Wearable Seismocardiography. In Proceedings of the 29th Annual International Conference of the IEEE Engineering in Medicine and Biology Society, Lyon, France, 22–26 August 2007; pp. 3954–3957.
18. Jourand, P.; De Clercq, H.; Corthout, R.; Puers, R. Textile integrated breathing and ECG monitoring system. *Procedia Chem.* **2009**, *1*, 722–725. [CrossRef]

19. Di Rienzo, M.; Rizzo, F.; Parati, G.; Brambilla, G.; Ferratini, M.; Castiglioni, P. MagIC system: A new textile-based wearable device for biological signal monitoring. Applicability in daily life and clinical setting. In Proceedings of the 2005 IEEE Engineering in Medicine and Biology 27th Annual Conference, Shanghai, China, 17–18 January 2005; pp. 7167–7169.
20. Bifulco, P.; Cesarelli, M.; Fratini, A.; Ruffo, M.; Pasquariello, G.; Gargiulo, G. A wearable device for recording of biopotentials and body movements. In Proceedings of the Medical Measurements and Applications Proceedings, Bari, Italy, 30–31 May 2011; pp. 469–472.
21. Mittleman, M.A.; Maclure, M.; Tofler, G.H.; Sherwood, J.B.; Goldberg, R.J.; Muller, J.E. Triggering of acute myocardial infarction by heavy physical exertion–protection against triggering by regular exertion. *New Engl. J. Med.* **1993**, *329*, 1677–1683. [CrossRef] [PubMed]
22. Hansson, A.; Madsen-Härdig, B.; Olsson, S.B. Arrhythmia-provoking factors and symptoms at the onset of paroxysmal atrial fibrillation: A study based on interviews with 100 patients seeking hospital assistance. *BMC Cardiovasc. Disor.* **2004**, *4*, 13. [CrossRef] [PubMed]
23. Lampert, R.; Joska, T.; Burg, M.M.; Batsford, W.P.; McPherson, C.A.; Jain, D. Emotional and physical precipitants of ventricular arrhythmia. *Circulation* **2002**, *106*, 1800–1805. [CrossRef] [PubMed]
24. Kohl, H.W., III; Powell, K.E.; Gordon, N.F.; Blair, S.N.; Paffenbarger, R.S., Jr. Physical activity, physical fitness, and sudden cardiac death. *Epidemiol. Rev.* **1992**, *14*, 37–58. [CrossRef] [PubMed]
25. Benarroch, E.E. Postural tachycardia syndrome: A heterogeneous and multifactorial disorder. *Mayo Clin. Proc.* **2012**, *87*, 1214–1225. [CrossRef] [PubMed]
26. Cobb, L.A.; Weaver, W.D. Exercise: A risk for sudden death in patients with coronary heart disease. *J. Am. Coll. Cardiol.* **1986**, *7*, 215–219. [CrossRef]
27. Burke, A.P.; Farb, A.; Malcom, G.T.; Liang, Y.H.; Smialek, J.E.; Virmani, R. Plaque rupture and sudden death related to exertion in men with coronary artery disease. *J. Am. Med. Assoc.* **1999**, *281*, 921–926. [CrossRef]
28. Siscovick, D.S.; Weiss, N.S.; Fletcher, R.H.; Lasky, T. The incidence of primary cardiac arrest during vigorous exercise. *N. Engl. J. Med.* **1984**, *311*, 874–877. [CrossRef] [PubMed]
29. Webster, J.G. *Medical Instrumentation-Application and Design*; John Wiley & Sons: Hoboken, NJ, USA, 2009.
30. Geddes, L.A.; Valentinuzzi, M.E. Temporal changes in electrode impedance while recording the electrocardiogram with 'dry' electrodes. *Ann. Biomed. Eng.* **1973**, *1*, 356–367. [CrossRef] [PubMed]
31. Olson, W.; Schmincke, D.; Henley, B. Time and frequency dependence of disposable ECG electrode-skin impedance. *Med. Instrum.* **1978**, *13*, 269–272.
32. Tiller, J.C.; Liao, C.J.; Lewis, K.; Klibanov, A.M. Designing surfaces that kill bacteria on contact. *Proc. Natl. Acad. Sci. USA* **2001**, *98*, 5981–5985. [CrossRef] [PubMed]
33. Lansdown, A.B. A pharmacological and toxicological profile of silver as an antimicrobial agent in medical devices. *Adv. Pharmacol. Sci.* **2010**. [CrossRef] [PubMed]
34. Lee, H.J.; Jeong, S.H. Bacteriostasis and skin innoxiousness of nanosize silver colloids on textile fabrics. *Text. Res. J.* **2005**, *75*, 551–556. [CrossRef]
35. Samberg, M.E.; Oldenburg, S.J.; Monteiro-Riviere, N.A. Evaluation of silver nanoparticle toxicity in skin in vivo and keratinocytes in vitro. *Environ. Health Perspect.* **2009**, *118*, 407–413. [CrossRef] [PubMed]
36. Tregear, R.T. *Physical Functions of Skin*; Elsevier: New York, NY, USA, 1966; Volume 5.
37. Priniotakis, G.; Westbroek, P.; Van Langenhove, L.; Kiekens, P. An experimental simulation of human body behaviour during sweat production measured at textile electrodes. *Int. J. Cloth. Sci. Technol.* **2005**, *17*, 232–241. [CrossRef]
38. Westbroek, P.; Priniotakis, G.; Palovuori, E.; De Clerck, K.; Van Langenhove, L.; Kiekens, P. Quality control of textile electrodes by electrochemical impedance spectroscopy. *Text. Res. J.* **2006**, *76*, 152–159. [CrossRef]
39. Xu, P.J.; Zhang, H.; Tao, X.M. Textile-structured electrodes for electrocardiogram. *Text. Prog.* **2008**, *40*, 183–213. [CrossRef]
40. Liu, H.; Tao, X.; Xu, P.; Zhang, H.; Bai, Z. A dynamic measurement system for evaluating dry bio-potential surface electrodes. *Measurement* **2013**, *46*, 1904–1913. [CrossRef]
41. Hearle, J.W.; Grosberg, P.; Backer, S. *Structural Mechanics of Fibers, Yarns and Fabrics*; Wiley Interscience: Hoboken, NJ, USA, 1969; Volume I.
42. Peirce, F.T. Geometrical principles applicable to the design of functional fabrics. *Text. Res. J.* **1947**, *17*, 123–147. [CrossRef]
43. Peirce, F.T. 5—The geometry of cloth structure. *J. Text. Inst. Trans.* **1937**, *28*, T45–T96. [CrossRef]

44. Xiang, A.; Orathai, T.; George, K.S. Investigating the Performance of Dry Textile Electrode for wearable End-Uses. *J. Text. Inst* **2018**, in press.
45. Puurtinen, M.M.; Komulainen, S.M.; Kauppinen, P.K.; Malmivuo, J.A.; Hyttinen, J.A. Measurement of noise and impedance of dry and wet textile electrodes, and textile electrodes with hydrogel. In Proceedings of the Engineering in Medicine and Biology Society, New York, NY, USA, 30 August–3 September 2006; pp. 6012–6015.

 © 2018 by the authors. Licensee MDPI, Basel, Switzerland. This article is an open access article distributed under the terms and conditions of the Creative Commons Attribution (CC BY) license (http://creativecommons.org/licenses/by/4.0/).

Article

Carbon Nanomaterials Based Smart Fabrics with Selectable Characteristics for In-Line Monitoring of High-Performance Composites

Guantao Wang [1,2], Yong Wang [1], Yun Luo [2] and Sida Luo [1,*]

1. Department of Material Processing and Controlling, School of Mechanical Engineering & Automation, Beihang University, Beijing 100191, China; wangtgt@cugb.edu.cn (G.W.); yongw@buaa.edu.cn (Y.W.)
2. School of Engineering and Technology, China University of Geosciences (Beijing), Beijing 100083, China; luoyun@cugb.edu.cn
* Correspondence: s.luo@buaa.edu.cn; Tel.: +86-186-0020-0787

Received: 31 July 2018; Accepted: 5 September 2018; Published: 11 September 2018

Abstract: Carbon nanomaterials have gradually demonstrated their superiority for in-line process monitoring of high-performance composites. To explore the advantages of structures, properties, as well as sensing mechanisms, three types of carbon nanomaterials-based fiber sensors, namely, carbon nanotube-coated fibers, reduced graphene oxide-coated fibers, and carbon fibers, were produced and used as key sensing elements embedded in fabrics for monitoring the manufacturing process of fiber-reinforced polymeric composites. Detailed microstructural characterizations were performed through SEM and Raman analyses. The resistance change of the smart fabric was monitored in the real-time process of composite manufacturing. By systematically analyzing the piezoresistive performance, a three-stage sensing behavior has been achieved for registering resin infiltration, gelation, cross-linking, and post-curing. In the first stage, the incorporation of resin expands the packing structure of various sensing media and introduces different levels of increases in the resistance. In the second stage, the concomitant resin shrinkage dominates the resistance attenuation after reaching the maximum level. In the last stage, the diminished shrinkage effect competes with the disruption of the conducting network, resulting in continuous rising or depressing of the resistance.

Keywords: carbon nanomaterials; smart fabrics; in-line monitoring; polymeric composites; carbon nanotubes; reduced graphene oxide

1. Introduction

By virtue of remarkable features, including high specific modulus and strength, adiabaticity, corrosion resistance, and shock absorption [1], fiber-reinforced polymeric composites (FRPs) have been applied in widespread fields, such as aerospace technology, automotive industry, shipbuilding, and civil engineering. The booming applications of FRPs have inevitably caused people to worry about their quality and life assurance [2]. Under such circumstances, structural health monitoring (SHM) has emerged as a scientific and necessary tool, playing an important role in identifying, quantifying, and deciding the health states of the high-performance composites. In addition to traditional methods, such as strain gages [3], optical fibers [4], metal oxide films [5], guided waves [6–8], and piezoelectric sensors [9], carbon nanomaterials are considered to be novel materials for establishing embeddable, built-in, lightweight, versatile, and flexible self-sensing technology for composites [10]. Because of their extraordinary mechanical robustness, structural non-invasiveness, and piezoresistive sensitivity [11], it is important to investigate the great potentiality of novel carbon nanomaterials-based sensors to improve the structural health monitoring performance of composites. In addition, the key carbon nanomaterials, such as carbon nanotubes (CNTs) and graphene, are being adopted for multiple SHM purposes for FRPs. For examples, CNTs and graphene powders have been respectively mixed in resin matrix to monitor the failures and damages of the FRPs under mechanical deformations [12,13]; CNT/graphene-based films [14–16] and papers [17] have been placed on top of or embedded between lamination layers as strain sensors for deformation detection. Graphene-coated fibers have also been utilized for monitoring tensions and/or compressions [18].

Although most of the current research on carbon nanomaterials-based sensing technology mainly focuses on SHM during the service stages of the composites, it is equally important to monitor the manufacturing stage to detect and guide resin infiltration and curing, which are significant for assuring the quality and performance of the final products [19]. Considering the off-line limitations of differential scanning calorimetry (DSC), rheological, spectroscopic [20], and simulation methods [21], carbon nanomaterials could also serve as novel in-line monitoring strategies to meet the standards and requirements of composite manufacturing. In this respect, Lu et al. [2,19] utilized CNT buckypapers to monitor the real-time cure behavior of FRPs; for reducing part-to-part variations, Gnidakouong et al. [13,22] monitored and assessed the resin flow and curing levels of CNT/fiber-enabled polyester composites in a vacuum-assisted resin transfer molding (VARTM) process. Ali et al. [23] exploited graphene-coated piezo-resistive fabrics for monitoring the process of liquid composite molding. Our group recently enabled CNT and graphene thin films coated on reinforcement fibers to monitor and quantitatively analyze the composite manufacturing under both dry [24,25] and liquid [26] molding processes.

Summarizing the sensing performance of all the carbon nanomaterials-based sensors, it is surprising that the maximum resistance change has varied substantially from ~30% for graphene-based sensors [23] to ~1600% for CNT-based sensors [26]. To investigate the possible structure–property relationship, this paper focuses on the systematic analysis and comparison of three types of carbon nanomaterials-based sensors for monitoring the VARTM process of FRPs, namely, CNT-enabled fabrics (CNTF), reduced graphene oxide (RGO)-enabled fabrics (RGOF), and carbon fiber (CF)-enabled fabrics (CFF). Specifically, a high-efficient fiber winding and coating system was established for coating CNTs or graphene on fiber substrates. Through scanning electron microscopy (SEM) and Raman analysis, the microstructures of various carbon nanomaterials-based fibers were characterized in detail. For instance, a CNT-coated fiber has a loose and disordered packing structure with visible pores. In comparison, RGO coating is ultrathin and conformal with large lateral dimensions. CF itself is composed of continuous and densely packed graphites. The as-produced fiber sensors were respectively braided into a fiber fabric as the smart fabric layer of the composites. Through complete monitoring of the resin infiltration and curing, the piezoresistive behavior of the smart fabrics was divided into three stages (i.e., the gelation stage, hardening stage, and post-curing stage). Significant dissimilarities in sensing performance can be found at each stage after comprehensive analysis of various fabrics, which leads us to be more convinced of the existence of selectable structure-dependent characteristics driving the generation of various sensing responses. Briefly, in the gelation stage, the packing structure of the embedded fiber sensors dominates the increasing level of resistance through the resin infiltration; in the hardening stage, the resistance attenuation is closely linked to the resin shrinkage, causing the densified conducting network; and in the post-curing stage, the diminished shrinkage effect competes with the defects of the packing structure, resulting in the continuous rising or depressing of the resistance.

2. Materials and Methods

2.1. Materials

In this work, multi-walled carbon nanotubes (MWCNTs, General Nano LLC., Cincinnati, OH, USA) were used as coating materials for the fabrication of CNTF. Graphene oxide (GO) was synthesized via the Hummers's method and used for assembling RGOF [27]. Triton™ X-100 (CAS # 9002-93-1, Sigma-Aldrich, Beijing, China) was selected as surfactant for dispersing CNTs in the aqueous solution.

A hydroiodic acid solution (CAS # 10034-85-2, Sigma-Aldrich, Beijing, China) was utilized for the GO reduction. Fiber bundles with lengths of 15 cm, drawn from plain-woven glass fabrics (Part # GF-PL-290-100, Easy Composites Ltd., Beijing, China), were used as substrates for fiber coatings. CF rovings (Part # CF-PL-210-100) with lengths of 15 cm were also acquired from the Easy Composites Ltd.(Beijing, China). The above-mentioned glass fabrics were also used for reinforcing composites.

2.2. Preparation of Smart Fabrics

Following established strategies of materials' exfoliation [28,29], MWCNTs (300 mg) were sonicated in deionized water (100 mL) with 5 mL of Triton X-100 surfactant for 120 min, using a Ultrasonics FS-600N probe sonicator operated in a pulse mode (on 10 s, off 10 s), with the power fixed at 480 W. Following the same conditions as in the sonication process, the GO dispersion was prepared with 300 mg of GO powders in 100 mL of deionized water. Based on our previous works [10,26], a modified fiber winding and coating system was established, as shown in Figure 1a, in which the fiber powertrain was made up of a stepping motor and multiple standing pulleys for CNT/GO bathing, aqueous cleaning, and thermal drying. After the coating process, the GO-coated fibers required an additional reduction procedure to form the RGO-coated fibers by dipping the fibers into a hydroiodic acid solution at 85 °C for 30 min. With a clear contrast of colors, Figure 1b visually compares the appearance of the RGO-coated fibers at three fabrication stages, including pre-coating (white), post-coating (dark yellow), and post-reduction (black). Then, to obtain the smart fabrics, the as-produced fiber sensor was embedded in a plain-woven fabric (15 cm × 15 cm) by manual extracting and weaving, as shown in Figure 1c.

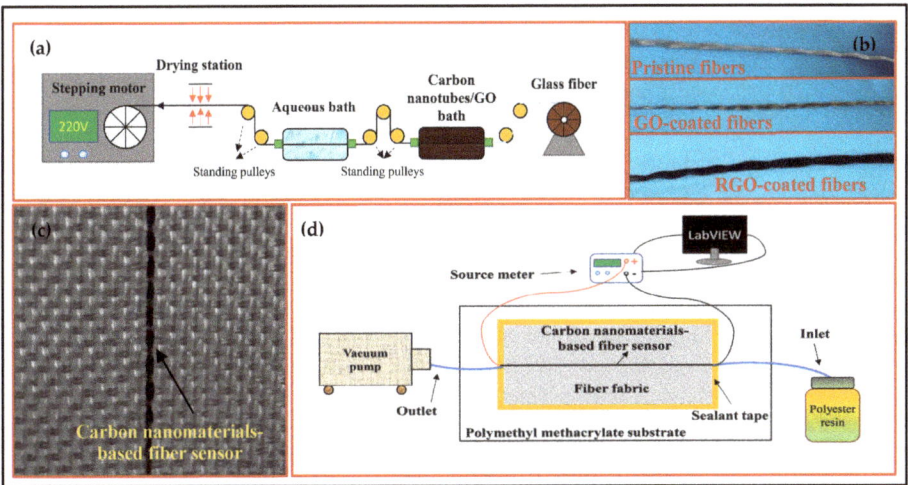

Figure 1. (**a**) Schematic diagram of the fiber winding and coating system. (**b**) Photographs of the pristine, graphene oxide (GO) and reduced graphene oxide (RGO)-coated fibers. (**c**) A representative smart fabric with an embedded carbon nanomaterials-based fiber sensor. (**d**) The schematic diagram of the setup for the monitoring of the vacuum-assisted resin transfer molding (VARTM) process.

2.3. In-Line Process Monitoring of Composites

A VARTM process was performed with the embedded smart fabrics for the in-line monitoring. Figure 1d schematically shows the experimental setup. Together with 2 layers of pristine fabrics, the smart fabric layer was first stacked on a polymethyl methacrylate substrate. To facilitate the electrical measurements, copper tapes were used as electrodes connecting both ends of the fiber sensor with a silver paste. Then, a peel ply (ELS60100, Airtech Ltd., Tianjin, China) and a flow mesh

(ELS60100, Airtech Ltd., Tianjin, China) were laid sequentially on the fabric layers for guiding the resin fluid flowing through the two nylon tubes fixed as the inlet and outlet. Assisted by a double-sided sealant tape (AT200Y1/250, Airtech Ltd., Tianjin, China), a vacuum bagging film (WL5400, Airtech Ltd., Tianjin, China) was used for sealing all the preforms.

After setting up the preforms, a vacuum pump was continuously running to introduce the polyester resin (IP2, Easy Composites Ltd., Beijing, China) mixed with 1.5 wt.% hardener (methyl ethyl ketone peroxide, MEKP, Easy Composites Ltd., Beijing, China) to infuse, infiltrate, and finally, fill the above-created vacuum space. Based on the guidance of the resin product, the manufacturing process lasted 24 h. During this period, the resistance of the embedded fiber sensor was recorded in real time by a source meter (2450, Keithley Ltd., Shanghai, China) interfaced with a homemade LabVIEW program. For each type of fiber sensor, at least 5 samples assembled by the same recipe and condition were tested to study reproducibility of the performance.

2.4. Structural Characterization and Performance Evaluation

To characterize the microstructure of the fiber sensor, scanning electron microscopy (SEM) and Raman spectroscopy were employed. The SEM was operated by a JEOL-JSM-7001F at 20 kV. The light source of the Raman microscope (Horiba-HR800, Horiba Ltd., Kyoto, Japan) is a 532-nm excitation laser controlled at 5 mW. In this work, the relative resistance change ($dR/R0$) was utilized to describe the sensitivity of the various smart fabrics [19], where $R0$ is the initial resistance. The initial resistance of all the CNT- and RGO-coated fibers and the CFs with lengths of 15 cm in the experiments were respectively stabilized at ~200 kΩ, ~400 kΩ, and ~20 Ω.

3. Results and Discussion

3.1. Microstructure of Carbon Nanomaterials-Coated Fibers

A detailed structural characterization of the embedded carbon nanomaterial-coated fibers is essential for exploring the structure-dependent characteristics of the smart fabrics. SEM was therefore requisitioned to examine the morphology and microstructure of the various sensors. Figure 2a–f show the SEM micrographs of the CNT- and RGO-coated fibers from multiple perspectives, as well as CF. The images at low magnification displayed the overall packing structures and were used for viewing the coating effectiveness. To specify, with respect to the CNT-coated fibers, large numbers of disordered CNT particles were intertwined and agglomerated together, forming a uniformly distributed coating layer on the filament substrate (Figure 2a). For the RGO-coated fibers, flexible graphene flakes were stacked in a staggered manner to splice a soft sheet tightly wrapping around the fiber (Figure 2b). As for CF, its highly consistent and continuous graphite structure was clearly visualized in Figure 2c [30]. More detailed structural information can be found in Figure 2d–f under high magnifications. With one-dimensional tubular structures, the fluffy CNTs were randomly entangled, introducing abundant porous structures throughout the network (Figure 2d). In contrast, the RGO possessed two-dimensional plate-like structures, making it easy to spread out on fiber surfaces [31]. Caused by layer-by-layer assembly, the wrinkles and small gaps between the neighboring flakes can be observed in Figure 2e. In Figure 2f, the graphite crystallites were seamlessly packed along the axial direction of the fiber and exhibited clear ridges.

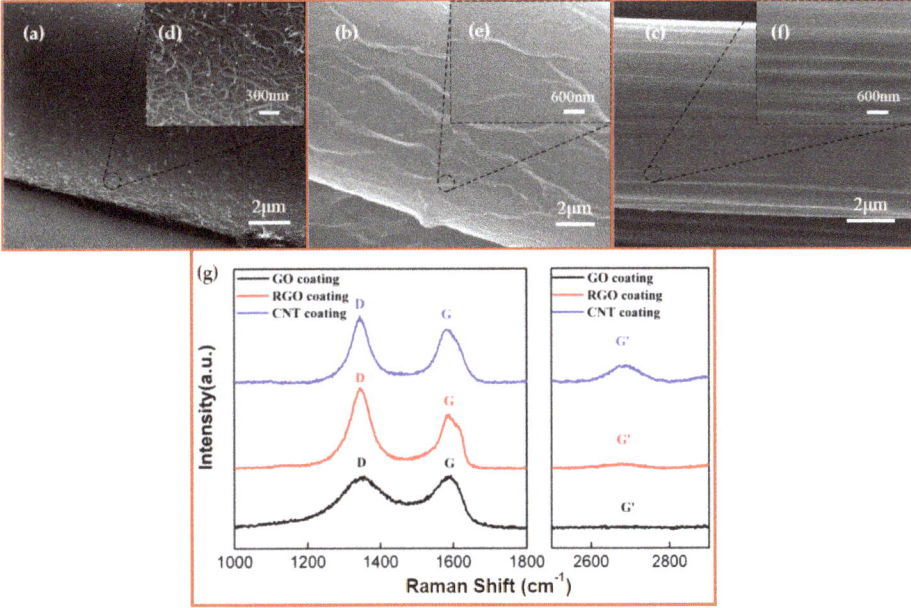

Figure 2. (a–c) SEM images of the CNT- and RGO-coated fibers, as well as CF, at a low magnification. (d–f) SEM images of CNT- and RGO-coated fibers, as well as CF, at a high magnification. (g) Raman spectrum of GO, RGO, and CNT coatings.

To further study the structural properties, Raman spectroscopy was explored to characterize the crystal structures of the carbon nanomaterials [32,33]. Figure 2g shows the Raman spectra of the CNT, RGO, and GO coatings. It was observed that all the samples presented similar Raman features with three characteristic peaks, confirming the existence of CNT and graphitic structures. Namely, the G peak appeared in the range of 1585–1595 cm^{-1} and was caused by the stretching of the C–C bonds in the graphitic structures corresponding to the first-order scattering of the E_{2g} mode [34,35]. The D peak, located around 1350 cm^{-1}, was a defect-induced Raman feature [36], which was closely related to the defects and crystal distortion [37]. The G' peak, located around 2700 cm^{-1}, originated from a double resonance Raman process [38], providing related information on the energy band structure of the carbon-based materials. As clearly exhibited in Figure 2g, the widths of the D and G peaks in the spectrum of the GO were broader than that of the RGO and CNTs, possibly due to the introduction of the disordered sp^3 carbon by breaking the symmetrical and ordered sp^2 structures in the graphite during the extensive oxidation. From the GO to the RGO, a clear enhancement in the intensity of the D peak can be observed. As stated above, this suggests the increased number of defects on the graphene flakes by the massive removal of the oxygen-containing groups through reduction. Accordingly, we have observed the increase in the intensity ratio between the D and G peaks (I_D/I_G). In accordance with other studies [35,39,40], this may be caused by the decrease in the average crystallite size of the sp^2 carbon domains in the RGO, because large numbers of sp^3 carbons are deoxygenated to form new sp^2 domains. Additionally, the intensity of the G' peak in the spectrum of the RGO is much weaker than that of the CNTs, indicating the formation of considerable defects in the double resonance process and disclosing the multi-layered structure of the RGO [37,41,42].

3.2. Piezoresistivity of Fiber Sensors for In-Line Monitoring

The conducting network created by the carbon nanoparticles attached to the fiber surface endows the fabrics with sensorial capabilities [18]. To explore the process monitoring characteristics and understand the structure-dependent mechanisms, each of the fabric sensors was applied to an in situ monitor to track the complete VARTM by recording the electrical resistance in real time. To demonstrate this idea, the real-time resistance change (dR/R0) of a representative CNT-enabled fabric (CNTF) is shown in Figure 3. To better understand the sensing behavior, we have divided the signal curve into three parts according to the key stages of the resin curing process.

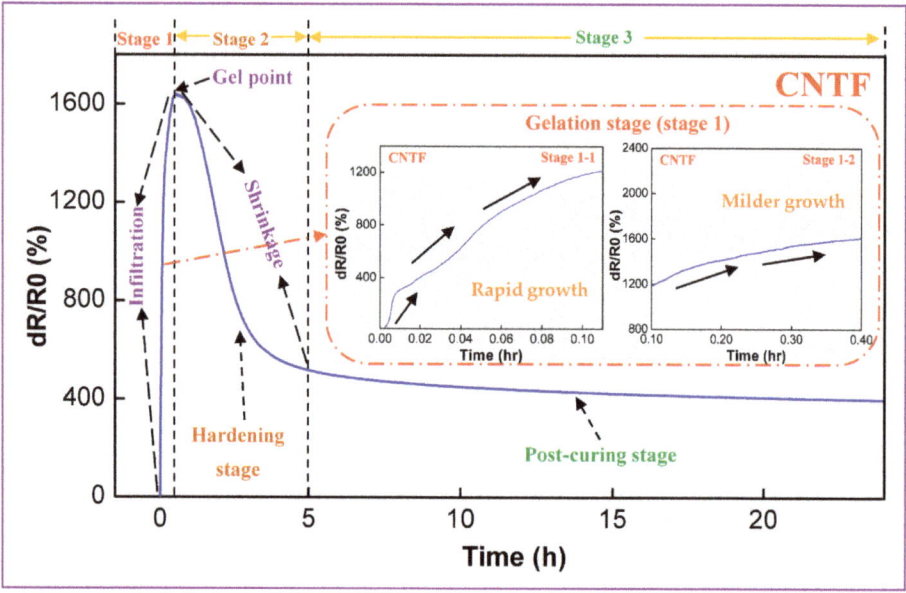

Figure 3. Representative sensing performance of a CNT-enabled fabric (CNTF) for the in-line monitoring of fiber-reinforced polymeric composites (FRP) manufacturing.

As shown in Figure 3, dR/R0 keeps changing throughout the whole process of composite molding. In the first 0.4 h, a significant increase of dR/R0 can be clearly observed, correlating closely with the resin infiltration by rapidly entering the dry reinforcement to gradually fill the entire space under the vacuumed condition. Interestingly, this time interval matches exactly the gel time provided by the resin supplier (~25 min) and thus is named the gelation stage (stage 1). In addition, the piezoresistivity of this stage can be further subdivided into two small stages, namely, stage 1-1 (0–0.1 h) with rapid resistance growth and stage 1-2 (0.1–0.4 h) with a milder increase of dR/R0. As the molding process proceeds, the manufacturing transitions into the hardening stage (stage 2) due to the occurrence of a drastic crosslinking reaction. In this stage, dR/R0 begins to decrease substantially until the time reaches about 5 h, indicating the gradually weakened reaction of crosslinking. Subsequently, the final stage starts, which lasts until the end of the process (5–24 h). It is called the post-curing stage (stage 3), because the degree of curing does not change drastically and dR/R0 also gradually approaches a stabilized value.

The abovementioned piezoresistive effect may be strongly associated with the phase change of polyester resin during the curing process. As investigated in the research of Chiacchiarelli et al. [43], the piezoresistivity of the carbon nanomaterials is dominated mainly by the numbers of conductive pathways and the variation of contact resistance between adjacent particles. In the gelation stage, with the elapse of time, the resin viscosity gradually changes from low to high and eventually loses its fluidity. In this context, the flow rate of the resin determines the increasing rate of dR/R0. At the beginning of the infusion, the lowest viscosity along with almost unobstructed space makes the resin flow swiftly to fill the vacuumed space. This brings about a rapid rise in dR/R0 (0–~1200%)

by expanding the distance between adjacent CNTs to deteriorate and even completely damage the interconnection of the conductive particles (stage 1-1). After filling the majority of the voids, the resin could infiltrate to the internal space of the fiber bundles to further improve the fiber wetting. At this point, the increased viscosity, as well as the reduced space, simultaneously slows down the resin flow [13]. Thus, the effect of the electrical disturbance is significantly reduced, leading to a gentle rise in $dR/R0$ (~1200%–~1600%, stage 1-2). By meeting the gelation time, the resin flow almost stops, making the value of $dR/R0$ stay temporarily in a stabilized state. In the hardening stage, the resin evolves from the gel state to the deep crosslinking state, causing certain levels of volume shrinkage [44]. Suffering from the matrix shrinkage, as well as internal stress, the impregnated nanoparticles become closer, re-connected, or even tightened. As a result, the attenuation of $dR/R0$ (from ~1600% to ~517%, stage 2) has been observed. Once the resin matrix can maintain its shape and hardness, it will enter the post-curing stage, in which the resin slowly evolves to a finalized product with optimized structural properties. Resultingly, the structure of the electrical network is merely disrupted, and its $dR/R0$ decays slowly and stabilizes eventually (between ~517% and ~401%, stage 3).

3.3. Comparison of Sensing Performance of Various Smart Fabrics

Taking CNTF as an example, we have discussed the typical piezoresistive behavior of the in-line monitoring in the previous section. To further investigate whether other carbon nanomaterials-enabled fabrics also have similar sensing performance and reveal the mechanisms contained, a comparative analysis of the piezoresistivity of CNTF, reduced graphene oxide-enabled fabrics (RGOF), and carbon fiber-enabled fabrics (CFF) was carried out. As shown in Figure 4, the three types of fabrics all exhibited obvious three-stage piezoresistive behavior, and the outline of each signal curve was also roughly similar. Nevertheless, there were still visible dissimilarities at each stage. To specify, in stage 1, the maximum resistance change of the CNTF (~1600%) was much higher than that of the RGOF (~77%) and the CFF (~24%); in stage 2, the degree of resistance attenuation varied substantially and its order from strong to weak is the CNTF (67.7%, $dR/R0$: from ~1600% to ~517%), the CFF (50%, $dR/R0$: from ~24% to ~12%), and the RGOF (28.6%, $dR/R0$: from ~77% to ~55%) by using the decrement rate of $dR/R0$ as a measure; and in stage 3, the $dR/R0$ value of the RGOF showed a tendency to slowly rise (from ~55% to ~69%), while the CNTF and the CFF almost stayed in a stabilized state.

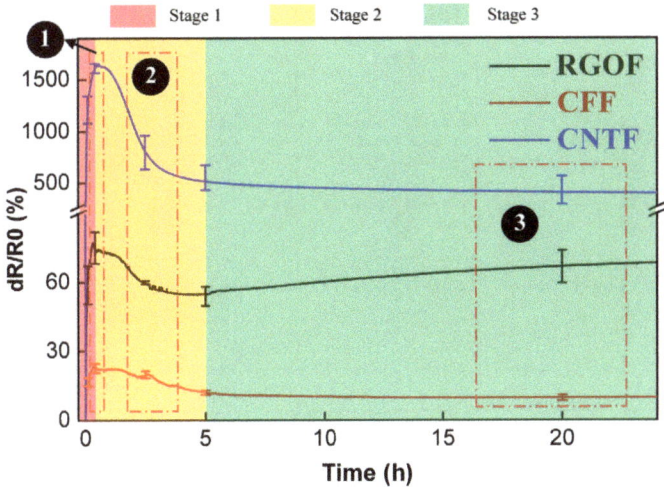

Figure 4. Comparison of the sensing performance of the CNTF, reduced graphene oxide-enabled fabrics (RGOF), and carbon fiber-enabled fabrics (CFF) for the in-line monitoring of FRP manufacturing.

3.4. Structure-Dependent Mechanism of In-line Monitoring

As mentioned above, although the three smart fabrics possessed a substantially uniform piezoresistive behavior, there were still discrepancies in certain details that cannot be ignored, because they may play an important role in taking advantage of the various carbon nanomaterials for the purposes of structural health monitoring..To better interpret the discrepancies with valuable insights, a structure-dependent mechanism has been suggested based on the structural properties of the carbon nanomaterials. It is because the conducting network of the fiber sensors has been formed by carbon nanoparticles and its diversified structure may strongly affect the level of the resin infiltration, as well as the deformation during the resin shrinkage.

To vividly describe our advocated mechanism, a schematic diagram is shown in Figure 5 to demonstrate the complete process from the resin infusion to the full solidification. In stage 1, due to the differences in the microstructures of the various carbon nanomaterial coatings, the level of the resin infiltration to the three types of fiber sensors varies obviously. Because of its fluffy and porous structure, the resin molecules can easily infiltrate and merge into the CNT coating [10]. The adequate flowing space allows them to be in full contact with the CNTs, causing a huge increase in $dR/R0$ and improving the upper limit of its growth. Comparatively, the RGO coating is stacked with flake-like graphene particles with large lateral dimensions. The poreless structure leads to a small amount of resin infiltration, and only the surface or upper layers of the conducting network are disturbed by resin. Thus, the improvement of $dR/R0$ for the RGOF is weakened. While for the CFF, almost no resin molecules penetrate the highly consistent and densely packed graphite structures with negligible pores. The resin flow just causes a slight volumetric change in the conductive structure and limited interruption of the contact points between the adjacent filaments by squeezing, impacting, and rubbing the carbon fibers. As a result, the sensitivity is reasonably the lowest.

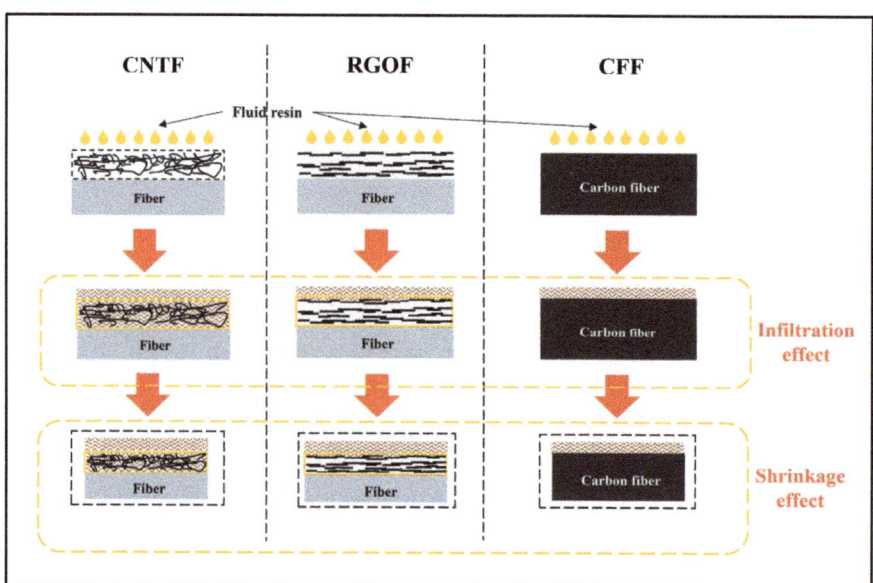

Figure 5. Schematic diagram of the comparison of the CNTF, RGOF, and CFF in the process of composite manufacturing.

The impact of the varied resin infiltration in stage 1 also extends to the second stage. For the CNTF whose conductive particles are completely impregnated, the volume shrinkage derived from the drastic resin crosslinking will generate strong multidirectional stress to promote tight junctions between the adjacent CNTs. In contrast, there are few resin molecules inside the RGO coating or CFF, such that most conductive particles are not in direct contact with resin matrix. As a result, the shrinkage effect mainly comes from the surface of the structure whose strength, as well as the triggered deformation, is less than that of the CNTF. Thus, the degree of resistance change attenuation of the CNTF is inevitably stronger than that of the other two fabrics. As for why the $dR/R0$ value of the CFF has more attenuation than that of the RGOF, we suppose that it may be closely related to the numerous tiny defects on the graphene flakes based on recent representative studies [10,43,45]. Specifically, during the process of the resin shrinkage, a competing behavior in the resistance change of the RGOF is proceeded, due to the combined effect of the compaction resulting from the drastic crosslinking and concentrated internal stress around the defects, which can cause an increase in the tunnel resistance by strengthening the stress intensity and even introducing new defects and voids [23]. Although the piezoresistive behavior indicates that the shrinking effect dominates this competition, defects from the graphene itself have somewhat curbed the momentum of the resistance attenuation. For the CFF, the CF's consistent and densely packed graphite structure with invisible defects cannot achieve the competing behavior. Its resistance change attenuation is, therefore, more than that of the RGOF.

In stage 3, the shrinkage effect from last stage has almost faded away. For the CNTF and the CFF, because the continuous improvement in the hardness and strength of the upcoming composite cannot further interrupt the configuration of the conducting network, the $dR/R0$ value levels off in the whole process. However, for the RGOF, even though the stress concentration effect is overshadowed in the hardening stage, it finally works here. Its unique tiny defects appear to be particularly sensitive to the internal structural stress. By enlarging the stress intensity to disturb the sensing structure, they can help to achieve a continuous rising electrical signal.

Based on all the results and analysis above, we finally conclude that there is a structure-dependent sensing mechanism of the CNTF, RGOF, and CFF in each stage of the composite manufacturing. In the gelation stage, the packing structure of the various fiber sensors dominates the level of the resin infiltration, leading to an increased resistance change with different upper limits. Benefitted by the great advantage of a fluffy and porous conducting network, which can be easily impregnated, the CNTF possesses the highest sensitivity. Because of its impenetrable continuous graphite structure, the CFF exhibits the lowest sensing response. In the hardening stage, the different levels of the resin infiltration result in different intensities of the shrinkage effect. The CNTF, whose sensing element is completely imprisoned by the resin, has more resistance attenuation than the RGOF and CFF. Compared with the CFF, a competitive relationship between the vacuum compaction and stress concentration originated from many tiny defects, making the RGOF's resistance decay less. In the post-curing stage, the diminished shrinkage effect competes with the disruption of the conducting network, resulting in the continuous rising (RGOF) or depressing (CNTF, CF) of the resistance.

4. Conclusions

In summary, the CNTF, RGOF, and CFF were fabricated by respectively weaving CNT- and RGO-coated fibers and carbon fibers into glass fiber fabrics for the in-line monitoring of FRPs. The structural properties of the various fiber sensors were systematically studied through SEM and Raman analysis. With the piezoresistive behavior divided into the gelation stage, hardening stage, and post-curing stage, a detailed comparison of the sensing performance of the CNTF, RGOF, and CFF was performed. Combining the significant piezoresistive dissimilarities found at each stage with the clear differences in the packing structure, we have been more convinced that there is a structure-dependent mechanism dominating the sensing performance of the varied smart fabrics. In the gelation stage, the different levels of the resin infiltration cause the different upside potentials of the resistance changes. In the hardening stage, the resin shrinkage effect results in the resistance attenuation. In the post-curing stage, massive defects on the conducting network can make smart fabrics more sensitive to internal stress, leading to continuous rise of the electrical signal.

Author Contributions: The study was conceived of and designed by S.L. and G.W.; the experiments were performed by G.W. and Y.W.; the data collection and analysis were performed by G.W.; the original draft was prepared by G.W.; the review and editing were performed by S.L. and Y.L.; and the funding was acquired by S.L.

Funding: This work was jointly supported by the National Nature Science Foundation of China (NSFC 61701015), Commercial Aircraft Corporation of China, Ltd. (COMAC-SFGS-2017-36740), and the Chinese Aeronautical Establishment (2017ZF51077).

Acknowledgments: S.L. gratefully acknowledges the Thousand Talents Plan funded by the Chinese government and Beihang University for young scientists.

Conflicts of Interest: The authors declare no conflict of interest.

References

1. Sureeyatanapas, P.; Hejda, M.; Eichhorn, S.J.; Young, R.J. Comparing singlewalled carbon nanotubes and samarium oxide as strain sensors for model glass-fibre/epoxy composites. *Compos. Sci. Technol.* **2010**, *70*, 88–93. [CrossRef]
2. Lu, S.; Chen, D.; Wang, X.; Shao, J.; Ma, K.; Zhang, L.; Araby, S.; Meng, Q. Real-time cure behaviour monitoring of polymer composites using a highly flexible and sensitive cnt buckypaper sensor. *Compos. Sci. Technol.* **2017**, *152*, 181–189. [CrossRef]
3. Dally, J.W.; Sanford, R.J. Strain-gage methods for measuring the opening-mode stress-intensity factor. *Exp. Mech.* **1987**, *27*, 381–388. [CrossRef]
4. Güemes, A.; Sierra, J.; Grooteman, F.; Kanakis, T.; Michaelides, P.; Habas, D.; Tur, M.; Gorbatov, N.; Koimtzoglou, C.; Kontis, N. Methodologies for the Damage Detection Based on Fiber-Optic Sensors. Applications to the Fuselage Panel and Lower Wing Panel. In *Smart Intelligent Aircraft Structures (SARISTU)*; Springer: Berlin, Germany, 2016; pp. 401–431.
5. Moletn, L.; Aktepe, B. Review of fatigue monitoring of agile military aircraft. *Fatigue Fract. Eng. Matrl. Struct.* **2000**, *23*, 767–785.
6. Purekar, A.S.; Pines, D.J. Damage Detection in Thin Composite Laminates Using Piezoelectric Phased Sensor Arrays and Guided Lamb Wave Interrogation. *J. Intell. Mater. Syst. Struct.* **2010**, *21*, 995–1010. [CrossRef]
7. Hay, T.R.; Royer, R.L.; Gao, H.; Zhao, X.; Rose, J.L. A comparison of embedded sensor Lamb wave ultrasonic tomography approaches for material loss detection. *Smart Mater. Struct.* **2006**, *15*, 946. [CrossRef]
8. Michaels, J.E.; Michaels, T.E. Guided wave signal processing and image fusion for in situ damage localization in plates. *Wave Motion* **2007**, *44*, 482–492. [CrossRef]
9. Giurgiutiu, V.; Zagrai, A.; Jing, J.B. Piezoelectric Wafer Embedded Active Sensors for Aging Aircraft Structural Health Monitoring. *Struct. Health Monit.* **2002**, *1*, 41–61. [CrossRef]
10. Wang, G.; Wang, Y.; Zhang, P.; Zhai, Y.; Luo, Y.; Li, L.; Luo, S. Structure dependent properties of carbon nanomaterials enabled fiber sensors for in situ monitoring of composites. *Compos. Struct.* **2018**, *195*, 36–44. [CrossRef]
11. Baughman, R.H.; Zakhidov, A.A.; Heer, W.A.D. Carbon nanotubes—The route toward applications. *Science* **2002**, *297*, 787–792. [CrossRef] [PubMed]
12. Gao, L.; Chou, T.; Thostenson, E.T.; Zhang, Z.; Coulaud, M. In situ sensing of impact damage in epoxy/glass fiber composites using percolating carbon nanotube networks. *Carbon* **2011**, *49*, 3382–3385. [CrossRef]
13. Gnidakouong, J.; Roh, H.; Kim, J.; Park, Y. In situ process monitoring of hierarchical micro-/nano-composites using percolated carbon nanotube networks. *Compos. Part A Appl. Sci. Manuf.* **2016**, *84*, 281–291. [CrossRef]
14. Luo, S.; Liu, T. SWCNT/graphite nanoplatelet hybrid thin films for self-temperature-compensated, highly sensitive, and extensible piezoresistive sensors. *Adv. Mater.* **2013**, *25*, 5650–5657. [CrossRef] [PubMed]
15. Luo, S.; Hoang, P.T.; Liu, T. Direct laser writing for creating porous graphitic structures and their use for flexible and highly sensitive sensor and sensor arrays. *Carbon* **2016**, *96*, 522–531. [CrossRef]
16. Luo, S.; Wang, Y.; Wang, G.; Liu, F.; Zhai, Y.; Luo, Y. Hybrid spray-coating, laser-scribing and ink-dispensing of graphene sensors/arrays with tunable piezoresistivity for in situ monitoring of composites. *Carbon* **2018**, *139*, 437–444. [CrossRef]
17. Wang, Y.; Wang, Y.; Zhang, P.; Liu, F.; Luo, S. Laser-Induced Freestanding Graphene Papers: A New Route of Scalable Fabrication with Tunable Morphologies and Properties for Multifunctional Devices and Structures. *Small* **2018**, 1802350. [CrossRef] [PubMed]

18. Moriche, R.; Jiménez-Suárez, A.; Sánchez, M.; Prolongo, S.G.; Ureña, A. Graphene nanoplatelets coated glass fibre fabrics as strain sensors. *Compos. Sci. Technol.* **2017**, *146*, 59–64. [CrossRef]
19. Lu, S.; Chen, D.; Wang, X.; Xiong, X.; Ma, K.; Zhang, L.; Meng, Q. Monitoring the manufacturing process of glass fiber reinforced composites with carbon nanotube buckypaper sensor. *Polym. Test.* **2016**, *52*, 79–84. [CrossRef]
20. Skrifvars, M.; Niemelä, P.; Koskinen, R.; Hormi, O. Process cure monitoring of unsaturated polyester resins, vinyl ester resins, and gel coats by raman spectroscopy. *J. Appl. Polym. Sci.* **2010**, *93*, 1285–1292. [CrossRef]
21. Carlone, P.; Rubino, F.; Paradiso, V.; Tucci, F. Multi-scale modeling and online monitoring of resin flow through dual-scale textiles in liquid composite molding processes. *Int. J. Adv. Manuf. Technol.* **2018**, *96*, 2215–2230. [CrossRef]
22. Gnidakouong, J.; Roh, H.; Kim, J.; Park, Y. In situ assessment of carbon nanotube flow and filtration monitoring through glass fabric using electrical resistance measurement. *Compos. Part A Appl. Sci. Manuf.* **2016**, *90*, 137–146. [CrossRef]
23. Ali, M.A.; Umer, R.; Khan, K.A.; Samad, Y.A.; Liao, K.; Cantwell, W. Graphene coated piezo-resistive fabrics for liquid composite molding process monitoring. *Compos. Sci. Technol.* **2017**, *148*, 106–114. [CrossRef]
24. Luo, S.; Obitayo, W.; Liu, T. SWCNT-thin-film-enabled fiber sensors for lifelong structural health monitoring of polymeric composites—From manufacturing to utilization to failure. *Carbon* **2014**, *76*, 321–329. [CrossRef]
25. Luo, S.; Liu, T. Graphite nanoplatelet enabled embeddable fiber sensor for in situ curing monitoring and structural health monitoring of polymeric composites. *ACS Appl. Mater. Interfaces* **2014**, *6*, 9314–9320. [CrossRef] [PubMed]
26. Luo, S.; Wang, Y.; Wang, G.; Wang, K.; Wang, Z.; Zhang, C.; Wang, B.; Luo, Y.; Li, L.; Liu, T. CNT enabled co-braided smart fabrics: A new route for non-invasive, highly sensitive & large-area monitoring of composites. *Sci. Rep.* **2017**, *7*, 44056. [PubMed]
27. Hummers, W.S., Jr.; Offeman, R.E. Preparation of Graphitic Oxide. *J. Am. Chem. Soc.* **1958**, *80*, 1339. [CrossRef]
28. Luo, S.; Liu, T. Structure-property-processing relationships of single-wall carbon nanotube thin film piezoresistive sensors. *Carbon* **2013**, *59*, 315–324. [CrossRef]
29. Li, Y.; Luo, S.; Yang, M.; Liang, R.; Zeng, C. Poisson ratio and piezoresistive sensing: A new route to high-performance 3d flexible and stretchable sensors of multimodal sensing capability. *Adv. Funct. Mater.* **2016**, *26*, 2900–2908. [CrossRef]
30. Harris, P.J.F. New perspectives on the structure of graphitic carbons. *Crit. Rev. Solid State Mater. Sci.* **2005**, *30*, 235–253. [CrossRef]
31. Hao, B.; Ma, Q.; Yang, S.; Mäder, E.; Ma, P.C. Comparative study on monitoring structural damage in fiber-reinforced polymers using glass fibers with carbon nanotubes and graphene coating. *Compos. Sci. Technol.* **2016**, *129*, 38–45. [CrossRef]
32. Kuzmany, H.; Pfeiffer, R.; Hulman, M.; Kramberger, C. Raman spectroscopy of fullerenes and fullerene-nanotube composites. *Philos. Trans. R. Soc. A Math Phys. Eng. Sci.* **2004**, *362*, 2375–2406. [CrossRef] [PubMed]
33. Borowiak-Palen, E.; Bachmatiuk, A.; Rümmeli, M.H.; Gemming, T.; Kruszynska, M.; Kalenczuk, R.J. Modifying cvd synthesised carbon nanotubes via the carbon feed rate. *Phys. E Low-dimens. Syst. Nanostruct.* **2008**, *40*, 2227–2230. [CrossRef]
34. Dresselhaus, M.S.; Jorio, A.; Hofmann, M.; Dresselhaus, G.; Saito, R. Perspectives on carbon nanotubes and graphene Raman spectroscopy. *Nano Lett.* **2010**, *10*, 751–758. [CrossRef] [PubMed]
35. Tuinstra, F.; Koenig, J.L. Raman spectrum of graphite. *J. Chem. Phys.* **1970**, *53*, 1126–1130. [CrossRef]
36. Pimenta, M.A.; Dresselhaus, G.; Dresselhaus, M.S.; Cançado, L.G.; Jorio, A.; Saito, R. Studying disorder in graphite-based systems by raman spectroscopy. *Phys. Chem. Chem. Phys.* **2007**, *9*, 1276–1291. [CrossRef] [PubMed]
37. Marquez, C.; Rodriguez, N.; Ruiz, R.; Gamiz, F. Electrical characterization and conductivity optimization of laser reduced graphene oxide on insulator using point-contact methods. *RSC Adv.* **2016**, *6*, 46231–46237. [CrossRef]
38. Baranov, A.V.; Bekhterev, A.N.; Bobovich, Y.S.; Petrov, V.I. Interpretation of certain characteristics in raman spectra of graphite and glassy carbon. *Opt. Spektrosk.* **1987**, *62*, 612–616.

39. Stankovich, S.; Dikin, D.A.; Piner, R.D.; Kohlhaas, K.A.; Kleinhammes, A.; Jia, Y.; Wu, Y.; Nguyen, S.T.; Ruoff, R.S. Synthesis of graphene-based nanosheets via chemical reduction of exfoliated graphite oxide. *Carbon* **2007**, *45*, 1558–1565. [CrossRef]
40. Ferrari, A.C. Raman spectroscopy of graphene and graphite: Disorder, electron–phonon coupling, doping and nonadiabatic effects. *Solid State Commun.* **2007**, *143*, 47–57. [CrossRef]
41. Kumar, R.; Avasthi, D.K.; Kaur, A. Fabrication of chemiresistive gas sensors based on multistep reduced graphene oxide for low parts per million monitoring of sulfur dioxide at room temperature. *Sens. Actuators B Chem.* **2017**, *242*, 461–468. [CrossRef]
42. Ferrari, A.C.; Basko, D.M. Raman spectroscopy as a versatile tool for studying the properties of graphene. *Nat. Nanotechnol.* **2013**, *8*, 235–246. [CrossRef] [PubMed]
43. Chiacchiarelli, L.M.; Rallini, M.; Monti, M.; Puglia, D.; Kenny, J.M.; Torre, L. The role of irreversible and reversible phenomena in the piezoresistive behavior of graphene epoxy nanocomposites applied to structural health monitoring. *Compos. Sci. Technol.* **2013**, *80*, 73–79. [CrossRef]
44. Haider, M.; Hubert, P.; Lessard, L. Cure shrinkage characterization and modeling of a polyester resin containing low profile additives. *Compos. Part A Appl. Sci. Manuf.* **2007**, *38*, 994–1009. [CrossRef]
45. Tran, H.P.; Salazar, N.; Porkka, T.N.; Joshi, K.; Liu, T.; Dickens, T.J.; Yu, Z. Engineering crack formation in carbon nanotube-silver nanoparticle composite films for sensitive and durable piezoresistive sensors. *Nanoscale Res. Lett.* **2016**, *11*, 422. [CrossRef] [PubMed]

© 2018 by the authors. Licensee MDPI, Basel, Switzerland. This article is an open access article distributed under the terms and conditions of the Creative Commons Attribution (CC BY) license (http://creativecommons.org/licenses/by/4.0/).

Article

Electromagnetic Shielding Effectiveness of Woven Fabrics with High Electrical Conductivity: Complete Derivation and Verification of Analytical Model

Marek Neruda * and Lukas Vojtech

Department of Telecommunication Engineering, Faculty of Electrical Engineering, Czech Technical University in Prague, Technicka 2, 166 27 Prague, Czech Republic; lukas.vojtech@fel.cvut.cz
* Correspondence: marek.neruda@fel.cvut.cz; Tel.: +420-224-355-824

Received: 31 July 2018; Accepted: 5 September 2018; Published: 7 September 2018

Abstract: In this paper, electromagnetic shielding effectiveness of woven fabrics with high electrical conductivity is investigated. Electromagnetic interference-shielding woven-textile composite materials were developed from a highly electrically conductive blend of polyester and the coated yarns of Au on a polyamide base. A complete analytical model of the electromagnetic shielding effectiveness of the materials with apertures is derived in detail, including foil, material with one aperture, and material with multiple apertures (fabrics). The derived analytical model is compared for fabrics with measurement of real samples. The key finding of the research is that the presented analytical model expands the shielding theory and is valid for woven fabrics manufactured from mixed and coated yarns with a value of electrical conductivity equal to and/or higher than σ = 244 S/m and an excellent electromagnetic shielding effectiveness value of 25–50 dB at 0.03–1.5 GHz, which makes it a promising candidate for application in electromagnetic interference (EMI) shielding.

Keywords: analytical model; electromagnetic shielding effectiveness; electric properties; fabric; woven textiles

1. Introduction

Electromagnetic compatibility is the branch of electrical engineering focused on generation, propagation, and reception of electromagnetic energy that can affect the proper function of electronic systems. One of the methods for ensuring proper function of these systems is a shielding, expressed by a quantity called shielding effectiveness (SE), electromagnetic shielding, or electromagnetic shielding effectiveness. Primarily, the shielding of electronic systems is performed by metals. Nowadays, the metals can be replaced by electrically conductive textiles in order to obtain a relevant value of the SE, which has been a highly discussed topic in recent years. The structure of these textile materials can be in the form of coated/metallized fabric, which can be categorized as a multi-layered "stack-up" system of composite shielding materials, or particulate-blended shielding textile composites, which are made up by metallic inclusions like aluminum, copper, silver, or nickel particles heterogeneously mixed in a host medium such as polymer/plastic. The main benefits include lower consumption of metals, flexibility of the textile materials, mechanical properties, and/or lower weight of the shielding. Woven fabrics with high electrical conductivity are being increasingly utilized in the shielding of electromagnetic interference (EMI) and in electrostatic protection in various applications such as the shields for equipment cases, the protective clothing for personnel working under high-voltage magnetic fields and/or in radiofrequency/microwave environments, shielding and grounding curtains, electrostatic discharge wipers, flexible shielded shrouds, smocks, stockings, boots, etc.

Many research papers describe SE evaluation from different perspectives, i.e., measurement techniques [1–7], composition of materials [8–17], influence of washing/drying cycles on values of

SE of fabrics [18–20], or calculation of SE [4,21–31]. SE measurement is commonly performed by a coaxial transmission line method specified in ASTM 4935-10 [1–4,6,8–10,12,14,17,18] by measuring the insertion loss with a dual transverse electromagnetic (TEM) cell [3,5], or by measurement in a free space, shielding box, or shielding room with receiving and transmitting antennas [7,15,19,20]. The papers which focus on measurement techniques of various electrically conductive fabrics usually present basic equations for SE calculation [4,21–28] (SI units are used in all formulas unless otherwise stated) as shown in Equation (1).

$$SE = 20 \cdot \log_{10} \left| \frac{E_i}{E_t} \right| = 20 \cdot \log_{10} \left| \frac{H_i}{H_t} \right| = 10 \cdot \log_{10} \left| \frac{P_i}{P_t} \right| = R + A + B \tag{1}$$

where E_i, H_i, and P_i are the electric field intensity, magnetic field intensity, and power without the presence of tested material (incident electromagnetic field on the tested material), respectively, E_t, H_t, and P_t are the same physical quantities with the presence of tested material (transmitted electromagnetic field measured behind the tested material), R is the reflection loss, A is the absorption loss, and B is multiple reflections.

Reflection loss R (also called return attenuation) is a consequence of the electromagnetic wave reflection on the interface. The absorption loss A (also called absorption attenuation) is produced if the electromagnetic wave is transferred through the shielding barrier. A portion of energy is absorbed in the shielding barrier due to heat loss. Attenuation caused by multiple reflections, B, is physically caused by electromagnetic wave propagation in the conducted shielding barrier. The electromagnetic wave is repetitively reflected on the "inner" interfaces of the material.

The handbook of electromagnetic materials [28] describes an expression for SE calculation of metallized fabrics based on transmission line theory, i.e., an analysis of the leakage through apertures in the fabric, as shown in Equation (2):

$$SE = A_a + R_a + B_a + K_1 + K_2 + K_3 \tag{2}$$

where A_a is the attenuation introduced by a particular discontinuity, R_a is a fabric aperture with single reflection loss, B_a is the multiple reflection correction coefficient, K_1 is the correction coefficient to account for the number of like discontinuities, K_2 is the low-frequency correction coefficient to account for skin depth, and K_3 is the correction coefficient to account for a coupling between adjacent holes.

The authors in [4] adopted this formula without description of its derivation for hybrid fabrics, i.e., fabrics composed of hybrid yarns containing polypropylene and different content. This formula is also compared with the wave-transmission-matrix (WTM) method in [8]. The authors evaluated the SE of the laminated and anisotropic composites for single-layer and multi-layer fabrics. The same formula is also presented in [23] for evaluation of copper core-woven fabrics in order to identify dependencies of the SE on the material structure. None of the papers [4,8,23] nor the handbook [28] present a derivation of this formula. The influence of seaming stitches on the SE fabric is described in [29]. That paper presents a computation model of the SE based on the equivalent seaming gap. Analytical formulation for the SE of enclosures with apertures is described in [30]. The paper presents an extended theory to account for electromagnetic losses, circular apertures, and multiple apertures. Formulas for the apertures (especially multiple apertures) are key to the analytical modeling of fabrics. A calculation method of SE for woven fabric containing metal fiber yarns is deduced through the transfer matrix of the electromagnetic field numerical calculation in [31].

A semi-empirical model describing the plane wave SE for fabrics is presented in [25]. The authors focus only on coated fabrics and derivation of the SE formula based on electrical properties (especially electrical conductivity). The same formula is also described in the handbook [28] without a formula derivation, Equation (3).

$$SE_{fabric} = e^{-0.129 \cdot \ell \cdot \sqrt{f}} \cdot SE_{foil} + \left(1 - e^{-0.129 \cdot \ell \cdot \sqrt{f}}\right) \cdot SE_{aperture} \tag{3}$$

where SE_{foil} and $SE_{aperture}$ are the SE values for metallic foil (of the same thickness as the fabric) and for the same foil with aperture (s), l is the aperture size of the fabric, and f is the frequency.

Calculation of the SE_{foil} is well-known from shielding theory [24,26–28,32,33] as:

$$SE_{foil} = \underbrace{20 \cdot \log\left|\frac{(Z_0 + Z_M)^2}{4 Z_0 Z_M}\right|}_{R_{foil}} + \underbrace{20 \cdot \log\left|e^{t/\delta} e^{-j\beta_0 t} e^{j\beta t}\right|}_{A_{foil}} + \underbrace{20 \cdot \log\left|1 - e^{-2t/\delta} e^{-j2\beta t}\left(\frac{Z_0 - Z_M}{Z_0 + Z_M}\right)^2\right|}_{B_{foil}} \quad (4)$$

where Z_0 is the impedance of free space, Z_M is the impedance of shielding barrier, t is the thickness, δ is the penetration depth, β_0 is the vacuum phase constant, and β is the phase constant.

A complete derivation of Equation (4) is published in our previous research papers [26,27]. Calculation of $SE_{aperture}$ is usually expressed similarly to Equation (1), and it is expressed only for metallized fabric shields as [25,28], Equation (5).

$$SE_{aperture} = R_{aperture} + A_{aperture} + K_{aperture} = 100 - 20 \cdot \log(L \cdot f) + 20 \cdot \log\left(1 + \ln\left(\frac{L}{s}\right)\right) + 30\frac{D}{L} \quad (5)$$

where L is the maximum aperture size, f is the frequency of operation, s is the minimum aperture size, and D is the depth of the aperture.

Calculation and derivation of $SE_{aperture}$ is not present in the scientific literature for particulate-blended shielding electrically conductive textile composites. Therefore, the main contribution of this this paper is that the research performed a complete derivation of an analytical model of SE for woven textile materials manufactured from electrically conductive mixed and coated yarns, i.e., for particulate-blended shielding electrically conductive textile composites. Basic simplifications, which are valid for metals, were also evaluated for these textile materials. A complete derivation of SE evaluation was also performed for SE_{fabric} (Equation (3)) and $SE_{aperture}$ of metallized fabric shields (Equation (5)). A general equation for SE evaluation for particulate-blended shielding electrically conductive textile composites with electrical conductivity bigger than 244 S/m was derived and compared with measurement of real samples according to ASTM 4935-10. The maximal difference between modeling and measurement results was in the range of 2–6 dB, which is within the random error of the used measurement method, i.e., ±5 dB.

2. Experimental Materials

The samples are particulate-blended shielding electrically conductive textile composites manufactured from two types of yarns that are mixed and coated with a plain weave fabric structure, Table 1. The coated yarns SilveR.STAT® (samples #1–#2) contain a very pure silver layer on the polymer base (Polyamide). The mixed yarns (samples #3–#7) are blended from the non-conductive textile material, i.e., Polyester (PES), and the conductive material, i.e., silver in the form of the SilveR.STAT® coated yarns. The plain weave is chosen because of its simple and regular structure.

The samples #1 and #2 and #3–#5 are made of the same material (the same ratio of conductive and non-conductive textile material in the case of #3–#5) with the same fabric structure and differ from each other mainly by the warp and weft density used in a production process. The samples #6 and #7 are characterized by the same warp and weft density, the same fabric structure, and differ from each other by the ratio of conductive and non-conductive textile material, i.e., 40%/60% and vice versa. The selected parameters result in a different value of mass per unit area and, more importantly, in the different electrical conductivity value. As a result, the three groups of electrically conductive textile materials can be distinguished by the value of the order of the electrical conductivity, i.e., #1–#2, #3–#5 and #6–#7, which is an important parameter in the SE calculation as shown in the equations, e.g., Equations (4) and (6).

Measurement of the electrical conductivity of samples #1–#7 is based on a four electrode test method described in BS EN 16812:2016 [34] and conclusions presented in [35], i.e., measurement of surface and bulk resistance is equal for high electrically conductive textile materials and, therefore, thickness of the sample can be taken into account in electrical conductivity evaluation. Mean value (evaluated for five different lengths and five different areas of the sample, 65% RH, 20 °C) and standard deviation of electrical conductivity are depicted in Table 1.

Table 1. Fabric specifications.

No.	Composition	Fabric Structure	Mass per Unit Area [g/m^2]	Warp/Weft Density d_w [t/cm]	Linear Density [tex]	Electrical Conductivity [S/m]	Standard Deviation [S/m]
1	SilveR.STAT® 240dtex/10F	Plain weave	75	13/13	24	1.71×10^4	9.72×10^2
2	SilveR.STAT® 240dtex/10F	Plain weave	95	16/16	24	1.77×10^4	5.69×10^2
3	60% PES/40% SilveR.STAT® 3.3dtex	Plain weave	92	13/13	29.5	1.07×10^3	2.32×10^1
4	60% PES/40% SilveR.STAT® 3.3dtex	Plain weave	115	16/16	29.5	1.00×10^3	3.29×10^1
5	60% PES/40% SilveR.STAT® 3.3dtex	Plain weave	135	19/19	29.5	1.37×10^3	5.14×10^1
6	40% PES/60% SilveR.STAT® 1.7dtex	Plain weave	115	16/16	29.5	2.44×10^2	8.99×10^0
7	60% PES/40% SilveR.STAT® 1.7dtex	Plain weave	117	16/16	29.5	3.9×10^1	2.13×10^0

3. Evaluation of Reflection Loss of Foil

Reflection loss R_{foil} is generally expressed in Equations (4) and (6). It can be simplified for metals because of its good electrical conductivity, i.e., the inequality $Z_M \ll Z_0$ is valid. Moreover, the impedance of the material Z_M is further simplified because of the validity $\sigma \gg \omega\varepsilon$. The R_{foil} is then calculated as [26,27], Equation (6).

$$R_{foil} = 20 \cdot \log\left|\frac{(Z_0+Z_M)^2}{4Z_0Z_M}\right| \approx 20 \cdot \log\left|\frac{Z_0}{4Z_M}\right| = 20 \cdot \log\left|\frac{\sqrt{\frac{\mu_0}{\varepsilon_0}}}{4\sqrt{\frac{j\omega\mu}{\sigma+j\omega\varepsilon}}}\right| \approx 20 \cdot \log\left|\frac{\sqrt{\frac{\mu_0}{\varepsilon_0}}}{4\sqrt{\frac{\omega\mu}{\sigma}}}\right| = 20 \cdot \log\left(\frac{1}{4}\sqrt{\frac{\sigma}{\omega\mu_r\varepsilon_0}}\right) \quad (6)$$

where μ_0 is the vacuum permeability, μ_r is the relative permeability, μ is the permeability of a specific medium, ω is the angular speed, σ is the conductivity, ε_0 is the vacuum permittivity, and ε is the absolute permittivity.

The conductivity of a material can be expressed as conductivity relative to copper [28]. The value of copper conductivity is equal to $\sigma_{Cu} = 5.8 \times 10^7$ S/m [33,36]. Material conductivity is described as $\sigma = \sigma_r\sigma_{Cu}$, and R_{foil} is expressed as shown in Equation (7).

$$R_{foil} = 20 \cdot \log\left(\frac{1}{4}\sqrt{\frac{\sigma_{Cu}}{2\pi\varepsilon_0}}\right) + 20 \cdot \log\left(\sqrt{\frac{\sigma_r}{f\mu_r}}\right) = 168.14 + 20 \cdot \log\left(\sqrt{\frac{\sigma_r}{f\mu_r}}\right) \quad (7)$$

where f is the frequency of the operation.

A similar equation can be also found in [4,21,22,24]. The calculation of reflection loss R_{foil} corresponds to the copper conductivity value, e.g., $\sigma_{Cu} = 5.82 \times 10^7$ S/m [32], 5.7×10^7 S/m [37], or 5.85×10^7 S/m [38], which depends on the purity and the production method of copper. Nevertheless, Equation (7) is presented for fabrics, i.e., the value $\sigma_{Cu} = 5.8 \times 10^7$ S/m is used. This means the authors presume the validity of presented inequalities, i.e., $Z_M \ll Z_0$ and $\sigma \gg \omega\varepsilon$. This presumption is furthermore verified. Definitions of the impedances Z_M and Z_0 and their simplified versions are described in Equation (8).

$$Z_0 = \sqrt{\frac{\mu_0}{\varepsilon_0}} = \sqrt{\frac{4 \cdot \pi \cdot 10^{-7}}{8,854 \cdot 10^{-12}}} = 120\,\pi \approx 377 \quad Z_M = \sqrt{\frac{j\omega\mu}{\sigma+j\omega\varepsilon}} \approx \sqrt{\frac{\omega\mu}{\sigma}} \quad (8)$$

The validity of the $\sigma \gg \omega\varepsilon$ can be easily verified for the lowest values of the electrical conductivity of the samples, i.e., #6 and #7. The value of relative permittivity of the used electrically conductive material is considered to be $\varepsilon_r = 1$, because the non-conductive textile material is blended with a conductive material, i.e., silver, Table 1. As a consequence, the resultant textile material is categorized as lossy conductive material, which can be characterized as $\varepsilon_r = 1$. The value of relative permeability is considered to be equal to $\mu_r = 1$. Results for different frequencies are presented in Table 2, and Figures 1 and 2. The condition of $\sigma \gg \omega\varepsilon$ is fulfilled for sample #6 in the entire analyzed frequency range, i.e., 30 MHz–10 GHz because of the difference between the values of σ and $\omega\varepsilon$ in at least two orders of magnitude. Sample #7 fulfills the condition up to approximately 6.9 GHz. As a consequence, a simplified version of the Z_M and Equation (7) can be used for #1–#6 up to 10 GHz and for #7 up to approximately 6.9 GHz. Materials with lower electrical conductivity than #7, i.e., 39 S/m, have to be analyzed in order to obtain the frequency limit of validity $\sigma \gg \omega\varepsilon$ and Equation (7). It can also be noted the limit of #6 is found to be 43.85 GHz, and the SE measurement is usually performed by a coaxial transmission line method specified in ASTM 4935-10 in the range of 30 MHz–1.5 GHz [39].

Table 2. Results of ratio of the σ and $\omega\varepsilon$ evaluation.

Frequency [GHz]	Sample #6 $\sigma/\omega\varepsilon$	Sample #7 $\sigma/\omega\varepsilon$
1.5	2924	461.4
3	1462	230.7
4	1097	173
10	438.6	69.21

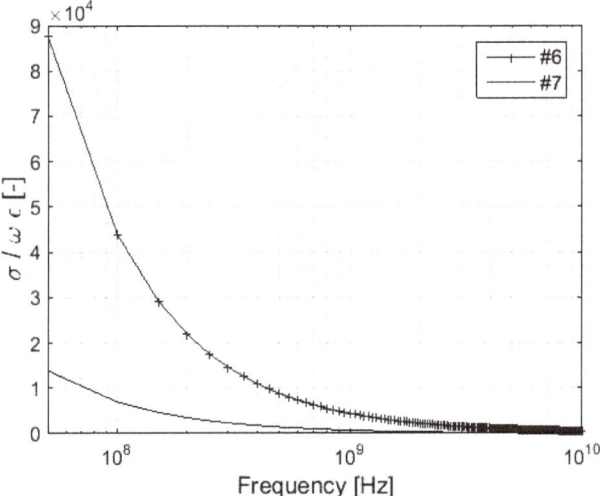

Figure 1. Ratio $\sigma/\omega\varepsilon$ for samples #6 and #7.

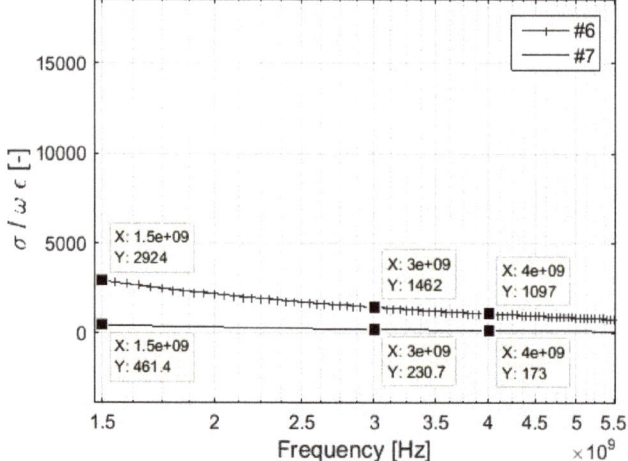

Figure 2. Details of the ratio $\sigma/\omega\varepsilon$ for samples #6 and #7 in 1.5–5.5 GHz.

The validity of $Z_M \ll Z_0$ is verified for samples #7, #6, and #4, which are characterized by lower values of electrical conductivity from all described samples, Figures 3 and 4. The validity of $\sigma \gg \omega\varepsilon$ is assumed, i.e., a simplified version of Z_M is considered. A difference in values of two orders of magnitude in the frequencies 70 MHz, 439 MHz, and 1.8 GHz can be seen for #7, #6, and #4, respectively. If the validity of $\sigma \gg \omega\varepsilon$ is not assumed, the values of two orders of magnitude are in the frequencies 34.6 MHz, 219.5 MHz, and 899.9 MHz for #7, #6, and #4, respectively.

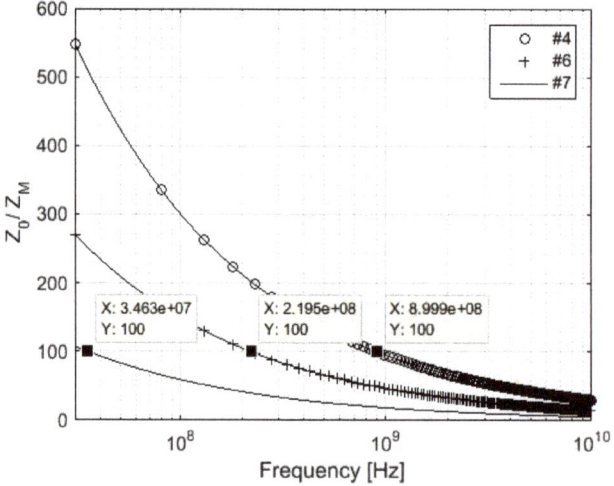

Figure 3. Ratio Z_0/Z_M for samples #4, #6, and #7; $\sigma \gg \omega\varepsilon$ is not assumed.

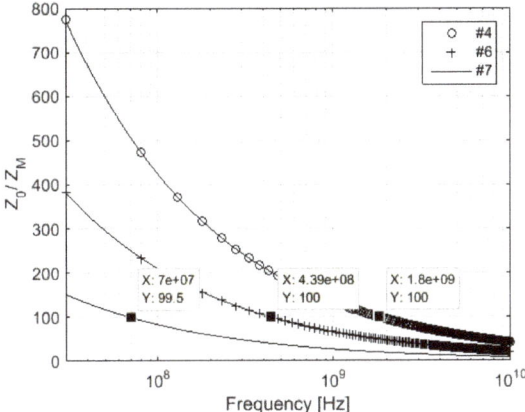

Figure 4. Ratio Z_0/Z_M for samples #4, #6, and #7; $\sigma \gg \omega\varepsilon$ is assumed.

The clarity of application of the validity $Z_M \ll Z_0$ and $\sigma \gg \omega\varepsilon$ can be also seen in Figure 5. It compares the R_{foil} parameter of the four cases, i.e., $Z_M \ll Z_0$ and $\sigma \gg \omega\varepsilon$ are/are not considered (in all four cases), and also for all samples #1–#7. All four cases are almost identical to #1–#5, i.e., the greatest difference is reached for the sample with the lowest electrical conductivity (#4), and it is equal to about 0.5 dB in 10 GHz. The results also show an insignificant difference, i.e., the greatest difference is about a thousandth of a dB, for cases $Z_M \ll Z_0$ with (dash-dot line) and without (dotted line) consideration of $\sigma \gg \omega\varepsilon$ validity for the sample with the lowest electrical conductivity (#7) (case 3 and case 4). The greatest difference between these four cases is obtained for #7 in the frequency 10 GHz. It is about 2 dB for the cases where $Z_M \ll Z_0$ is not considered and $\sigma \gg \omega\varepsilon$ is considered (solid line) and where $Z_M \ll Z_0$ is considered and $\sigma \gg \omega\varepsilon$ is (dash-dot line) / is not (dotted line) considered (both cases of $\sigma \gg \omega\varepsilon$ show similar results as previously mentioned) (case 1 and case 3). The same situation is also valid for #6 with a difference not exceeding 1 dB. The results also show higher values of R_{foil} for #7 at 10 GHz, i.e., about 0.5 dB, for the case where $Z_M \ll Z_0$ is not considered and $\sigma \gg \omega\varepsilon$ is considered (solid line) in comparison with the case where $Z_M \ll Z_0$ and $\sigma \gg \omega\varepsilon$ are not considered (dashed line) (case 1 and case 2), i.e., no simplification is performed.

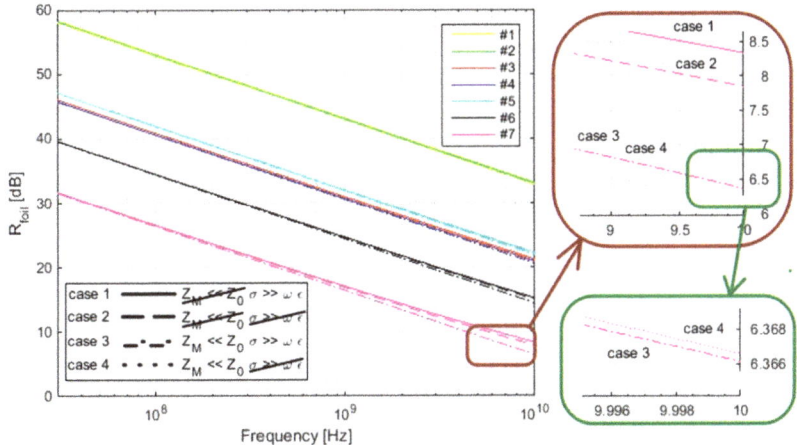

Figure 5. R_{foil} evaluation for #1–#7 and $Z_M \ll Z_0$ and $\sigma \gg \omega\varepsilon$ are/are not considered (four cases).

As a consequence, a simplification of SE evaluation by $Z_M \ll Z_0$ and $\sigma \gg \omega\varepsilon$ in frequency range up to 10 GHz is valid for samples with electrical conductivity values higher than 1000 S/m with an error up to 0.5 dB, for samples with an electrical conductivity value of 244 S/m with an error not exceeding 1 dB and for samples with an electrical conductivity value of 39 S/m and an error up to 2 dB.

4. Evaluation of Reflection Loss of Aperture

Reflection loss of aperture $R_{aperture}$ is generally derived from the basic relation between the gain and effective aperture of the antenna. It is described as [33], Equation (9).

$$A_e = \frac{\lambda^2}{4\pi} G \qquad (9)$$

where λ is wavelength, A_e is effective aperture and, G is the gain.

The parameter A_e differs for different shapes of antenna loops [40]. Woven textile materials are manufactured by interlacing the yarns at right angles and, therefore, a rectangular or square shape of the apertures can be considered. Presented samples are manufactured from the same yarn and by the same sett in the warp and weft directions. As a consequence, the aperture is square shaped. For the square shape of the antenna loop, the effective aperture is calculated as [40], Equation (10).

$$A_e = l^2 \qquad (10)$$

where l is the length of the aperture.

The equality of Equations (9) and (10) describes the gain in Equation (11).

$$G_{aperture\_square} = \frac{4\pi l^2}{\lambda^2} = \left(\frac{2l\sqrt{\pi}}{\lambda}\right)^2 \qquad (11)$$

The equation for G calculation for the circular loop antenna and for a slot is commonly mentioned in the literature as [33], Equations (12) and (13).

$$G_{aperture\_circular} = \left(\frac{2\pi r}{\lambda}\right)^2 \qquad (12)$$

where r is the radius of the circular loop.

$$G_{aperture\_slot} = \left(\frac{2l}{\lambda}\right)^2 \qquad (13)$$

where l is the length of the slot.

Reflection loss $R_{aperture}$ is then calculated [33,37] as:

$$R_{aperture\_square} = 10 \cdot \log\left(\frac{1}{G_{aperture\_square}}\right) = 20 \cdot \log\left(\frac{\lambda}{2l\sqrt{\pi}}\right) \qquad (14)$$

$$R_{aperture\_circular} = 10 \cdot \log\left(\frac{1}{G_{aperture\_circular}}\right) = 20 \cdot \log\left(\frac{\lambda}{2\pi r}\right) \qquad (15)$$

$$R_{aperture\_slot} = 10 \cdot \log\left(\frac{1}{G_{aperture\_slot}}\right) = 20 \cdot \log\left(\frac{\lambda}{2l}\right) \qquad (16)$$

Equations (14)–(16) can be also described as:

$$R_{aperture\_square} = 20 \cdot \log\left(\frac{c}{2\sqrt{\pi}}\right) - 20 \cdot \log(l \cdot f) = 158.55 - 20 \cdot \log(l \cdot f) \qquad (17)$$

$$R_{aperture\_circular} = 20 \cdot \log\left(\frac{c}{2\pi}\right) - 20 \cdot \log(r \cdot f) = 153.58 - 20 \cdot \log(r \cdot f) \tag{18}$$

$$R_{aperture\_slot} = 20 \cdot \log\left(\frac{c}{2}\right) - 20 \cdot \log(l \cdot f) = 163.52 - 20 \cdot \log(l \cdot f) \tag{19}$$

where c is the speed of light.

Equations (17)–(19) are derived for one aperture in the foil. It can be seen that they are valid for any material with an aperture, because reflection loss of aperture $R_{aperture}$ is dependent on the frequency and dimensions of an aperture.

5. Evaluation of Reflection Loss of Multiple Apertures

Multiple apertures are discussed in [32,36,37]. The equation for multiple apertures is described as:

$$R_{aperture\_multiple} = -20 \cdot \log \sqrt{n} \tag{20}$$

where n is the number of apertures.

Nevertheless, calculation of the number of apertures n is not unified in [32,36,37]. The conditions for validity of Equation (20) follow:

- Reference [32]: linear array of apertures, equal sizes, closely spaced apertures, and the total length of linear array of apertures is less than 1/2 of the wavelength. If the two-dimensional array of holes is considered, Equation (20) can be directly applied only for the first row of apertures (the rest of the apertures are not included in parameter n). This means, if the two-dimensional array is given by 7 × 12 holes, then n = 12. This approximation is motivated by experience.
- Reference [36]: equally sized perforations, hole spacing < $\lambda/2$, hole spacing > thickness, n is the number of all apertures.
- Reference [37]: thin material, equally sized apertures, n is the number of all apertures.

The minimal value of the wavelength can be easily found as depicted in Figure 6.

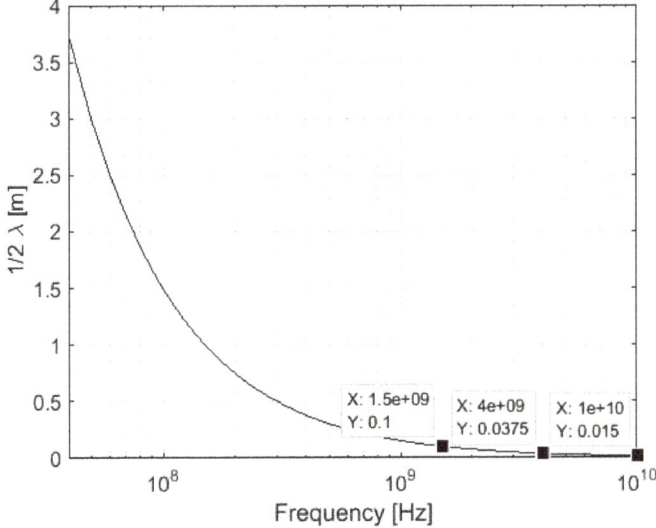

Figure 6. Calculation of the 1/2 wavelength.

The longest linear array of apertures can be determined with respect to ASTM D4935-10. The standard ASTM D4935-99 was withdrawn in 2005, because the committee could not maintain a standard for which the expertise did not lie within the current committee membership [41]. It also describes the dimensions of the measured samples. The longest linear array of apertures that can be found on the sample is the tangent of the inner circle limited by the middle circle. It is indicated by the double arrow with the parameter l_c in Figure 7a. It shows the shape of the reference sample, which matches the size of the sample holder, i.e., the measured part of the sample corresponds to the white annulus in Figure 7a. The distance is equal to $l_c = 0.069$ m, i.e., the total length of apertures is less than 1/2 of the wavelength at 0.03–1.5 GHz. Figure 7b shows the load sample.

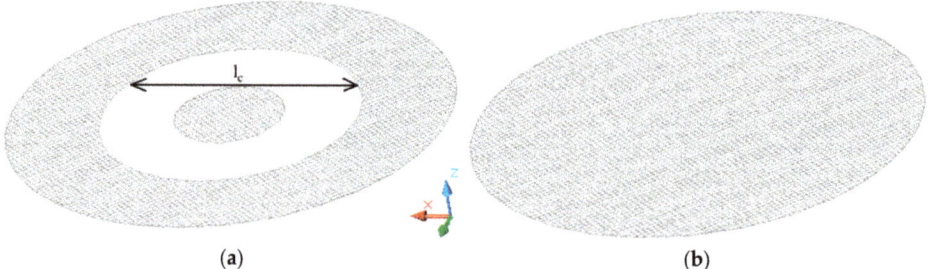

Figure 7. The longest linear array of apertures found in the reference (**a**) and load sample (**b**) according to ASTM 4935-10 (the white annulus of (**a**) matches the size of the measured sample).

The apertures are equally sized, closely spaced, and form a linear array of apertures because of the production process of textiles and parameters used during the production of samples.

The textile structure forms the two-dimensional array of holes and the longest linear array of apertures is equal to $l_c = 0.069$ m.

Hole spacing is equal to the yarn diameter, which is in the range of 0.220–0.251×10^{-3} m. It is less than 1/2 of the wavelength, and it is less than the thickness of the material, which is at a minimum equal to 0.295×10^{-3} m, Table 3.

Table 3. Material characterization and thin material evaluation.

#	Warp/Weft Density d_w [t/cm]	Fabric Thickness [μm]	Yarn Diameter [μm]	Aperture Length [μm]	Thin Material 7δ [GHz]	Thin Material 5δ [GHz]
1	13/13	295	251	517	0.03–8.34	0.03–4.25
2	16/16	300	251	373	0.03–7.79	0.03–3.98
3	13/13	469	239	530	0.03–52	0.03–27
4	16/16	537	239	386	0.03–43	0.03–22
5	19/19	533	239	287	0.03–31	0.03–16
6	16/16	476	220	405	0.03–228	0.03–110
7	16/16	491	240	385	0.03–1316	0.03–680

A difference of the thin and thick material is presented in [36]. The material is considered to be thick when there is no reflection from the "far" interface of the material. This definition can be verified by the equivalent depth of penetration δ, Equation (21), which defines a distance of wave penetration to amplitude wave degradation to the value e^{-1}, i.e., amplitude wave degradation of about 36.8% in comparison with the thickness of the material. If we consider 3δ, amplitude wave degradation is about 95%, i.e., 95% of the current flows within a material. This is the point beyond which current flow is negligible in a material [25]. Nevertheless, comparison for almost 100% of amplitude wave degradation can be performed. The penetration depth 4δ decreases the amplitude wave to about 98.2%, 5δ to about 99.3%, 6δ to about 99.8%, and 7δ to about 99.9%. If the penetration depths 3δ, 5δ, and 7δ are calculated for sample #2 (the sample with the highest value of electrical conductivity, i.e., the value of penetration depth is the lowest), the dependence for the frequency band 30 MHz–10 GHz is obtained, Figure 8. The results show the penetration depths 3δ, 5δ, and 7δ are lower than the thickness of #2 in the frequency range 1.43–10 GHz, 3.98–10 GHz, and 7.79–10 GHz, respectively. This means that in this frequency range there is no reflection from the "far" interface of the material. In other words, the material is considered to be thick. In the frequency ranges 30 MHz–1.43 GHz for 3δ, 30 MHz–3.98 GHz for 5δ, and 30 MHz–7.79 GHz for 7δ, there are reflections from the "far" interface of the material and therefore the material is considered to be thin, Table 3.

$$\delta = \sqrt{\frac{2}{\omega\mu\sigma}} = \frac{1}{\sqrt{\pi f \mu \sigma}} \tag{21}$$

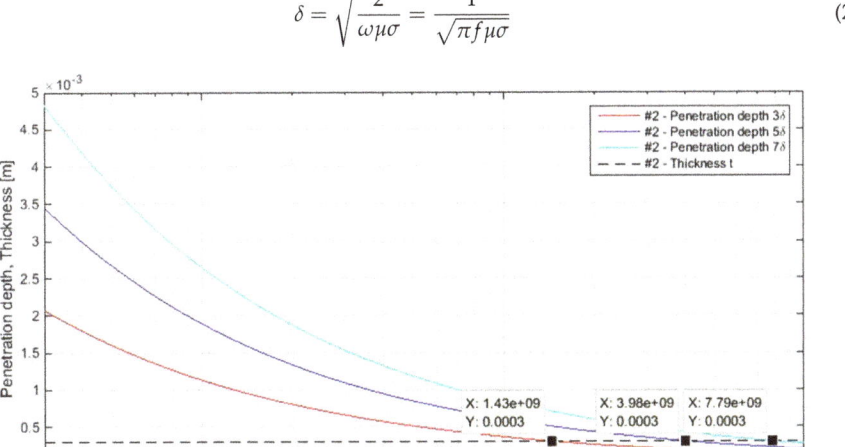

Figure 8. Penetration depths 3δ, 5δ, and 7δ in comparison with thickness for #2.

As a consequence of the reflection loss of multiple apertures, Equation (3) has to be specified for the electrically conductive textile samples described, i.e., Equation (20) is added and the values of electrical conductivity of the described samples are considered.

6. Evaluation of SE Fabric

An expression for the SE calculation of fabric has been developed on the basis of plane wave shielding theory [25,28]. It is based on a linear combination of the SE of the compact material $f_1(l, \lambda)$ (in lower frequency ranges) and the SE of the apertures $f_2(l, \lambda)$ (in higher frequency ranges) as shown in Equation (3). It can be written as shown in Equations (22) and (23):

$$f_1(\ell, \lambda) + f_2(\ell, \lambda) = 1; \quad f_2(\ell, \lambda) = 1 - f_1(\ell, \lambda) \tag{22}$$

$$SE_{fabric} = f_1(\ell, \lambda) \cdot SE_{foil} + (1 - f_1(\ell, \lambda)) \cdot SE_{aperture} \tag{23}$$

A one-dimensional base of the solution $f_1(l, \lambda)$, i.e., $f_1(l, f)$ with respect to λ calculation λ = c/f, can be written as [25], Equation (24).

$$f_1(\ell, f) = e^{-C \cdot \ell \cdot \sqrt{f}} \tag{24}$$

where C is the constant.

An assumption of the equality of components, which corresponds to reflection loss R of compact material R_{foil} and material with apertures $R_{aperture}$, is used for $f_1(l, f)$ derivation, Equation (25).

$$R_{foil} = R_{aperture} \quad (25)$$

The parameter R_{foil} can be used in its simplified version because the R_{foil} evaluation shows the difference is not significant for samples #1–#6, especially for the frequency range up to 3 GHz, i.e., error does not exceed 0.5 dB, Figure 5. Then Equations (6) and (7) for materials with electrical conductivity σ are valid. $R_{aperture}$ is used from Equations (17) and (20). It is written as:

$$20 \cdot \log\left(\frac{1}{4}\sqrt{\frac{\sigma}{\omega\mu_r\varepsilon_0}}\right) = 20 \cdot \log\left(\frac{c}{2\sqrt{\pi}}\right) - 20 \cdot \log(\ell f) - 20 \cdot \log(\sqrt{n}) \quad (26)$$

Equation (26) can be modified as shown in Equations (27)–(30).

$$\log\left(\frac{\sigma}{16\omega\mu_r\varepsilon_0}\right) = \log\left(\frac{c^2}{4\pi}\right) - \log(\ell^2 f^2) - \log(n) \quad (27)$$

$$\log\left(\frac{\sigma}{32\pi f \mu_r\varepsilon_0}\right) = \log\left(\frac{c^2}{4\pi\ell^2 f^2 n}\right) \quad (28)$$

$$\frac{\sigma n}{8c^2\mu_r\varepsilon_0} = \frac{1}{\ell^2 f} \quad (29)$$

$$\ell\sqrt{f} = \sqrt{\frac{8c^2\mu_r\varepsilon_0}{\sigma n}} \quad (30)$$

A boundary condition, which defines a decrease of the amplitude about 95% in specific material [25], i.e., equivalent of 3 depth of penetration (3δ decrease of the amplitude on the multiple $e^{-1}e^{-1}e^{-1} = e^{-3}$ of original value) can be used in Equations (31) and (32).

$$C\ell\sqrt{f} = e^{-3} = C\sqrt{\frac{8c^2\mu_r\varepsilon_0}{\sigma n}} \quad (31)$$

$$C = \frac{e^{-3}}{\ell\sqrt{f}} = e^{-3}\sqrt{\frac{\sigma n}{8c^2\mu_r\varepsilon_0}} = 1.972 \cdot 10^{-5}\sqrt{\sigma n} \quad (32)$$

Evaluation of the constant C and n are shown for samples #1–#7 in Table 4. The number of apertures n is calculated with respect to the longest linear array of apertures of ASTM 4935-10, i.e., $l_c = 0.069$ m and the sett of each sample d_w, Equation (33).

$$n = l_c \cdot d_w \quad (33)$$

Table 4. Calculation of the C constant.

#	σ [S/m]	d_w [t/cm]	n	C
1	1.71×10^4	13	89	2.44×10^{-2}
2	1.77×10^4	16	110	2.76×10^{-2}
3	1.07×10^3	13	89	6.1×10^{-3}
4	1.00×10^3	16	110	6.6×10^{-3}
5	1.37×10^2	19	131	8.4×10^{-3}
6	2.44×10^2	16	110	3.2×10^{-3}
7	3.9×10^1	16	110	1.3×10^{-3}

As a consequence, Equation (23) is specified for sample #1 as shown in Equation (34).

$$SE_{fabric} = e^{-0.0244 \cdot \ell \cdot \sqrt{f}} \cdot SE_{foil} + \left(1 - e^{-0.0244 \cdot \ell \cdot \sqrt{f}}\right) \cdot SE_{aperture} \quad (34)$$

It is obvious the SE calculation has to be specified for each sample with regards to its electrical conductivity, sett, and number of apertures. Therefore, an equation for SE calculation of woven fabrics manufactured from the electrically conductive mixed and coated yarns with square apertures can be written generally with respect to the C constant calculation Equation (32) as shown in Equation (35).

$$SE_{fabric} = e^{-C \cdot \ell \cdot \sqrt{f}} \cdot SE_{foil} + \left(1 - e^{-C \cdot \ell \cdot \sqrt{f}}\right) \cdot SE_{aperture} \qquad (35)$$

The calculation of SE_{foil} is performed according to Equations (4), (7), and (8), and it is also described in depth in [26–28,32,33] as shown in Equation (36).

$$SE_{foil} = 168.14 + 20 \cdot \log\left(\sqrt{\frac{\sigma_r}{f\mu_r}}\right) + 8.6859\frac{t}{\delta} + 20 \cdot \log\left|1 - e^{-2t/\delta}e^{-j2\beta t}\left(\frac{Z_0 - Z_M}{Z_0 + Z_M}\right)^2\right| \qquad (36)$$

6.1. Evaluation of SE of Apertures

$SE_{aperture}$ is calculated as a sum of $R_{aperture}$, $A_{aperture}$, and $K_{aperture}$. $R_{aperture}$ is derived in this paper and expressed in Equations (17) and (20) as shown in Equation (37).

$$R_{aperture} = 158.55 - 20 \cdot \log(\ell \cdot f) - 20 \cdot \log(\sqrt{n}) \qquad (37)$$

The absorption loss of $A_{aperture}$ is included in $SE_{aperture}$ if the fabric is considered to be a thick material, Table 3. It is calculated for a subcritical rectangular waveguide as [32,37], Equation (38).

$$A_{aperture} = 27.2\frac{t_a}{l_a} \qquad (38)$$

where l_a is the largest linear dimension of the cross-section of the aperture and t_a is the depth of the aperture (length of "waveguide").

As shown in Table 3, samples #1–#7 are considered to be thin in a specific analyzed frequency range 30 MHz–1.5 GHz for 7δ, 5δ, and also 3δ (with the exception of the most electrically conductive sample #2 in the frequency range 1.43–1.5 GHz), and therefore $A_{aperture}$ is not included in the $SE_{aperture}$ calculation. $K_{aperture}$ takes into account the geometrical dimensions of the aperture in a shielding barrier. It is described as [28], Equation (39).

$$K_{aperture} = 20 \cdot \log\left(1 + \ln\left(\frac{\ell}{s}\right)\right) \qquad (39)$$

Equation (39) clearly shows the square apertures, i.e., $l = s$, do not influence $SE_{aperture}$. Therefore, the resultant $SE_{aperture}$ is calculated for #1–#7, characterized as thin material, as shown in Equation (40).

$$SE_{aperture} = R_{aperture} = 158.55 - 20 \cdot \log(\ell \cdot f) - 20 \cdot \log(\sqrt{n}) \qquad (40)$$

6.2. Comparison of Equations for SE Fabric

As previously mentioned, the SE_{fabric} is calculated as Equation (35) or Equation (34) for #1. A similar equation was also previously mentioned as Equation (3) for metallized fabric shields [25,28]. Equation (3) is furthermore derived in order to compare Equations (3) and (34) for specific samples. The constant C = 0.129 is obtained in Equation (41) as:

$$Cl\sqrt{f} = e^{-3} => C = \frac{e^{-3}}{l\sqrt{f}} = \frac{2.71^{-3}}{0.389} = 0.129 \qquad (41)$$

The authors in [25] use the value 2.71 for the mathematical constant e, which is approximately equal to $e = 2.718\,828$. Moreover, the value of $l\sqrt{f}$ is equal to $l\sqrt{f} = 0.389$, which is written as $l\sqrt{f} = 0.398$ in [25] (and obviously calculated as $l\sqrt{f} = 0.389$). The value of $l\sqrt{f} = 0.389$ is calculated with respect to the description of Equation (26) from Equations (42)–(44).

$$20 \cdot \log\left(\frac{1}{4}\sqrt{\frac{\sigma}{\omega\mu_r\varepsilon_0}}\right) = 100 - 20 \cdot \log(\ell f) \tag{42}$$

$$\log\left(\frac{\sigma}{32\pi f \mu_r \varepsilon_0}\right) = 10 - \log\left(\ell^2 f^2\right) \tag{43}$$

$$\log\left(\frac{\sigma}{32\pi \mu_r \varepsilon_0}\right) - 10 = -\log\left(\frac{1}{f}\right) - \log\left(\ell^2 f^2\right) \tag{44}$$

Considering the electrical conductivity of copper, i.e., $\sigma = 5.85 \times 10^7$ S/m [38], the material used for electrically conductive textile material production in [25], Equation (44) is rewritten as shown in Equations (45)–(47).

$$16.818 - 10 = \log\left(\frac{1}{\ell^2 f}\right) \tag{45}$$

$$10^{-6.818} = \ell^2 f \tag{46}$$

$$\ell\sqrt{f} = 10^{\frac{-6.818}{2}} = 10^{-3.409} = 389.9 \cdot 10^{-6} \tag{47}$$

The order of the value $l\sqrt{f} = 389.9 \times 10^{-6}$ is multiplied by 1000 because of the units [mm] and [MHz] that are used in [25], i.e., 10^{-3} [m] and 10^6 [Hz].

Equations (42)–(44) with no apertures are considered, and the condition $Z_M \ll Z_0$ is applied. The value 100 is derived from the $R_{aperture}$ equation, i.e., Equations (14)–(19), as shown in Equations (48)–(50).

$$R_{aperture} = 20 \cdot \log\left(\frac{c}{x}\right) - 20 \cdot \log(\ell \cdot f) = 100 - 20 \cdot \log(\ell \cdot f) \tag{48}$$

$$20 \cdot \log\left(\frac{c}{x}\right) = 100 \Longrightarrow 10^5 = \frac{c}{x}, \tag{49}$$

$$x = \frac{c}{10^5}, \tag{50}$$

The Equations (17)–(19) show the value of parameter x is equal to 2 (slot aperture), 2π (circular aperture), or $2\sqrt{\pi}$ (square aperture). If the speed of light $c = 3 \times 10^8$ m/s is considered, the x is equal to $x = 3000$. Nevertheless, if the speed of light is equal to $c = 186{,}000$ miles/s, the $x = 1.86$. This result is close to the value, which is valid for the slot aperture. If the value $x = 2$ is used, Equation (48) is described as:

$$R_{aperture} = 20 \cdot \log\left(\frac{186000}{2}\right) - 20 \cdot \log(\ell \cdot f) = 99.4 - 20 \cdot \log(\ell \cdot f), \tag{51}$$

The value 99.4 presented in Equation (51) is further rounded to the value 100.

The derivation of Equation (3) clearly shows the used equation is valid for copper metallized fabric, i.e., fabric without apertures, as the authors present in [25,28], and not valid for electrically conductive woven textile materials manufactured from the electrically conductive mixed and coated yarns.

7. Results and Discussion

The presumption of validity of $Z_M \ll Z_0$ and $\sigma \gg \omega\varepsilon$ for reflection loss of foil evaluation is presented in detail in chapter 3. It is also shown in Figures 1–5, and Table 2. The validity of presented inequalities is based on a ratio of magnitudes of individual values, i.e., at least two orders of values of magnitude are required. As a result, a simplified version of the Z_M, i.e., $\sigma \gg \omega\varepsilon$ is valid, and Equation (7) can be used for #1–#6 up to 10 GHz and for #7 up to approximately 6.9 GHz. The presumption of $Z_M \ll Z_0$ is valid for #7, #6, and #4 up to 70 MHz, 439 MHz, and 1.8 GHz, respectively, if the validity of $\sigma \gg \omega\varepsilon$ is assumed and up to 34.6 MHz, 219.5 MHz, and 899.9 MHz, respectively, if the validity of $\sigma \gg \omega\varepsilon$ is not assumed. It can be seen that the greater the value of electrical conductivity of the samples is, the greater is the frequency limit that can be obtained. As a result, #1–#3 and #5 fulfill this validity up to the frequency limit, which is greater than 1.8 GHz (#4, i.e., $\sigma = 1000$ S/m). This frequency limit is chosen with respect to the limits of ASTM D4935-10, i.e., 0.03–1.5 GHz. The presumption of validity of $Z_M \ll Z_0$ and $\sigma \gg \omega\varepsilon$ is also verified for the R_{foil} parameter, i.e., $Z_M \ll Z_0$ and $\sigma \gg \omega\varepsilon$ are/are not considered (in all four combinations), and also for all samples #1–#7, Figure 5. It shows the greatest difference is reached for the sample with the lowest electrical conductivity from #1–#5, i.e., #4, and it is equal to about 0.5 dB in 10 GHz. Similar results are obtained for #6 and #7, i.e., 1 dB and 2 dB, respectively, in 10 GHz. As a consequence, the presented limits for $Z_M \ll Z_0$, e.g., 70 MHz, 439 MHz, and 1.8 GHz for #7, #6, and #4, respectively, can be ignored and the relevant error has to be taken into account.

Derivation of reflection loss of one aperture shows an importance of determination of the effective aperture A_e for different shapes of apertures. It is clear that knitted fabrics require a different calculation of reflection loss for one aperture in comparison with woven fabrics.

Calculation of reflection loss of multiple apertures is not unified in the scientific literature [32,36,37] because different conditions for calculation of the number of apertures n are presented. It is, for instance, the calculation of the total length of linear array of apertures l_c. Obviously, different sizes of samples result in different values of total length of linear arrays of apertures l_c. We consider standard ASTM D4935-10, which is one of the most used standards for SE evaluation, and $l_c = 0.069$ m, Figure 7. This parameter has to be less than 1/2 of the wavelength, and it fulfills this condition in the frequency range defined in ASTM D4935-10, i.e., 0.03–1.5 GHz, Figure 6. One of the conditions is also that the material has to be thin, which is verified by comparison of the equivalent depth of penetration δ, usually 3δ, and the thickness of the material t, i.e., material is considered to be thin if inequality $3\delta > t$ is valid. The results show the penetration depth 3δ is greater than the thickness of #2 (the sample with the highest value of electrical conductivity, i.e., the value of penetration depth is the lowest from all samples) in the frequency range 0.03–1.43 GHz, Figure 8 and Table 3. The penetration depth for 5δ and 7δ is also analyzed in order to verify whether there are any reflections from the "far" interface of the material for the frequency beyond 1.43 GHz, i.e., the material can be considered to be thin. The results show it is valid for 5δ and 7δ in the frequency ranges 0.03–3.98 GHz, and 0.03–7.79 GHz, respectively, Figure 8 and Table 3. The results of reflection loss of multiple apertures evaluation show (20) has to be considered in reflection loss calculations and the values of electrical conductivity, and thickness of samples has to be considered because of thin/thick material evaluation.

Evaluation of the SE fabric considers a simplified version of the reflection loss of foil, i.e., $Z_M \ll Z_0$ and $\sigma \gg \omega\varepsilon$ are valid, a boundary condition, which defines a decrease of the amplitude by about 95% in specific materials, i.e., equivalent of 3 depth of penetration e^{-3}, and number of apertures, which is calculated with respect to the longest linear array of apertures of ASTM 4935-10, i.e., $l_c = 0.069$ m. As a result, constant C, the value in the exponent of Euler's number in the equation of the SE fabric calculation, is derived in Equation (32), Table 4. It clearly shows the SE fabric evaluation depends on sett, number of the longest linear array of apertures, and electric conductivity of each sample. As a consequence, the equation for SE calculation of woven fabrics manufactured from the electrically conductive mixed and coated yarns with square apertures is generally derived by Equation (35) with respect to Equation (32).

Individual components of SE fabric evaluation are SE_{foil}, i.e., the SE values for metallic foil of the same thickness as the fabric Equations (4) and (36), which is derived and described in many research papers [24,26–30], and $SE_{aperture}$, i.e., the SE values for metallic foil of the same thickness as the fabric with aperture(s), which is derived in this paper for particulate-blended shielding electrically conductive textile composites, i.e., woven fabrics manufactured from the electrically conductive mixed and coated yarns with square apertures, samples #1–#7. Calculation of the reflection loss of aperture

$R_{aperture}$ is a sum of reflection loss of one aperture $R_{aperture\_square}$ (Equation (17)) and reflection loss of multiple apertures $R_{aperture\_multiple}$ (Equation (20)), i.e., Equation (37). The absorption loss $A_{aperture}$ is neglected because the material is considered to be thin. A correction of geometrical dimensions of the aperture $K_{aperture}$ does not influence $SE_{aperture}$ because of square apertures. As a result, the resultant $SE_{aperture}$ is equal to $R_{aperture}$.

Derivation of $SE_{aperture}$ (Equation (40)) and SE_{fabric} (Equation (35)) clearly shows many factors have to be considered, i.e., shape of apertures, thickness of fabric in comparison with penetration depth (in order to determine conditions for thin/thick material), values of electrical conductivity, validation of $Z_M \ll Z_0$ and $\sigma \gg \omega\varepsilon$, total length of linear array of apertures, and sett of the fabric. It also shows (3) is valid for copper metallized fabric, i.e., fabric without apertures, and not valid for electrically conductive woven textile materials manufactured from the electrically conductive mixed and coated yarns.

Modeling of the SE_{fabric} (Equation (35)) with respect to used textile material, i.e., electrical conductivity of samples described in Table 1, evaluation of the constant C (Equation (32)) and n (Equation (33)) shown in Table 4, calculation of the SE_{foil} presented in Equation (4) and specified in Equation (7), and $SE_{aperture}$ derived in Equation (40) can be performed and compared with measurement results, Figure 9.

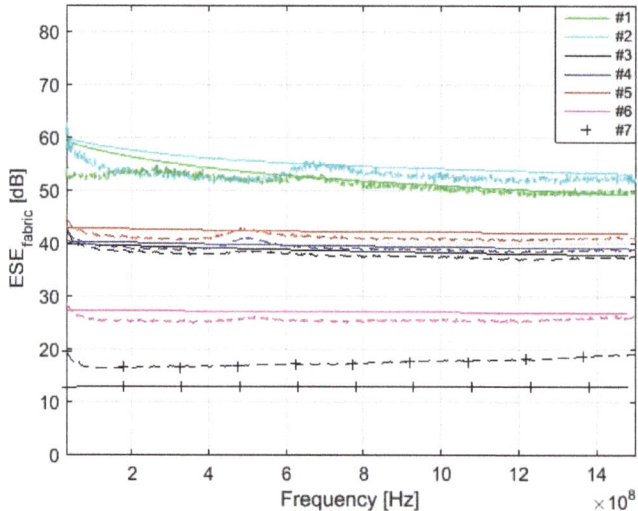

Figure 9. Comparison of the modeling (solid line) and measurement results for #1–#7 in 30 MHz–1.5 GHz.

Measurement is performed according to ASTM D4935-10 [39] (22 °C, RH 48%). A schematic block diagram of the experimental setup is shown in Figure 10 and a cross section of the sample holder with reference sample is shown in Figure 11. The sample holder is an enlarged coaxial transmission line with special taper sections to maintain a characteristic impedance of 50 Ω throughout the entire length of the sample holder. The reference sample is intended for calibration of the measurement setup. The load sample causes the loss of the passing high-frequency signal, which can be recorded by spectral analyzer. The results show the presented equations are valid for electrically conductive textile materials with a value of electrical conductivity equal to and higher than σ = 244 S/m, i.e., samples #1–#6. The maximal difference between modeling and measurement results was obtained for #1 and #2 in the frequency range 30–280 MHz. This is in the range of 2–6 dB. Nevertheless, it is within the random error of the used measurement method, which is defined in ASTM D4935-10 as ±5 dB. It is also within an observed standard deviation based on measurements by five laboratories on five samples presented in ASTM D4935-10 as 6 dB [39]. It is important to note that the measurement results are evaluated with respect to ASTM D4935-10, which defines a test procedure in the frequency range 0.03–1.5 GHz. The measurement results are therefore only informative in the frequency range 1.5 GHz–3 GHz, i.e., increasing value of the measured SE is caused by the excitation of modes other than the transverse electromagnetic mode (TEM), Figure 12. The results for sample #7 show these equations have to be modified for other materials (σ = 39 S/m). The frequency range of the model can also be extended, Figure 13. It shows both an increasing and decreasing trend of the SE_{fabric} of samples.

Figure 10. Schematic block diagram of experimental setup according to ASTM D4935-10.

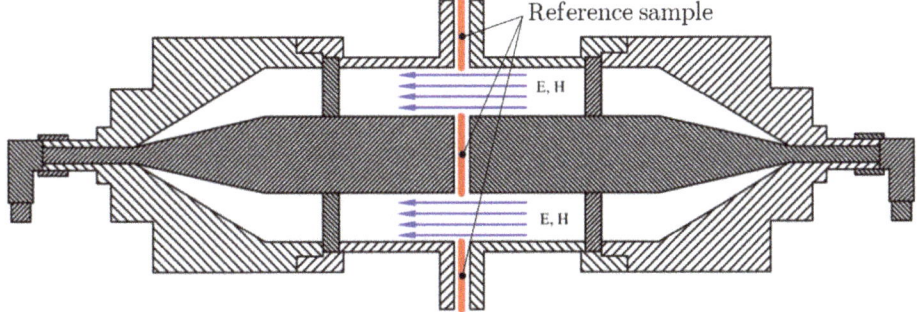

Figure 11. Cross section of sample holder with reference sample according to ASTM D4935-10.

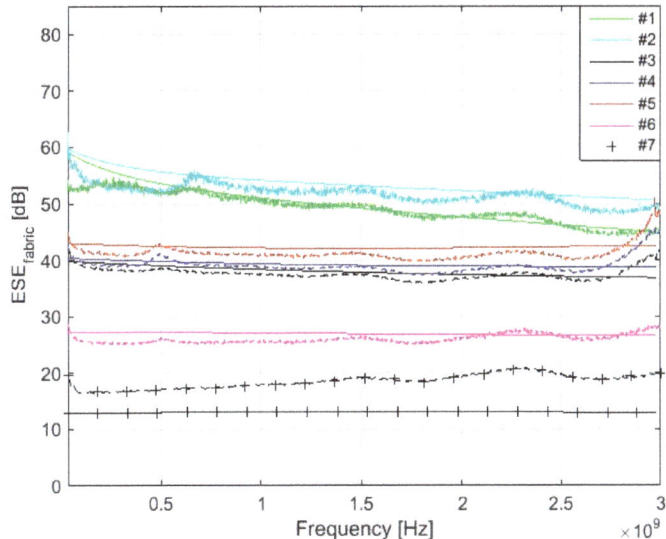

Figure 12. Comparison of the modeling (solid line) and measurement results for #1–#7 in 30 MHz–3 GHz.

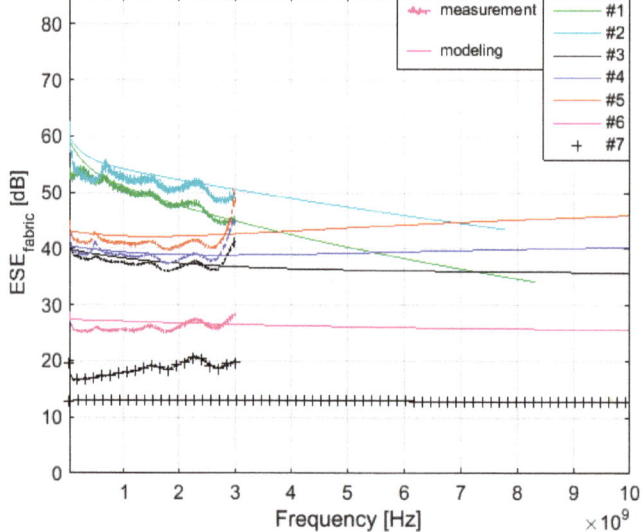

Figure 13. Comparison of the modeling (solid line) and measurement results for #1–#7 in 30 MHz–3 GHz (7δ) and modeling (solid line) results in 3 GHz–10 GHz.

8. Conclusions

This paper is focused on a derivation of a numerical model of electromagnetic shielding effectiveness for woven fabrics manufactured from electrically conductive mixed and coated yarns. Commonly used measurement techniques are mentioned. Basic equations of electromagnetic shielding effectiveness calculations are presented.

An evaluation of reflection loss of foil is described in detail and verifies the assumption of $Z_M \ll Z_0$ and $\sigma \gg \omega\varepsilon$ in a frequency range up to 10 GHz is valid for samples with electrical conductivity higher than 1000 S/m with an error up to 0.5 dB, for a sample with electrical conductivity 244 S/m with an error not exceeding 1 dB, and for a sample with electrical conductivity 39 S/m and an error up to 2 dB.

A derivation of reflection loss of one aperture is performed for slot, square, and circular apertures. The evaluation of reflection loss of multiple apertures describes the different calculation of the number of apertures in a shielding barrier and verifies presented conditions for its calculation. The longest linear array of apertures is used for the numerical model.

A complete derivation of electromagnetic shielding effectiveness of woven fabrics manufactured from the electrically conductive mixed and coated yarns is presented in detail. It shows the equations for electromagnetic shielding effectiveness evaluation differ for materials that are considered to be thin or thick (based on penetration depth and thickness comparison), for different values of electrical conductivity, and for different setts used in the manufacturing process.

A comparison of modeling and measurement results of electromagnetic shielding effectiveness fabric is performed in the frequency range 0.03–1.5 GHz according to ASTM D4935-10. The results clearly show a numerical model is valid for electrically conductive woven textile materials with a value of electrical conductivity equal to and higher than σ = 244 S/m. The results of the numerical mode are also extended up to 10 GHz in order to show the trend of electromagnetic shielding effectiveness.

Author Contributions: M.N. investigated and wrote the original manuscript. L.V. led the research project, obtained funding, and conceptualized as well as reviewed and edited the manuscript.

Funding: This research was funded by the Czech Ministry of Industry and Trade grant number FR-TI4/202.

Conflicts of Interest: The authors declare no conflict of interest.

References

1. Safarova, V.; Tunak, M.; Truhlar, M.; Militky, J. A new method and apparatus for evaluating the electromagnetic shielding effectiveness of textiles. *Text. Res. J.* **2016**, *86*, 44–56. [CrossRef]
2. Çeven, E.K.; Karaküçük, A.; Dirik, A.E.; Yalçin, U. Evaluation of electromagnetic shielding effectiveness of fabrics produced from yarns containing metal wire with a mobile based measurement system. *Ind. Text.* **2017**, *68*, 289–295.
3. Chen, H.C.; Lee, K.C.; Lin, J.H.; Koch, M. Comparison of electromagnetic shielding effectiveness properties of diverse conductive textiles via various measurement techniques. *J. Mater. Process. Techol.* **2007**, *192*, 549–554. [CrossRef]
4. Safarova, V.; Militky, J. Comparison of methods for evaluating the electromagnetic shielding of textiles. *Fibers Text.* **2012**, *19*, 50–55.
5. Avloni, J.; Lau, R.; Ouyang, M.; Florio, L.; Henn, A.R.; Sparavigna, A. Shielding Effectiveness Evaluation of Metallized and Polypyrrole-Coated Fabrics. *J. Thermoplast. Comp. Mater.* **2007**, *20*, 241–254. [CrossRef]
6. Ozen, M.S.; Usta, I.; Yuksek, M.; Sancak, E.; Soin, N. Investigation of the Electromagnetic Shielding Effectiveness of Needle Punched Nonwoven Fabrics Produced from Stainless Steel and Carbon Fibres. *Fibers Text. East. Eur.* **2018**, *26*, 94–100. [CrossRef]
7. Hakan, Ö.; Uğurlu, Ş.S.; Özkurt, A. The electromagnetic shielding of textured steel yarn based woven fabrics used for clothing. *J. Ind. Text.* **2015**, *45*, 416–436. [CrossRef]
8. Chen, H.C.; Lee, K.C.; Lin, J.H.; Koch, M. Fabrication of conductive woven fabric and analysis of electromagnetic shielding via measurement and empirical equation. *J. Mater. Process. Technol.* **2007**, *184*, 124–130. [CrossRef]
9. Tugirumubano, A.; Vijay, S.J.; Go, S.H.; Kwac, L.K.; Kim, H.G. Investigation of Mechanical and Electromagnetic Interference Shielding Properties of Nickel-CFRP Textile Composites. *J. Mater. Eng. Perform.* **2018**, *27*, 1–8. [CrossRef]
10. King, J.A.; Pisani, W.A.; Klimek-McDonald, D.R.; Perger, W.F.; Odegard, G.M.; Turpeinen, D.G. Shielding effectiveness of carbon-filled polypropylene composites. *J. Compos. Mater.* **2015**, *50*, 2177–2189. [CrossRef]
11. Pothupitiya Gamage, S.J.; Yang, K.; Braveenth, R.; Raagulan, K.; Kim, H.S.; Lee, Y.S.; Yang, C.M.; Moon, J.J.; Chai, K.Y. MWCNT Coated Free-Standing Carbon Fiber Fabric for Enhanced Performance in EMI Shielding with a Higher Absolute EMI SE. *Materials* **2017**, *10*, 1350. [CrossRef]
12. Lu, Y.; Xue, L. Electromagnetic interference shielding, mechanical properties and water absorption of copper/bamboo fabric (Cu/BF) composites. *Compos. Sci. Technol.* **2012**, *72*, 828–834. [CrossRef]
13. Maity, S.; Chatterjee, A. Conductive polymer-based electro-conductive textile composites for electromagnetic interference shielding: A review. *J. Ind. Text.* **2018**, *47*, 2228–2252. [CrossRef]
14. Tunáková, V.; Grégr, J.; Tunák, M. Functional polyester fabric/polypyrrole polymer composites for electromagnetic shielding: Optimization of process parameters. *J. Ind. Text.* **2016**, *47*, 686–711. [CrossRef]
15. Duran, D.; Kadoğlu, H. Electromagnetic shielding characterization of conductive woven fabrics produced with silver-containing yarns. *Text. Res. J.* **2015**, *85*, 1009–1021. [CrossRef]
16. Ortlek, H.G.; Alpyildiz, T.; Kilic, G. Determination of electromagnetic shielding performance of hybrid yarn knitted fabrics with anechoic chamber method. *Text. Res. J.* **2013**, *83*, 90–99. [CrossRef]
17. Safarova, V.; Militky, J. Electromagnetic shielding properties of woven fabrics made from high-performance fibers. *Text. Res. J.* **2014**, *84*, 1255–1267. [CrossRef]
18. Tunáková, V.; Techniková, L.; Militký, J. Influence of washing/drying cycles on fundamental properties of metal fiber-containing fabrics designed for electromagnetic shielding purposes. *Text. Res. J.* **2017**, *87*, 175–192. [CrossRef]
19. Šaravanja, B.; Malarić, K.; Pušić, T.; Ujević, D. Impact of dry cleaning on the electromagnetic shield characteristics of interlining fabric. *Fibers Text. East. Eur.* **2015**, *23*, 104–108.
20. Kayacan, Ö. The Effect of Washing Process on the Electromagnetic Shielding of Knitted Fabrics. *J. Text. Apparel* **2014**, *24*, 356–362.
21. Wang, X.; Liu, Z.; Jiao, M. Computation model of shielding effectiveness of symmetric partial for anti-electromagnetic radiation garment. *Prog. Electromagn. Res. B* **2013**, *47*, 19–35. [CrossRef]

22. Liu, Z.; Wang, X.C.; Zhou, Z. Computation of shielding effectiveness for electromagnetic shielding blended fabric. *Przeg. Elektrotechnol.* **2013**, *89*, 228–230.
23. Perumalraj, R.; Dasaradan, B.S.; Anbarasu, R.; Arokiaraj, P.; Harish, S.L. Electromagnetic shielding effectiveness of copper core-woven fabrics. *J. Text. Inst.* **2009**, *100*, 512–524. [CrossRef]
24. Hong, X.; Mei, W.S.; Qun, W.; Qian, L. The electromagnetic shielding and reflective properties of electromagnetic textiles with pores, planar periodic units and space structures. *Text. Res. J.* **2014**, *84*, 1679–1691. [CrossRef]
25. Henn, A.R.; Crib, R.M. Modeling the shielding effectiveness of metallized fabrics. Presented at the International Symposium on Electromagnetic Compatibility, Anaheim, CA, USA, 17–21 August 1992.
26. Vojtech, L.; Neruda, M.; Hajek, J. Planar Materials Electromagnetic Shielding Effectiveness Derivation. *Int. J. Com. Antenna Propag.* **2011**, *1*, 21–28.
27. Vojtech, L.; Neruda, M.; Hajek, J. Modelling of Electromagnetic Parameters of Planar Textile Materials. In *New Trends in the Field of Materials and Technologies Engineering*; Beneš, L., Ulewicz, R.M., Eds.; Oficyna Wydawnicza Stowarzyszenia Menedżerów Jakości i Produkcji: Czestochowa, Poland, 2012; pp. 123–159, ISBN 978-83-934225-2-4.
28. Neelakanta, P.S. *Handbook of Electromagnetic Materials: Monolithic and Composite Versions and Their Applications*, 1st ed.; CRC Press: Boca Raton, FL, USA, 1995; pp. 447–490, ISBN 0-8493-2500-5.
29. Wang, X.C.; Su, Y.; Li, Y.; Liu, Z. Computation Model of Shielding Effectiveness of Electromagnetic Shielding Fabrics with Seaming Stitch. *Prog. Electromagn. Res.* **2017**, *55*, 85–93. [CrossRef]
30. Robinson, M.P.; Benson, T.M.; Christopoulos, C.; Dawson, J.F.; Ganley, M.D.; Marvin, A.C.; Porter, S.J.; Thomas, D.W. Analytical formulation for the shielding effectiveness of enclosures with apertures. *IEEE Trans. Electromagn. Compat.* **1998**, *40*, 240–248. [CrossRef]
31. Liang, R.; Cheng, W.; Xiao, H.; Shi, M.; Tang, Z.; Wang, N. A calculating method for the electromagnetic shielding effectiveness of metal fiber blended fabric. *Text. Res. J.* **2017**, *88*, 973–986. [CrossRef]
32. Ott, H.W. *Electromagnetic Compatibility Engineering*, 1st ed.; John Wiley & Sons: Hoboken, NJ, USA, 2009; pp. 238–300, ISBN 978-0470189306.
33. Christopoulos, C. *Principles and Techniques of Electromagnetic Compatibility*, 2nd ed.; CRC Press: Boca Raton, FL, USA, 2007; ISBN 978-0-8493-7035-9.
34. BS EN 16812:2016. *Textiles and Textile Products; Electrically Conductive Textiles; Determination of the Linear Electrical Resistance of Conductive Tracks*; British Standard European Norm: London, UK, 2016.
35. Vojtech, L.; Neruda, M. Modelling of surface and bulk resistance for wearable textile antenna design. *Przeg. Elektrotech.* **2013**, *89*, 217–222.
36. Tim, W. *EMC for Product Designers*, 3rd ed.; Newnes: Oxford, UK, 2001; pp. 293–308, ISBN 978-0-7506-4930-8.
37. Electromagnetic Compatibility. Available online: www.urel.feec.vutbr.cz/~drinovsky/?download=Skripta_EMC.pdf (accessed on 30 July 2018).
38. Properties Table of Stainless Steel, Metals and Other Conductive Materials. Available online: www.tibtech.com/conductivity.php (accessed on 8 May 2018).
39. ASTM D4935-10:2010. *Standard Test Method for Measuring the Electromagnetic Shielding Effectiveness of Planar Materials*; ASTM International: West Conshohocken, PA, USA, 2010.
40. Sensitivity of Multi Turn Receiving Loops. Available online: www.vlf.it/octoloop/rlt-n4ywk.htm (accessed on 30 July 2018).
41. Wieckowski, T.W.; Jankukiewicz, M.J. Methods for evaluating the shielding effectiveness of textiles. *Fibers Text. East. Eur.* **2006**, *14*, 18–22.

© 2018 by the authors. Licensee MDPI, Basel, Switzerland. This article is an open access article distributed under the terms and conditions of the Creative Commons Attribution (CC BY) license (http://creativecommons.org/licenses/by/4.0/).

Article

Spectral Analysis and Parameter Identification of Textile-Based Dye-Sensitized Solar Cells

László Juhász [1,*] and Irén Juhász Junger [2]

[1] Deggendorf Institute of Technology, Faculty of Electrical Engineering, Media Technology and Computer Science, 94469 Deggendorf, Germany
[2] Bielefeld University of Applied Sciences, Faculty of Engineering and Mathematics, 33619 Bielefeld, Germany; iren.juhas_junger@fh-bielefeld.de
* Correspondence: laszlo.juhasz@th-deg.de; Tel.: +49-991-3615-532

Received: 31 July 2018; Accepted: 2 September 2018; Published: 5 September 2018

Abstract: Linearized equivalent electrical-circuit representation of dye-sensitized solar cells is helpful both for the better understanding of the physical processes in the cell as well as for various optimizations of the cells. White-box and grey-box modelling approaches are well-known and they are widely used for standard cell types. However, in the case of new cell types or the lack of deep knowledge of the cell's physic such approaches may not be applicable immediately. In this article a black-box approach for such cases is presented applied together with spectral analysis. The spectral analysis and the black-box approach were as first validated with a standard glass-based dye-sensitized solar cell and thereafter applied for the characterization of a new type of textile-based dye-sensitized solar cells. Although there are still improvement potentials, the results are encouraging and the authors believe that the black-box method with spectral analysis may be used particularly for new types of dye-sensitized solar cells.

Keywords: dye-sensitized solar cell; half-textile; spectral analysis; parameter identification; equivalent circuit; black-box; grey-box; power spectral density; optimization

1. Introduction

In praxis, it is often helpful to have a linear model for the investigated object or system. This enables among others the use of analysis techniques and definitions which are available only for linear plants. In the case of dye-sensitized solar cells (DSSC), the linearized frequency-dependent impedance is from great interest. As DSSC are generally non-linear, an appropriate linearization in the steady-state operating point is necessary. Because the linearized model is only accurate in the neighborhood of the given steady-state, it is important to choose it according to the intended use-case, which is usually in the point of the maximal energy generation.

In scope of the identification procedure an experimental measurement followed by a spectral analysis is conducted as first. Instead of the impedance spectroscopy, we propose the spectral analysis method for DSSC. Spectral analysis is not yet applied in the context with DSSC. The proposed spectral analysis method is a reasonable alternative to the standard spectroscopy methods because of its efficiency regarding the measurement duration. The standard spectroscopy methods require a separate measurement for each discrete frequency, whereas the proposed spectral estimation method processes the complete frequency band of interest based on a single measurement. The result of the spectral analysis is the non-parametric frequency transfer function (TF) for the covered frequency range. In order to get a parameterized standard TF, parameter identification is performed by using least-squares optimization procedure while using quadratic goal function.

If our goal is the optimization a process or the plant, it is important to have an accurate model in the intended steady-state working-point over the broad frequency range of interest. This is straightforward if the physical system is known in detail. However, even if the physical system is not known in detail it is still possible to perform an identification process, which is called black-box identification (BBI, [1]). If some of the elements of the system are well-known, but there are also unknown parts, grey-box identification (GBI, [1]) methods can be used. White-box linearization and parameter identification methods (WBI) are applicable if deep knowledge of the object's physic and internal processes is available. Using the knowledge and the technological properties of the plant, in this case an equivalent model can be directly composed using known parameter values. In case of missing parameter values but known model structure, the GBI may be used. While WBI and GBI follow the bottom-up approach starting with known elements (or structure) and ending with the parameterized model of the complete system, the BBI approach is rather a top-down one. It starts with the experimental data taken from the whole system, estimates the system's structure and it ends with the identified parameter of the structure elements.

For standard glass-based DSSC several equivalent models are published [2]. Therefore, the GBI approach is for such cells indeed a reasonable choice. In this paper, we present the BBI approach with glass-based DSSC to prove the proposed BBI concept as first. This method consists of a special form of spectral analysis and the recognition of the dominant elements in the Nyquist- and Bode-diagram followed with the proposed parameter identification procedure. The intermediate result is the parametrized frequency-transfer function, whereas the final goal of the approach is to find and parametrize a suitable equivalent electrical-circuit. As the concept of frequency transfer functions is limited to linear and time-invariant (LTI) systems, the currently proposed approach is limited to linear elements as well. The resulting parametric TF is here always a rational function and the resulting equivalent models are composed from passive linear elements (particularly R, L, and C). The extension of the proposed approach to further elements used in the modeling of DSSC (like the Warburg-element) will be the subject of our future work, whereas the concept of fractional differential equations could be helpful [3,4].

Indeed, it is not clear if the published equivalent models of glass-based DSSC are also suitable for textile-based DSSCs. Therefore, for the determination of the linear frequency-dependent impedance of the half-textile DSSC, we apply the method following the proposed BBI approach. Particularly, the BBI Spectral identification approach is applied for the characterization of dye-sensitized solar cells with counter electrodes built from electrospun polyacrylonitrile (PAN) nanofiber mat coated by different number of poly(3,4-ethylenedioxythiopene) polystyrene sulfonate (PEDOT: PSS) layers. Among others, the linearized resistance of the cells in the steady-state is determined, which shows a clear correlation with the number of PEDOT: PSS layers.

2. Materials and Methods

2.1. Measurements and Sprectral Analysis

The spectral analysis was performed for each DSSC separately in the individual steady-state point, which yields maximal active power. A sine-sweep voltage $u_{AC}(t)$ with 10% amplitude of the constant steady-state voltage and the frequency range of 1 Hz–50 kHz is applied to the DSSC. The applied voltage is recorded together with the measured current $i_{AC}(t)$ and the measurement timestamps. The measurement frequency was 100 kHz (10 μs steps). A sine-sweep generator with square frequency distribution was created and used. For the signal generation and measurements, a dSPACE MicroLabBox (MLBX, [5]) is used, which was equipped with a specially designed signal-attenuator and -converter box to achieve the best possible A/D and D/A resolution. The programming of the MLBX was performed while using MathWorks Simulink [6] in a graphical model-based way. The sine-sweep generator was implemented in Simulink. A single measurement covered 200 s and contained 5 complete sine-sweep sequences.

2.2. Parameter Identification

The spectral identification was performed by the use of specially written Matlab-scripts that implemented the method of Welch for the non-parametric estimation of the TF and a quadratic goal function for the parameter identification. The periodogram size of 32768 elements with 50% overlapping is selected and Hanning window is used.

3. Spectral Analysis and Parameter Identification of the Linearized DSSC Model

3.1. Spectral Analysis Versus Impedance Spectroscopy

The widely used impedance spectroscopy of DSSCs requires a separate measurement for each discrete frequency. A sine-wave voltage input is applied on top of the selected steady-state voltage to the cell and after the transient effects are faded, the amplitude and phase of the same frequency component of the established current through the DSSC is observed. In this way, the linearized complex dynamic impedance of the DSSC for the single angular frequency ω may be determined. Through repeating this procedure with different frequencies, the shape of the TF $\hat{Z}(j\omega) \cong Z(j\omega) = u_{AC}(j\omega)/i_{AC}(j\omega)$ can be obtained. Commercially available equipment uses an automated method for the determination of the real and imaginary part of the single points of the TF while using the correlation analysis [1]. The action-diagram of such a system is shown in Figure 1.

In Figure 1, a signal generator outputs two signals, which are of the same, selectable frequency, and of identical amplitude but with a phase difference of $\pi/2$. The process output $y_U(t)$ is distorted by the measurement noise $n(t)$; thus the measured output $y(t)$ is also distorted. By choosing the measurement duration $T = k \cdot \frac{2\pi}{\omega}$ the integration yields following signals:

$$R = \frac{1}{2}U^2 \cdot k \cdot \frac{2\pi}{\omega} \cdot Re\{G(j\omega)\} + \int_0^{k \cdot \frac{2\pi}{\omega}} n(t) \cdot U \cdot \cos(\omega \cdot t) \cdot dt, \qquad (1)$$

$$I = -\frac{1}{2}U^2 \cdot k \cdot \frac{2\pi}{\omega} \cdot Im\{G(j\omega)\} + \int_0^{k \cdot \frac{2\pi}{\omega}} n(t) \cdot U \cdot \sin(\omega \cdot t) \cdot dt. \qquad (2)$$

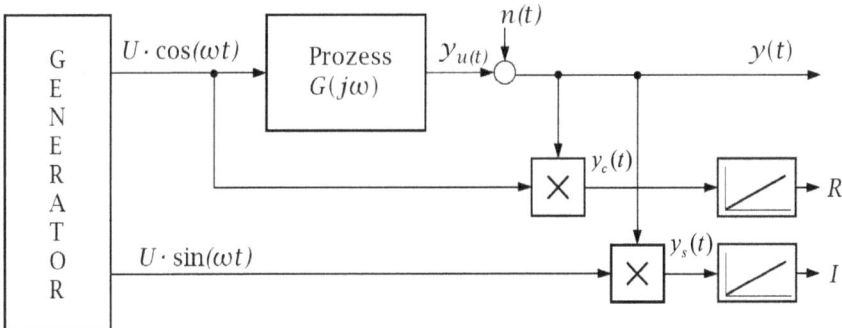

Figure 1. Action-diagram of correlation analysis used during impedance spectroscopy for determination of the real and imaginary parts of the transfer function (TF) for the discrete angular frequency ω.

In case when the disturbance $n(t)$ is independent on the excitation, the integrals remain limited, whereas the other terms are directly dependent on k. That means, for large k, the influence of the disturbance is minimized. The real and imaginary parts of the transfer function are directly obtained for the angular frequency ω. Knowing the real and imaginary part of the TF its amplitude and phase can be also calculated. Performing further measurements with different angular frequencies, the nonparametric transfer function of the linearized system can be reconstructed.

The obvious drawback of the impedance spectroscopy is that, for obtaining the TF of a single DSSC, hundreds of measurements are necessary. Even when these measurements are performed automatically, the complete sequence takes long time and requires expensive equipment. Spectroscopy was applied in field of electrical and mechanical engineering for long time [1], however due to available and inexpensive CPU-resources nowadays, mainly the spectral analysis is used instead.

The standard spectral analysis uses the Fourier-transformation of the input and the output signal of a stable process while using a single measurement. In the case of DSSC, this means applying of AC voltage of continuously changing frequency to the DSSC additionally to the steady-state DC-voltage and the measurement of the AC-component of the established current trough the cell. Knowing the applied AC components of a DSSC cell voltage u_{AC} and the measured current i_{AC}, the frequency-dependent dynamic cell impedance may be calculated, as follows:

$$\widetilde{Z}(j\omega) = \frac{\mathcal{F}(u_{AC}(t))}{\mathcal{F}(i_{AC}(t))} = \frac{u_{AC}(j\omega)}{i_{AC}(j\omega)}. \tag{3}$$

However, this method returns correct results only in case when the disturbances (e.g., measurement noise) is negligibly related to the system output. If the measurement noise $n(t)$ is considerable, the direct use of the Fourier-transformation yields erroneous results as shown in Equation (4).

$$\widetilde{Z}(j\omega) = \frac{\mathcal{F}(u_{AC}(t))}{\mathcal{F}(i_{AC}(t) + n(t))} \neq \frac{u_{AC}(j\omega)}{i_{AC}(j\omega)}. \tag{4}$$

Furthermore, the leaking-effect causes a "blurred" spectrum, which further deteriorates the result. In order to determine the frequency-dependent impedance with satisfying accuracy by the direct use of the Fourier-transformation, high-end equipment, and long measurements are necessary.

3.2. Spectral Analysis Using the Correlation Functions

The calculation of the nonparametric frequency-response using common measurement equipment and shorter measurement periods is still possible if the periodogram-method, according to Welch [7], is applied. Using the theorem of Wiener-Khintchine [8], the non-parametric estimation of the DSSC impedance in form of a TF $\hat{Z}(j\omega) \cong Z(j\omega) = u_{AC}(j\omega)/i_{AC}(j\omega)$ may be obtained using the auto-correlation of the AC current component $\phi_{i_{AC}i_{AC}}$ which is here formally the input signal and the cross-correlation of the AC voltage component (which here plays formally the role of the system output) with the AC current component $\phi_{i_{AC}u_{AC}}$.

On the one hand, the Fourier transformation of the correlation function yields the spectral power density (SPD). On the other hand, the SPD can be represented as a product of the individual signals represented in the frequency domain, see Equation (5). Thus, the non-parametric estimation of the impedance can be calculated through division of the SPD of the auto-correlation function with the SPD of the cross-correlation function.

$$\begin{aligned} \mathcal{F}(\phi_{i_{AC}i_{AC}}) &= S_{i_{AC}i_{AC}}(j\omega) = i^*_{AC}(j\omega) \cdot i_{AC}(j\omega) \\ \mathcal{F}(\phi_{i_{AC}u_{AC}}) &= S_{i_{AC}u_{AC}}(j\omega) = i^*_{AC}(j\omega) \cdot u_{AC}(j\omega) \\ \hat{Z}(j\omega) &= \frac{S_{i_{AC}u_{AC}}(j\omega)}{S_{i_{AC}i_{AC}}(j\omega)}. \end{aligned} \tag{5}$$

The impact of the measurement noise to the calculated non-parametric impedance estimation is now significantly reduced if the measured signal is not correlated with the measurement noise. To reduce the variance of the spectral analysis, the periodogram method may be further applied [9] with an appropriate windowing of the individual periodogram data in the time domain. The information about the quality of the estimation over the used frequency range is given by the coherence-function:

$$\gamma^2_{i_{AC}u_{AC}}(j\omega) = \frac{|S_{i_{AC}u_{AC}}(j\omega)|^2}{S_{i_{AC}i_{AC}}(j\omega) \cdot S_{i_{AC}u_{AC}}(j\omega)}, \quad 0 \leq \gamma^2 \leq 1. \tag{6}$$

By coherence-function values near to one, the linear dependence between the input and output is dominant, thus the non-parametric estimation is good. If the coherence-function tends to zero, there is no linear dependence between the input and output and the non-parametric estimation is poor. As one may notice, this is useful information for the parameter estimation procedure.

The result of the non-parametric spectral estimation is the frequency transfer function, which is defined with complex values related to discrete frequencies. Using these values graphical representation using Nyquist or Bode diagrams can be plotted. Based on the diagrams and their characteristic elements, the appropriate TF form and finally the elements of the equivalent electrical circuits of a DSSC may be chosen.

3.3. Parameter Identification Procedure

The non-parametric frequency transfer-function $\hat{Z}(j\omega)$ is the basis for the estimation of the parametric TF $Z(j\omega)$. During this process, we suppose the parametric TF structure and have to solve a quadratic optimization problem with some imposed restrictions.

The choice of the TF structure $Z(j\omega)$ is from crucial importance, because it determines whether the optimization task succeeds or fails. The number of parameter is here also important because in case too many parameters are supposed, the system is overdetermined, and the optimization may fail because of numeric issues. In case of a too small parameter number, the parametric TF cannot describe the system behavior and the optimization task fails as well.

In case when resonance peaks are observed in the Bode-diagram, a general n-th order TF should be supposed depending on the number of the resonant peaks. For example, in case of one resonant peak, a second order transfer-function may be eligible as shown in Equation (7). Thus, n resonant peaks would require a $2n^{th}$ order TF, which can be decomposed into a product (or alternatively through partial fraction decomposition into a sum) of second-order TFs.

$$Z(j\omega) = R \cdot \frac{b_2 \cdot (j\omega)^2 + b_1 \cdot j\omega + 1}{a_2 \cdot (j\omega)^2 + a_1 \cdot j\omega + 1} \quad (7)$$

Due to physical circumstances, only positive values should be allowed for the static gain R during the optimization process and furthermore only such TF parameters should be allowed which results in a stable plant. Such a second-order transfer function matches an equivalent electrical circuit consisting of RLC-elements.

If no resonant peaks are observed in the Bode-diagram, a TF consisting of simple real-valued pole-zero elements can be supposed. The number of zeros and poles may be supposed according to the number of observed "half-circles" observed in the Nyquist plot. For example, in the case when two half-circles are observed, a second-order TF with parameters $s_{p1}, s_{p2}, s_{z1}, s_{z2}$ having the static gain R in Equation (8) can be supposed. The stability condition and the positivity of R were here also enforced from the optimization algorithm.

$$Z(s) = R \cdot \frac{\left(1 - \frac{s}{s_{p1}}\right) \cdot \left(1 - \frac{s}{s_{p2}}\right)}{\left(1 - \frac{s}{s_{z1}}\right) \cdot \left(1 - \frac{s}{s_{z2}}\right)}, \ Z(j\omega) = R \cdot \frac{\left(1 - \frac{j\omega}{s_{p1}}\right) \cdot \left(1 - \frac{j\omega}{s_{p2}}\right)}{\left(1 - \frac{j\omega}{s_{z1}}\right) \cdot \left(1 - \frac{j\omega}{s_{z2}}\right)} \quad (8)$$

Both in the case of form Equations (7) and (8) it is obvious, that the parameter R represents the linearized thermal resistance of the DSSC in the steady-state working-point, whereas in Equation (8), the relative values of the poles and zeros determine the capacitive or inductive nature of the impedance for the given angular frequency ω_i. When vertical lines are observed in the Nyquist-diagram, they indicate the sum of static gain and pole or zero element. They can be substituted in the equivalent circuit with a series connection of R and L/C-element.

Finally, the possibility to fit the results to a specific equivalent circuit known from the literature should be checked. In cases where this was not possible, a hint for a possible equivalent circuit may be given, which fits the more general transfer function form.

The estimation of the parameter of the supposed TF-structure is performed by a quadratic optimization algorithm. One possibility is to compare the real- and imaginary-part of the non-parametric and the parametric TF for each discrete angular frequency ω_i and minimize the calculated quadratic error sum. The alternative way is to compare the phase and the gain differences for the building of the quadratic error sum [10]. This second option in our study showed significantly better results. The likelihood of the optimization can be improved by the multiplication of the individual error squares for discrete angular frequencies ω_i with the matching value of the coherence-function obtained for the same angular frequency ω_i in Equation (9) [10].

$$e_{Zmag} = \frac{\sqrt{\sum_{n=1}^{N}\left(\gamma^2_{i_{AC}u_{AC}}(j\omega_n)\cdot(|\hat{Z}(j\omega_n)|-|Z(j\omega_n)|)\right)^2}}{\sqrt{\sum_{n=1}^{N}(|\hat{Z}(j\omega_n)|)^2}}, e_{Zphase} = \frac{\sqrt{\sum_{n=1}^{N}\left(\gamma^2_{i_{AC}u_{AC}}(j\omega_n)\cdot(\arg(\hat{Z}(j\omega_n))-\arg(Z(j\omega_n)))\right)^2}}{\sqrt{\sum_{n=1}^{N}(\arg(\hat{Z}(j\omega_n)))^2}} \quad (9)$$

$$e_Z = e_{Zmag} + w_{phase}\cdot e_{Zphase}$$

In Equation (9), w_{phase} is a positive constant (weighting of the phase estimation), which gives a further possibility to improve the convergence of the optimization. The optimization goal is to find such parameter of the supposed TF, which results in a minimal value of the error-sum e_Z.

3.4. Parameter Identification Example with Glass-Based DSSC

The proposed identification method was tested while using a standard DSSC with glass electrodes prepared as described in [11]. The sine-sweep signal with range of (1 Hz–50,000 Hz) was added to the constant steady-state voltage. The lower frequency boundary was set to cover all the dynamic properties of the cell, whereas the higher boundary was imposed through the used CPU-based hardware platform. From the bode-diagram of the identified non-parametric TF, it was obvious that there are no resonant peaks. The Nyquist-diagram displayed two half-circles with negative imaginary part. As this is an obvious characteristic of second-order-systems with two time constants e.g., real poles and zeros, as first the parametric identification was performed while using a second-order zero-pole element. From the Nyquist plot of the non-parametric estimation, it is furthermore obvious that it agrees with two parallel RC circuits in series with a resistor [12], which is a published equivalent circuit for standard DSSC [2]. In the following step, the structure of the supposed TF for the glass based DSSC was calculated and displayed in Equation (10) while using the equivalent circuit from Figure 2 in accordance to Thévenin's theorem. The Bode and Nyquist diagrams of the non-parametric and parametric estimations are shown in Figures 3 and 4, together with the estimated parameter.

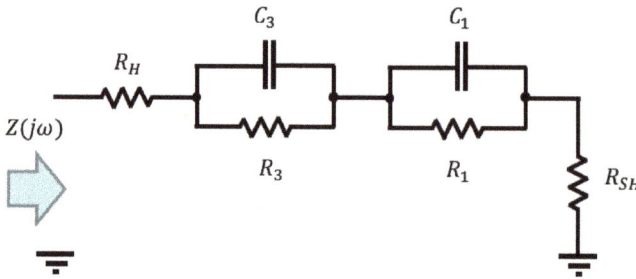

Figure 2. Equivalent circuit used for the parametric identification of the impedance of standard glass dye-sensitized solar cells (DSSC), according to [2].

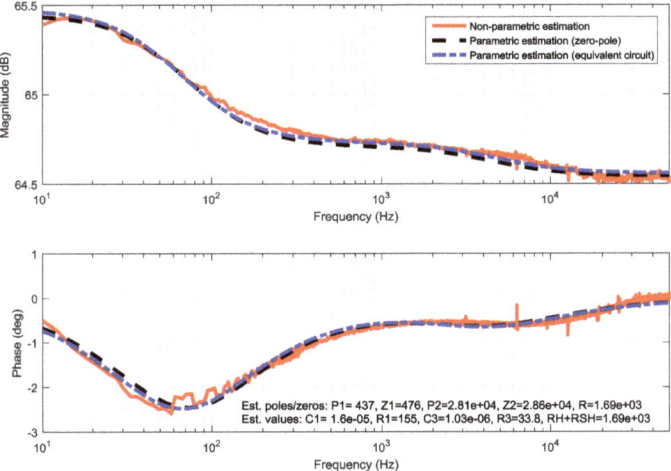

Figure 3. Comparison of the non-parametric and parametric TF identifications of a standard DSSC in Bode-diagram. Pole and zero values are given in (rad/s), resistance values in Ω and capacitances in F.

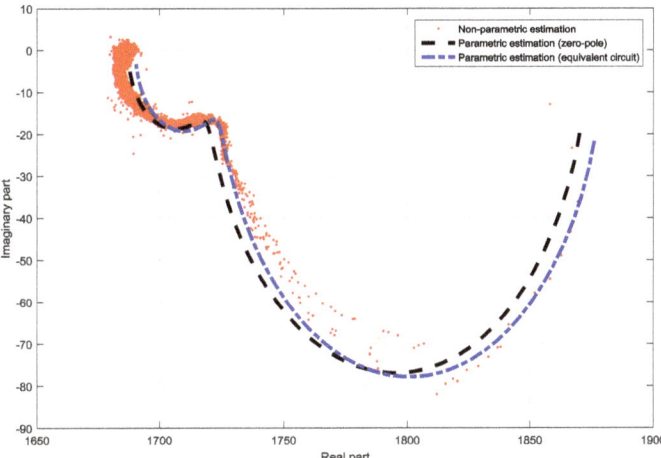

Figure 4. Comparison of the non-parametric and parametric TF identifications of a standard DSSC in Nyquist-diagram.

Generally, a very good agreement between the non-parametric and parametric estimations can be noticed. This indicates the validity of the proposed method for spectral analysis and parameter estimation. Indeed, as the resistors R_H and R_{SH} are connected in series only its sum appears in the parametric TF (but not the individual ones). For this reason, it is not possible to distinguish between them during the parametric identification: only their sum can be determined.

$$Z(j\omega) = \frac{(j\omega)^2 \cdot (R_H + R_{SH}) \cdot R_1 \cdot R_3 \cdot C_1 \cdot C_3 + j\omega \cdot ((R_H + R_{SH}) \cdot R_1 \cdot C_1 + (R_H + R_{SH}) \cdot R_3 \cdot C_3 + R_1 \cdot R_3 \cdot C_3) + (R_H + R_{SH}) + R_1 + R_3}{(j\omega)^2 \cdot R_1 \cdot R_3 \cdot C_1 \cdot C_3 + j\omega \cdot (R_1 \cdot C_1 + R_3 \cdot C_3) + 1} \quad (10)$$

3.5. Parameter Identification Examples with Half-Textile DSSC

In the next step, the proposed spectral analysis with black-box parameter identification procedure was applied to half-textile DSSCs with nanofiber-mat based counter electrodes. The working electrodes of the DSSCs were prepared on TiO_2-coated FTO (fluorine-doped thin oxide) glass plates (purchased from Man Solar) that were dyed with anthocyanins extracted from forest fruit tea (MAYFAIR). The counter electrodes were built on electrospun PAN nanofiber mat (prepared using the electrospinning machine "Nanospider Lab" (Elmarco, Czech Republic)) coated by one to five layers of PEDOT: PSS (CLEVIOS™ S V 4). As catalyst, a graphite layer performed by spraying was used (Graphit 33 by Kontakt Chemie). The both electrodes were put together and fixed by an adhesive tape, and the cells were filled with an iodine/triiodide based electrolyte (type 016 purchased from Man Solar). The detailed description of the cell preparation is given in [11].

The measurements were taken three days after cell preparation. In this section only few representative examples will be shown and discussed. In the case of the half-textile DSSC with one layer of PEDOT: PSS a qualitative difference in the Nyquist plot related to the standard glass based DSSC is noticeable. The Nyquist plot consists of a half-circle with negative imaginary part which can be described with a parallel RC circuit [12] and a half-circle with a positive imaginary part which can be described with a parallel RL circuit [12]. Here, instead of the pure capacitive nature of the impedance known from the standard DSSC, an inductive part is also observable; this becomes dominant in high-frequency area. As there were still no significant resonance peeks in the Bode plot noticeable, for the parametric identification the pole-zero form from Equation (8) was chosen. The comparison of the Nyquist plots of the non-parametric and parametric pole-zero estimation in the frequency range from 1 Hz to 50 kHz is shown in Figure 5. The estimated parameters are displayed in the Figure 5. The agreement between the non-parametric and the pole-zero estimation is excellent.

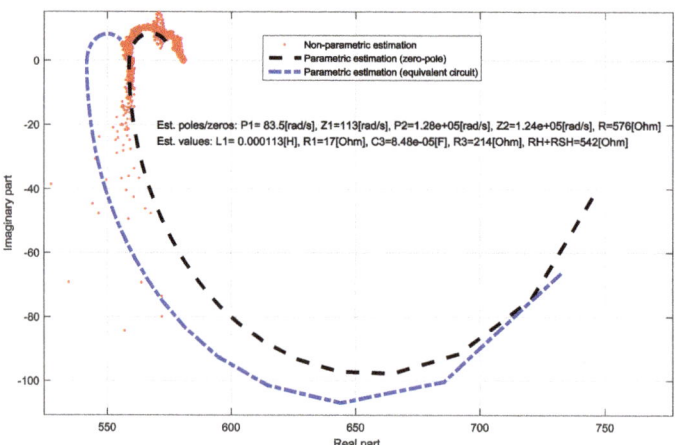

Figure 5. Comparison of the non-parametric and parametric TF identifications of a one-layer half-textile DSSC in Nyquist-diagram.

The effect of having inductive impedance at higher frequencies was observable by all the investigated half-textile DSSC. However, no significant resonance peaks were detected in the Bode-diagrams. Furthermore, the presence of only one R/C element was observed in all measurements for the utilized frequency range between 1 Hz and 50 kHz, e.g., only one half-circle under the ordinate in the Nyquist-plot was observable. A half-circle with positive imaginary part, which stays for inductive impedance, was always detectable. For this reason, according to the black-box method, a modified equivalent-model is supposed consisting of one parallel R/C and one parallel R/L element connected in series with the resistances, as shown in Figure 6. The equivalent impedance of the circuit is given in Equation (11).

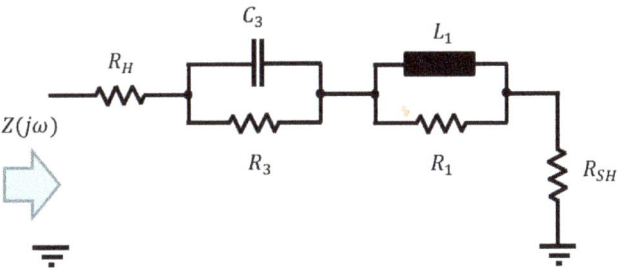

Figure 6. Proposed equivalent circuit for the parametric identification of the impedance of textile-based DSSC.

$$Z(j\omega) = (R_H + R_{SH}) + Z_3(j\omega) + Z_1(j\omega) = (R_H + R_{SH}) + \left(\frac{\frac{R_3}{j\omega C_3}}{R_3 + \frac{1}{j\omega C_3}}\right) + \frac{R_1 \cdot j\omega L_1}{R_1 + j\omega L_1},$$

$$Z(j\omega) = \frac{(j\omega)^2 \cdot ((R_H + R_{SH}) \cdot R_3 \cdot C_3 \cdot L_1 + R_1 \cdot R_3 \cdot L_1 \cdot C_3) + j\omega \cdot ((R_H + R_{SH}) \cdot (R_1 \cdot R_3 \cdot C_3 + L_1) + L_1 \cdot (R_3 + R_1)) + R_1 \cdot ((R_H + R_{SH}) + R_3)}{(j\omega)^2 \cdot R_3 \cdot L_1 \cdot C_3 + j\omega \cdot (R_1 \cdot R_3 \cdot C_3 + L_1) + R_1}$$

(11)

As it can be noticed from Figure 5, the parametric estimation of the proposed equivalent circuit elements shows a slight difference that is related to the parametric estimation while using zero-pole elements. Although this difference indicates that there are still secondary effects that are not modelled with the proposed equivalent circuit, the main properties of the textile-based DSSC, especially the inductivity are described with the proposed equivalent model.

Similar results are obtained for the textile DSSC with three PEDOT: PSS layers, where the differences between the non-parametric and parametric estimations are also small, despite of the obviously high measurement noise. The Nyquist-diagram of the identified impedance with the obtained parameter for this case is shown in Figure 7.

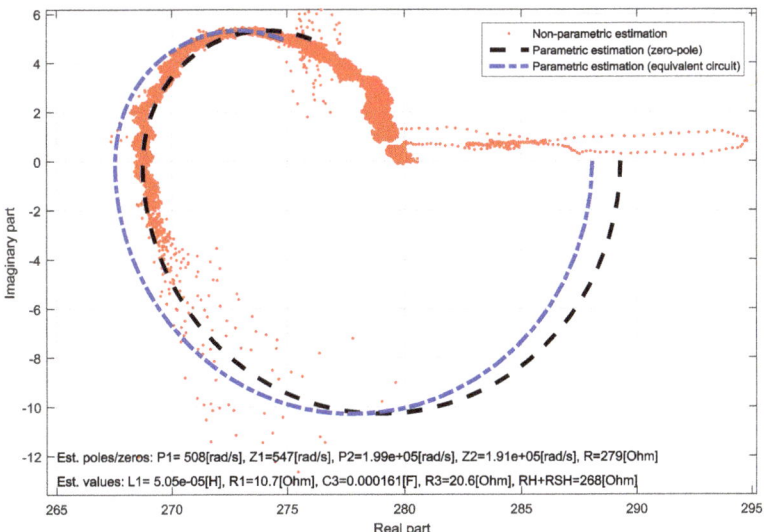

Figure 7. Comparison of the non-parametric and parametric TF identifications of a three-layer half-textile DSSC in Nyquist-diagram.

4. Identification Results Summary

For all the investigated cells, both pole-zero as well as equivalent-circuit based parameter identification are performed. The agreement between the two identification methods is good with slight differences. In Table 1 the summary of the parameter identification sequence is given, e.g., the obtained parameter for the different half-textile DSSCs. From the point of view of the potential user, the most important element may be the cell equivalent dynamic resistance at the steady-state working point, because it limits the generated energy flow-rate. This is the parameter R from Equation (8) or equivalently the sum $R_{eq} = R_H + R_{SH} + R_3$ from Equation (11).

Table 1. Obtained parameters for half-textile DSSC cells with one to five layers of PEDOT: PSS in the counter electrode by spectral identification.

Cell Type	R^1 (Ω)	$R_{eq}{}^2$ (Ω)	$L_1{}^2$ (mH)	$R_1{}^2$ (Ω)	$R_3{}^2$ (Ω)	$C_3{}^2$ (μF)
Single layer	575.91	755.53	0.112	16.95	213.65	84.83
Two layers	528.58	529.60	0.410	0.081	22.83	447.99
Three layers	279.14	288.14	0.050	10.72	20.59	161.31
Four layers	130.47	165.70	0.007	3.963	36.76	71.451
Five layers	242.43	253.66	0.016	6.744	17.13	144.27

[1] From the pole-zero TF estimation. [2] From the equivalent-circuit TF estimation.

According to the results that are shown in Table 1, there is a strong correlation noticeable between the number of layers and the linearized resistance of the half-textile DSSC in the steady-state working point. With the enlarging the number of layers up to four layers, the dynamic resistance (R or equivalently R_{eq}) decreases. The DSSC with five layers of PEDOT: PSS has, however an increased resistance related to the four-layer one. This is in agreement with the results in [11], where the efficiency of the cells with five layers of S V 4 is slightly decreased when compared to the efficiency of cells with four layers of S V 4, and furthermore the I-U curve of the cells with five layers of S V 4 lies lower than the I-U characteristics of cells with four layers. The reason for the increase of the resistance of cells with five S V 4 layers is, however, unclear. To exclude that it is a result of a systematic error, the experiment will be repeated in the future. An inverse tendency can be observed in the case of the identified cell inductivity L_1.

5. Discussion

Spectral analysis and parameter identification of half-textile dye-sensitized solar cells according the Black-Box approach are presented. The identification approach was first tested with a standard glass-based DSSC and thereafter applied to DSSCs with textile-based counter electrodes that were coated by a different number of PEDOT: PSS layers. A strong correlation between the number of layers and the linearized resistance of the half-textile DSSC was observed.

Although there are still improvement potentials left, the authors believe that the presented spectral analysis together with the proposed black-box method can be successfully applied by new types of DSSC where the deep knowledge of the processes is not established yet. In our future work, we will concentrate on further development and extensions of the proposed equivalent circuits in order to describe the higher-order effects. For this reason, on one hand, the extending of the frequency range for the spectral analysis up to 1 MHz by means of direct programming of the embedded FPGA of the MLBX is scheduled. On the other hand, the authors will continue the work toward supporting further elements which are used by modeling of DSSCs. The concept of TF based on fractional differential equations may be helpful for this task.

Author Contributions: Conceptualization, L.J. and I.J.J.; Methodology, L.J.; Software, L.J.; Validation, L.J. and I.J.J.; Writing-Original Draft Preparation, L.J. and I.J.J.

Funding: This research received no external funding.

Conflicts of Interest: The authors declare no conflict of interest.

References

1. Abel, D.; Bollig, A. *Rapid Control Prototyping: Methoden und Anwendungen*, 1st ed.; Springer: Berlin, Germany, 2006.
2. Naoki, K.; Ashraful, I.; Yasuo, C.; Liyuan, H. Improvement of efficiency of dye-sensitized solar cells based on analysis of equivalent circuit. *J. Photochem. Photobiol.* **2006**, *182*, 296–305. [CrossRef]
3. Kilbas, A.A.; Srivastava, H.M.; Trujillo, J.J. *Theory and Applications of Fractional Differential Equations*; Elsevier Science B.V.: Amsterdam, The Netherlands, 2006.
4. Jesus, I.S.; Machado, J.A.T. Development of fractional order capacitors based on electrolyte processes. *Nonlinear Dyn.* **2009**, *56*, 45–55. [CrossRef]
5. Available online: www.dspace.com.
6. Available online: www.mathworks.com.
7. Schoukens, J.; Pintelon, R. *Identification of Linear Systems—A Practical Guide to Accurate Modeling*; Pergamon press: Oxford, UK, 1991.
8. Cohen, L. Generalization of the Wiener-Khinchin Theorem. *IEEE Signal Process. Lett.* **1998**, *5*, 292–294. [CrossRef]
9. Welch, P. The use of fast Fourier transform for the estimation of power spectra: A method based on time averaging over short, modified periodograms. *IEEE Trans. Audio El. Acc.* **1967**, *15*, 70–73. [CrossRef]
10. Juhász, L. Control of Hybrid Micropositioning System for Use in Industry and Robotics. Ph.D. Thesis, University of Novi Sad, Novi Sad, Serbia, October 2011.
11. Juhász Junger, I.; Wehlage, D.; Böttjer, R.; Grothe, T.; Juhász, L.; Grassmann, C.; Blachowicz, T.; Ehrmann, A. Dye-sensitized solar cells with electrospun-nanofiber mat based counter electrodes. *Material* **2018**, *11*, 1604. [CrossRef]
12. Yuan, X.Z.; Song, C.; Wang, H.; Zhang, J. *Electrochemical Impedance Spectroscopy in PEM Fuel Cells*, 1st ed.; Springer-Verlag: London, UK, 2010.

© 2018 by the authors. Licensee MDPI, Basel, Switzerland. This article is an open access article distributed under the terms and conditions of the Creative Commons Attribution (CC BY) license (http://creativecommons.org/licenses/by/4.0/).

Article

Dye-Sensitized Solar Cells with Electrospun Nanofiber Mat-Based Counter Electrodes

Irén Juhász Junger [1], Daria Wehlage [1], Robin Böttjer [1], Timo Grothe [1], László Juhász [2], Carsten Grassmann [3], Tomasz Blachowicz [4] and Andrea Ehrmann [1,*]

[1] Faculty of Engineering and Mathematics, Bielefeld University of Applied Sciences, 33619 Bielefeld, Germany; iren.juhas_junger@fh-bielefeld.de (I.J.J.); daria.wehlage@fh-bielefeld.de (D.W.); robin.boettjer@fh-bielefeld.de (R.B.); timo.grothe@fh-bielefeld.de (T.G.)
[2] Faculty of Electrical Engineering, Media Technology and Computer Science, Deggendorf Institute of Technology, 94469 Deggendorf, Germany; laszlo.juhasz@th-deg.de
[3] Faculty of Textile and Clothing Technology, Niederrhein University of Applied Sciences, 41065 Mönchengladbach, Germany; carsten.grassmann@hsnr.de
[4] Institute of Physics—Center for Science and Education, Silesian University of Technology, 44-100 Gliwice, Poland; blachowicz@posls.pl
* Correspondence: andrea.ehrmann@fh-bielefeld.de; Tel.: +49-521-106-70254

Received: 31 July 2018; Accepted: 2 September 2018; Published: 4 September 2018

Abstract: Textile-based dye-sensitized solar cells (DSSCs) can be created by building the necessary layers on a textile fabric or around fibers which are afterwards used to prepare a textile layer, typically by weaving. Another approach is using electrospun nanofiber mats as one or more layers. In this work, electrospun polyacrylonitrile (PAN) nanofiber mats coated by a conductive polymer poly(3,4-ethylenedioxythiopene) polystyrene sulfonate (PEDOT:PSS) were used to produce the counter electrodes for half-textile DSSCs. The obtained efficiencies were comparable with the efficiencies of pure glass-based DSSCs and significantly higher than the efficiencies of DSSCs with cotton based counter electrodes. The efficiency could be further increased by increasing the number of PEDOT:PSS layers on the counter electrode. Additionally, the effect of the post treatment of the conductive layers by HCl, acetic acid, or dimethyl sulfoxide (DMSO) on the DSSC efficiencies was investigated. Only the treatment by HCl resulted in a slight improvement of the energy-conversion efficiency.

Keywords: dye-sensitized solar cell (DSSC); polyacrylonitrile (PAN); nanofiber mat; electrospinning; PEDOT:PSS

1. Introduction

Dye-sensitized solar cells (DSSCs) are being investigated intensively since their development in 1991 [1]. In the last years, typical conversion efficiencies in the lab reached about 11–29% [2–5], with commercially available cells showing efficiencies around 2–3%. Typically, ruthenium-based dyes are used as sensitizers although they are expensive and in most cases toxic [6,7]. Recently, DSSCs sensitized by porphyrin dyes were investigated [2,7–9] and even higher efficiencies than by the cells with Ru-based dyes were achieved [2]. Cells with metal-free organic dyes have slightly lower efficiencies, but since containing no noble metals, they are less cost intensive [7,10]. Even metal-complexes have been examined as sensitizers for DSSCs [11,12].

Natural dyes, on the other hand, are inexpensive and non-toxic, but show significantly reduced efficiencies [13–18]. This, however, is less important if large areas can be used—which is the case, e.g., for textile architecture and other outdoor textiles. Natural dyes are also used in this article.

Due to the possibility to work with such inexpensive, non-toxic materials outside a cleanroom, several approaches were used to create textile-based DSSCs [19–23]. Amongst them, however, only few concentrate on nanofiber mats produced by electrospinning [24,25]. This method allows for creation of nanofiber mats from diverse polymers or polymer blends, amongst them several biopolymers [26–29], man-made polymers [30,31], or blends with inorganic compounds [32–34]. Such electrospun nanofiber mats are used for a broad range of applications, from wastewater filters [35] to catalyzers [36] to medical wound dressings or cell growth substrates [37–40] to precursors for carbon nanofibers which may be included in polymers to enhance their mechanical properties [41–43].

However, only few reports can be found about integration of electrospun nanofiber mats in DSSCs. Such nanofiber mats were, e.g., used to create quasi-solid electrolytes for the integration in DSSCs [44,45]. Electrospun TiO_2 nanofibers were used to create the photo-anode [46,47]. Electrospun carbon nanofibers were applied with different coatings or blends as counter electrodes [48,49].

The simple approach of coating a non-conductive nanofiber mat—which can unambiguously be produced by electrospinning, opposite to conductive nanofibers which need a two-step procedure—by poly (3,4-ethylenedioxythiopene) polystyrene sulfonate (PEDOT:PSS), a conductive polymer, to combine the large surface–volume ratio of the nanofiber mat with the good conductivity of PEDOT:PSS, was nevertheless not yet found in the scientific literature. This article thus aims at showing in a first-principle study that PEDOT:PSS-coated nanofiber mats can be used as textile counter electrodes, investigating half-textile DSSCs in which the counter electrodes are replaced by textile substrates with conductive coatings (PEDOT:PSS), while the working electrodes are still made from glass and are prepared using a natural dye. DSSCs with counter electrodes prepared on PAN nanofiber mats and PAN nano-membranes are compared to reference cells with counter electrodes prepared on fluorine-doped tin oxide (FTO)-coated glass and on cotton as a macroscopic textile fabric. The effects of different numbers of PEDOT:PSS coating layers as well as of after-treatments of the PEDOT:PSS by dimethyl sulfoxide (DMSO), HCl, or acetic acid which is reported in the literature [50] were investigated to increase the conductivity of the counter electrode. The resulting efficiencies are similar to or even significantly higher than those achieved with common glass-based approaches.

2. Materials and Methods

2.1. Preparation of the Counter Electrodes (Textile Half-Cell)

Nanofiber mats were prepared using the electrospinning machine "Nanospider Lab" (Elmarco, Czech Republic), a needleless electrospinning machine based on the wire technology. For the spinning solution, 16% PAN were dissolved in DMSO (min 99.9%, from S3 Chemicals, Bad Oeynhausen, Germany). Spinning was performed for 1 h 25 min with the following spinning parameters: voltage 65 kV, electrode-substrate distance 240 mm, nozzle diameter 0.8 mm, carriage speed 50 mm/s, substrate speed 0 mm/min, relative humidity 32–33%, and temperature 22.0 °C. The resulting nanofiber mat was 0.105 mm thick with an areal weight of 21.08 g/m^2.

The PAN membranes were produced by treating the electrospun nanofiber mat with 11.25 µL DMSO per cm^2 and drying at 50 °C on a hot plate. The DMSO treatment dissolves the PAN transforming the original nanofiber mats into a membrane.

For comparison, a cotton fabric (thickness 0.19 mm, areal weight 84.49 g/m^2) and FTO-coated glass plates (purchased from Man Solar, Petten, The Netherlands) were used.

The nanofiber mats and nano-membranes as well as the cotton fabric were placed on microscopy slides and coated by doctor blading with 11.25 µL PEDOT:PSS per cm^2 area (1 to 5 times) to make them conductive. For this, CLEVIOS™ S V 4 (sheet resistance 500 Ohm) and Orgacon™ S305 (sheet resistance 200 Ohm) were used. The sheet resistances differ strongly, enabling the investigation of the influence of this parameter. After each layer the textiles were heated to 110 °C together with the microscopy slides for 50 min in an oven. The microscopy slides served as support for the textile electrodes.

To increase the conductivity, some of the samples with a single PEDOT:PSS layer were treated by 11.25 µL/cm^2 DMSO, 0.2 mol/L HCl, or 80% (vol.) acetic acid and heated for 10 min to 110 °C, 140 °C, or 160 °C, respectively, as proposed in Ref. [50]. After cooling to room temperature, the DMSO, the HCl and the acetic acid were removed by dipping the samples into distilled water. The textile and glass counter electrodes were coated with graphite spray (Graphit 33 by Kontakt Chemie), spraying for ca. 1 s from a distance of 25–30 cm, and heated to 90 °C for 60 min in an oven.

2.2. Preparation of the Working Electrodes (Glass Half-Cell) and Cell Assembly

The working electrodes were prepared using TiO_2-coated FTO glasses (Man Solar). For the dye, 25 g forest fruit tea (Mayfair, Wilken Tee GmbH, Fulda, Germany) were ground, mixed with 300 mL distilled water, and stirred with 200 min^{-1} at 40 °C for 30 min. Afterwards, the solid contents were filtered. The TiO_2 coated plates were inserted into the dye solution at room temperature for 2 days to avoid potential irregularities due to unequal dyeing times. The excess dye was washed off with distilled water, and the plates were dried at the air.

Both electrodes were pressed together and fixed with adhesive tape. A iodine/triiodide-based electrolyte (electrolyte type 016 purchased from Man Solar) was dropped on the gap between both electrodes and left to spread by capillary force. The cells were not sealed. The measurements of the I-U characteristics were taken with unmasked solar cells with an active area of 6 cm^2.

2.3. Measurements

Confocal laser scanning microscope (CLSM) images were taken with a VK-9000 (Keyence, Neu-Isenburg, Germany) with a nominal magnification of 2000×. All scale bars have dimensions of 10 μm.

Current-voltage (I-U) curves were measured with a source meter unit 2450 SourceMeter (Keithley Instruments, Solon, OH, USA) in the voltage range from 0.5 to 0 V. A halogen lamp with a color temperature of 3000 K was used for these investigations. The light intensity was measured by a solarimeter KIMO SL-200 and set to 1000 W/m^2 on the DSSC surface.

The spectral analysis and parameter identification of the linearized DSSC model was performed as described in detail in Ref. [51].

Three DSSCs are prepared with each type of counter electrodes and their current-voltage (I-U) characteristics were averaged.

3. Results and Discussion

3.1. Effect of the Textile Substrate

First, the effect of the textile substrate of the counter electrode on the photovoltaic characteristics of the DSSCs is investigated. For that purpose, the counter electrodes for the DSSCs are prepared from a PAN nanofiber mat, PAN nano-membrane and cotton, coated by a single layer of PEDOT:PSS (S305 or S V 4). For comparison, conventional cells with counter electrodes made from FTO glass were also prepared.

Table 1 shows the open-circuit voltages, short-circuit currents, fill factors, and efficiencies of these cells. The energy-conversion efficiency η of the cells is calculated from the I-U characteristics averaged for the three cells with the same counter electrode as $\eta/\% = 100 \cdot FF \cdot U_{OC} \cdot I_{SC}/(A \cdot I_0)$, where U_{OC} is the open-circuit voltage, I_{SC} is the short-circuit current, A is the cell area of 6 cm^2, I_0 is the incident irradiation, and FF is the fill factor defined as $FF = U_m \cdot I_m/(U_{OC} \cdot I_{SC})$. U_m and I_m are the photovoltage and photocurrent of the maximum power of the DSSC.

Table 1. Open-circuit voltage U_{OC}, short-circuit current I_{SC}, fill factor FF, and efficiency η for cells prepared with different textile fabrics and sorts of PEDOT:PSS in comparison to FTO glass cells.

PEDOT:PSS	Counter Electrode	U_{OC}/V	I_{SC}/mA	FF	η/%
	FTO glass	0.462	0.46	0.57	0.020
S305	Cotton	0.387	0.02	0.24	3×10^{-4}
	PAN nanof. mat	0.448	0.81	0.33	0.020
	PAN membrane	0.354	0.42	0.31	0.008
S V 4	Cotton	0.449	0.54	0.41	0.017
	PAN nanof. mat	0.434	1.06	0.43	0.033
	PAN membrane	0.408	1.72	0.28	0.033

PEDOT:PSS, poly(3,4-ethylenedioxythiopene) polystyrene sulfonate; FTO, fluorine-doped tin oxide; PAN, polyacrylonitrile.

In Figure 1, the corresponding current-voltage characteristics are depicted. The pure FTO glass cells show relatively rectangular I-U curves, verifying a higher fill factor than the other cells shown here. While the cotton cells prepared with S305 have currents near zero, the ones produced with S V 4 have even slightly higher short-circuit currents than the glass cells, but a reduced fill factor.

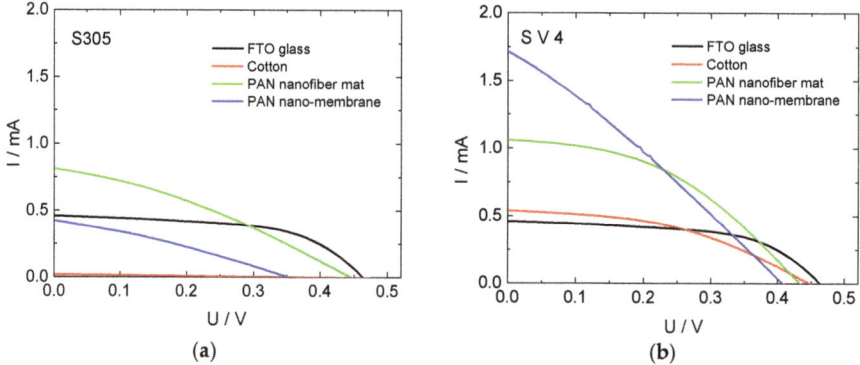

Figure 1. Current-voltage curves of dye-sensitized solar cells (DSSCs), prepared with different counter electrodes: (**a**) S305 as conductive coating; (**b**) S V 4 as conductive coating.

Comparing the results from both sorts of PEDOT:PSS used in this investigation, it can be recognized that the cells prepared with S V 4 show slightly higher currents although this PEDOT:PSS has a higher sheet resistance. This shows that the conductivity is not the only relevant factor.

Especially the nanofiber mat and the nano-membrane prepared with S V 4, although both with reduced fill factors, have significantly higher short circuit currents than the cells prepared on glass. For S305, the nanofiber mat still shows a higher short-circuit current than the FTO glass.

Figure 2 depicts a comparison of the efficiencies of DSSCs with counter electrodes built on different textile substrate with the efficiency of pure glass cell. As usual for DSSCs prepared from non-toxic natural substances, the efficiencies are relatively low. However, the best textile-based cells reach the efficiencies which are typically gained with glass-based cells using otherwise similar materials [15,17] and even clearly outperform the reference glass cell prepared with identical materials by more than 50%, underlining that nanofiber mats and nano-membranes coated with PEDOT:PSS are indeed a well-suited alternative for FTO glasses which are typically used for low-cost, non-toxic DSSCs. It should be mentioned that while with the lower-conductive PEDOT:PSS S V 4, nanofiber mat and nano-membrane show very similar efficiencies, the nano-membranes coated with S305 have clearly smaller efficiencies than the corresponding nanofiber mat. On the other hand, the efficiencies gained with cotton are lower than those obtained with pure glass cells, indicating that such macroscopic textile fabrics are not the ideal choice for the creation of textile-based solar cells. Since the measurements of the I-U characteristics were performed with unmasked DSSCs, the efficiencies of all cells may be slightly overestimated. Snaith et al. reported a decrease of the efficiency by about 30% when measured with masked cells compared to the measurements with unmasked cells [52]. The reason for the overestimation of the efficiency in measurements without mask is that in this case, not only the light falling on the active area can enter the cell, but the light falling on the glass boarder surrounding the active area or on the edges of the cell may also be trapped by the glass and enter the cell, contributing to the energy conversion. Our DSSCs have glass parts bordering the active area only at one edge. In half-textile DSSCs, extra light can enter the cells through these small glass areas and through the edges of the working electrodes, while the counter electrodes are not transparent. Therefore, we believe that in our case, the grade of the efficiency overestimation is reduced compared to that reported by Snaith et al.

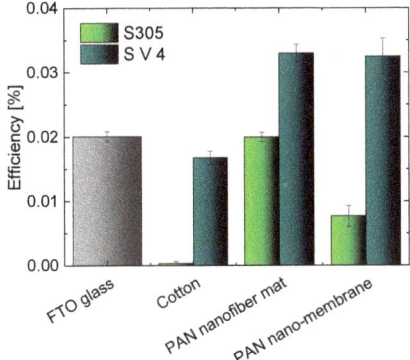

Figure 2. Efficiencies of DSSCs prepared with different textile-based counter electrodes and FTO glass as a reference.

The results of optical examination of the PEDOT:PSS-coated nanofiber mats and membranes are depicted in Figures 3 and 4. In both cases, coating cotton with PEDOT:PSS (Figures 3a and 4a) shows a thin coating on the fibers, with some interconnections due to the coating, as expected. The macroscopic fibers nevertheless impede creation of a closed conductive layer.

Coating the PAN nanofiber mats with PEDOT:PSS (Figures 3b and 4b) reveals the typical nanofiber mat structure with some small agglomerations which are typical for PEDOT:PSS. As expected, the membranes (Figures 3c and 4c) do not show any fibrous regions.

The sealing of the half-textile DSSCs is still an unsolved problem since the electrolyte evaporates not only on the edges of the cells, but through the whole surface of the textile electrode. This leads to a loss of efficiency in a relatively short period of time.

3.2. Effect of the Number of Conductive Layers

As we have seen in the previous subsection, the efficiency of some textile-based DSSCs reaches the efficiency of pure glass cells, but is still low. It could be increased by decreasing the inner resistance of the cell by enhancing the conductivity of the PEDOT:PSS coating on the counter electrode. One of the possibilities to reach this goal is applying more than one PEDOT:PSS layer, as proposed in the literature [53,54].

As a substrate for the counter electrodes, PAN nanofiber mats were used due to the above described findings that they are less sensitive to the choice of the PEDOT:PSS than nano-membranes, can be prepared in one production step (cf. Section 2.1) and show higher fill factors. The nanofiber mats were coated up to five times with PEDOT:PSS. The results are depicted in Figure 5. Coating the samples with PEDOT:PSS twice results in a significant increase of the short circuit currents for both PEDOT:PSS versions. This can be explained by monitoring the coating process carefully: The first PEDOT:PSS layer flows into the nanofiber mat where it is partly disconnected by the non-conductive PAN nanofibers. The second layer is placed on top of the first one and can form a more continuous conductive layer, in this way strongly increasing the current transport along the nanofiber mat.

Adding more PEDOT:PSS layers, the short circuit currents continuously increase (for S305) or increase until a maximum is reached for four layers (for S V 4), respectively. The open circuit voltages decrease slightly from one layer to two layers and stay constant afterwards.

Figure 6 depicts the corresponding efficiencies. While the value of the glass cells prepared as benchmark (0.02%) is even reached for one layer of the PEDOT:PSS S305, adding further layers clearly increases this value, until saturation is approached after 4 layers for both PEDOT:PSS versions, indicating that the cost–benefit ratio may be ideal for this number of layers.

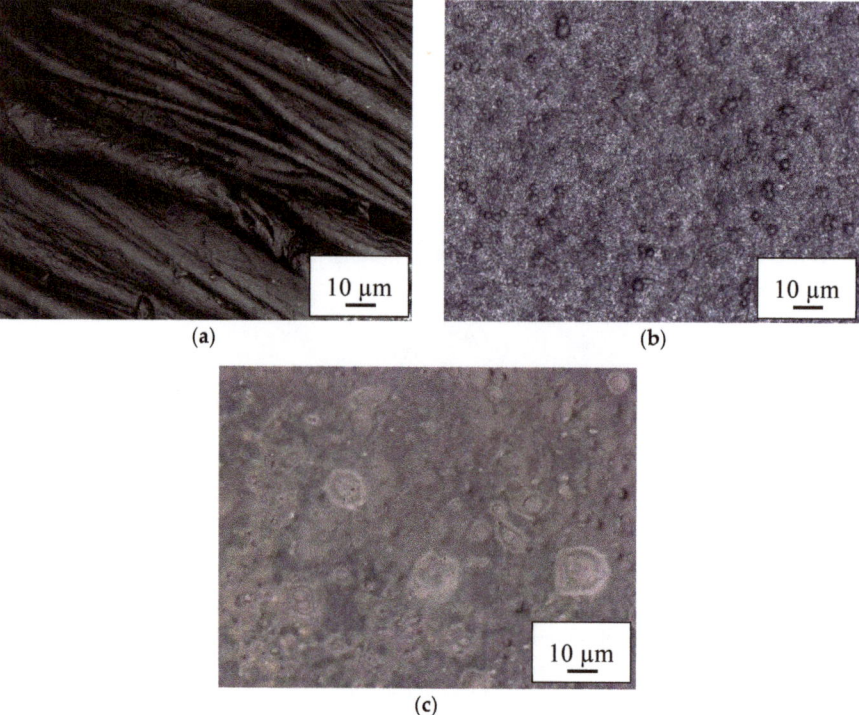

Figure 3. Confocal laser scanning microscope (CLSM) images of the textile-based counter electrodes prepared with S305 on (**a**) cotton; (**b**) PAN nanofiber mat; and (**c**) PAN nano-membrane taken with nominal magnification 2000×. The length of the scale bars is 10 μm in all subfigures.

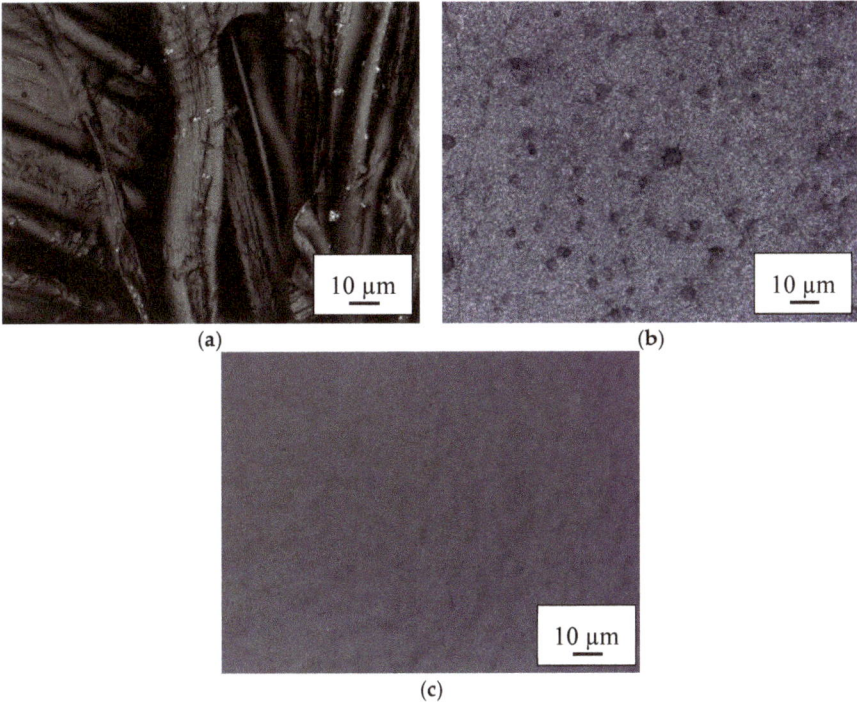

Figure 4. Confocal laser scanning microscope (CLSM) images of the textile-based counter electrodes prepared with S V 4 on (**a**) cotton; (**b**) PAN nanofiber mat; and (**c**) PAN nano-membrane taken with nominal magnification 2000×.

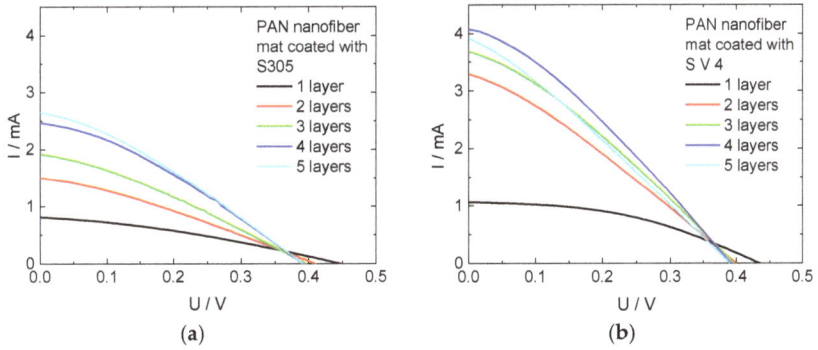

Figure 5. Current-voltage curves of DSSCs, prepared with different counter electrodes: (**a**) S305 as conductive coating; (**b**) S V 4 as conductive coating.

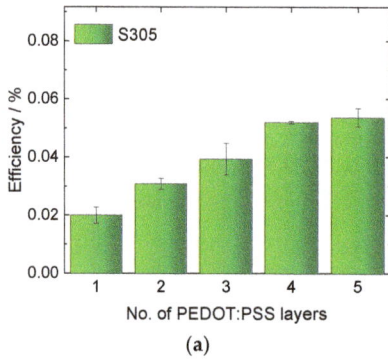 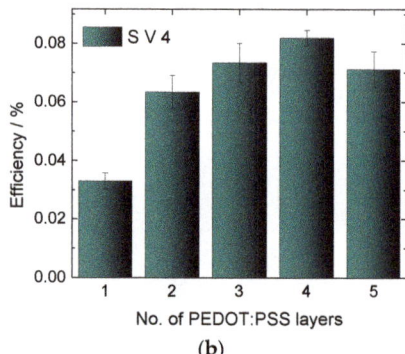

Figure 6. Efficiencies, obtained with PAN nanofibers mats and different numbers of PEDOT:PSS layers. (a) S305 as conductive coating; (b) S V 4 as conductive coating.

For S V 4, the maximum efficiency of 0.08%, i.e., four times the value of the glass reference, is reached for four layers. Unexpectedly, the efficiency of the cells with five layers of S V 4 is slightly decreased compared to the efficiency of cells with four layers of S V 4. A similar observation is made for the current, i.e., the I-U curve of the cells with five layers of S V 4 is lower than the I-U curve of cells with four layers. The reason may be an increased inner resistance of the cells with five S V 4 layers in comparison to cells with four layers. To test this assumption, AC measurements followed by spectral analysis were performed [51]. According to the obtained results (Table 2) there is a clear correlation noticeable between the number of layers and the linearized resistance of the textile-based DSSC. With increasing the number of layers up to four layers, the linearized dynamic resistance decreases. The DSSC with five layers of S V 4 has, however, an increased resistance related to the four-layer one. This justifies our assumption and explains the drop of the efficiency of five-layer cell. The difference between both systems under investigation may be attributed to S V 4 necessitating a smaller number of layers to reach the ideal combination of high conductivity and low thickness, both of which will reduce the inner resistance of the cells.

Table 2. Resistance R_1 of cells prepared with S V 4, calculated from pole-zero transfer function (TF) estimation, and resistance R_2, calculated by the equivalent-circuit TF estimation [51].

Cell Type	R_1/Ω	R_2/Ω
Single layer	575.91	755.53
Two layers	528.58	529.60
Three layers	279.14	288.14
Four layers	130.47	165.70
Five layers	242.43	253.66

3.3. Effect of Chemical Post-Treatment of the Conductive Layer

In the previous subsection, it was shown that the conductivity of PEDOT:PSS and therewith the efficiency of the cells can be increased by coating more than one conductive layer on the counter electrode. However, increasing the number of coating layers is cost consuming and results in losing the textile nanostructure after several coating layers. Therefore, an alternative solution should be found.

In Ref. [55] and the references therein, the possibility to enhance the conductivity of PEDOT:PSS layers on a glass substrate by more than two orders of magnitude by adding polar organic molecules such as DMSO, ethylene-glycol, diethylene glycol, or sorbitol to the aqueous solution of PEDOT:PSS was reported. According to Refs. [50,56], a chemical post-treatment of the PEDOT:PSS layer by inorganic acids (hydrochloric acid, sulfurous acids), organic acids (acetic acid, propionic acid etc.), and inorganic solvents (DMSO, ethylene glycol, etc.) results in an enhancement of the conductivity by a factor greater than 1000. Inspired by these encouraging results, the nanofiber mats coated with a single layer of S V 4 (the other PEDOT:PSS, S305, was omitted since the cells built with it yielded lower efficiencies in previous investigations) were treated with DMSO, HCl or acetic acid, as described in detail in Section 2.1, and the resulting counter electrodes were used to prepare DSSCs. The I-U characteristics and the efficiencies of the obtained cells are depicted in Figure 7.

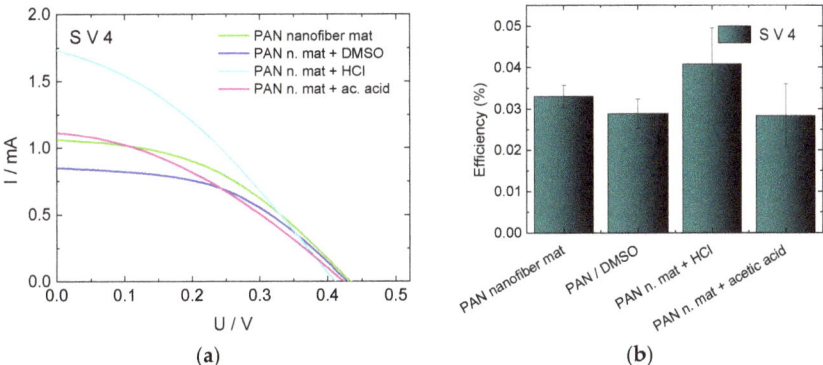

Figure 7. (a) Current-voltage characteristics and (b) efficiencies of DSSCs with counter electrodes prepared from PAN nanofiber mat coated by a single layer of S V 4, pure and with a post treatment by DMSO, HCl, or acetic acid.

While a clear change of the short-circuit current could be observed, the open circuit voltage remained nearly the same as for the cells without chemical treatment. Since a significant enhancement of the conductivity of PEDOT:PSS layer through the chemical treatment was reported in the literature, we expected also a significant increase of the efficiency of DSSCs compared to cells with untreated conductive layer. However, only the cells treated with HCl show a slight efficiency enhancement. This finding can be attributed to the much lower concentration of the HCl used here in comparison to the experiment described in the literature. In Ref. [50] it was shown that the conductivity increase depends on the concentration of the used HCl and reaches its maximum at a concentration of 9.6 mol/L. Since highly concentrated HCl could damage the nanofiber mat, the optimal concentration has to be found in future tests.

The DSSCs treated with DMSO or with acetic acid showed a slight decrease of the efficiency instead of an efficiency increase. It is known that DMSO dissolves PAN. Through the DMSO treatment of the PEDOT:PSS layer on a PAN nanofiber mat, the PAN mat is dissolved leading to mixing of PEDOT:PSS with PAN. The nanofiber structure is lost, and a membrane is formed. By forming poorly connected PEDOT:PSS islands within the isolating PAN membrane, the conductivity of the PEDOT:PSS layer is decreased. This results in a higher inner resistance and decreased efficiency of corresponding DSSCs. We assume that the efficiency drop of acetic acid treated cells may have the same reason.

4. Conclusions

In this paper, half-textile DSSCs with textile-based counter electrodes coated by a conducting polymer and working electrodes built on TiO_2 coated FTO glass dyed with a natural dye were investigated. The conductive coating was performed by two types of PEDOT:PSS: by CLEVIOS™ S V 4 and by Orgacon™ S305.

Coating of non-conductive electrospun PAN nanofiber mats with conductive PEDOT:PSS has been shown to result in high-quality counter-electrodes for half-textile DSSCs, with efficiencies comparable to those gained with pure FTO-glass cells even for the lower-conductive PEDOT:PSS used in this investigation and clearly superior to those of PEDOT:PSS coated cotton. The efficiency was dependent on the number of coating layers. Using 4 layers of S V 4, the efficiency could be increased by a factor of 4, as compared to the FTO-glass reference cells.

Additionally, a post-treatment of the conducting layer with DMSO, acetic acid or HCl was investigated. Only the treatment with HCl led to a slight increase of the efficiency, whereas the efficiency of cells treated with DMSO and acetic acid was slightly decreased compared to the untreated cells. This was ascribed to dissolving or damaging the nanofiber mat by treatment with both latter materials and forming of an only partially conducting layer.

Generally, this study has shown that electrospun nanofiber mats are a good alternative to common glass-based electrodes, offering a method to create a nanostructured electrode with a simple, textile-based technology. Aiming at increasing the efficiency, in the future the influence of the nanofiber mat morphology, as defined by the spinning and solution parameters, will be studied. The HCl-treatment will be optimized and performed on coatings with more than one conductive layer. The possibilities of sealing the half-textile cells will be investigated.

Author Contributions: Conceptualization, I.J.J. and A.E.; Methodology, I.J.J.; Validation, I.J.J. and A.E.; Formal Analysis, I.J.J., L.J., C.G., T.B., and A.E.; Investigation, I.J.J., D.W., R.B., T.G., and L.J.; Writing—Original Draft Preparation, I.J.J. and A.E.; Writing—Review & Editing, D.W., R.B., T.G., L.J., C.G., and T.B.; Supervision, T.B.

Funding: This research was funded by the Ph.D. fund and HiF fund of Bielefeld University of Applied Sciences, the SUT Rector Grants 14/990/RGJ17/0075-02 and 14/990/RGJ18/0099, the internal SUT project BK-229/RIF/2017 as well as by Volkswagen Foundation grant "Adaptive Computing with Electrospun Nanofiber Networks" No. 93679.

Acknowledgments: The authors acknowledge gratefully the program FH Basis of the German federal country North Rhine-Westphalia for funding the "Nanospider Lab".

Conflicts of Interest: The authors declare no conflict of interest.

References

1. O'Regan, B.; Grätzel, M. A low-cost, high-efficiency solar cell based on dye-sensitized colloidal TiO_2 films. *Nature* **1991**, *353*, 737–740. [CrossRef]
2. Yella, A.; Lee, H.-W.; Tsao, H.N.; Yi, C.; Chandiran, A.K.; Nazeeruddin, M.K.; Diau, E.W.-G.; Yeh, C.-Y.; Zakeeruddin, S.M.; Grätzel, M. Porphyrin-sensitized solar cells with cobalt (II/III)-based redox electrolyte exceed 12 percent efficiency. *Science* **2011**, *334*, 629–634. [CrossRef] [PubMed]
3. Mathew, S.; Yella, A.; Gao, P.; Humphry-Baker, R.; Curchod, B.F.E.; Ashari-Astani, N.; Tavernelli, I.; Rothlisberger, U.; Nazeeruddin, M.K.; Grätzel, M. Dye-sensitized solar cells with 13% efficiency achieved through the molecular engineering of porphyrin sensitizers. *Nat. Chem.* **2014**, *6*, 242–247. [CrossRef] [PubMed]
4. Kakiage, K.; Aoyama, Y.; Yano, T.; Oya, K.; Fujisawa, J.-I.; Hanaya, M. Highly-efficient dye-sensitized solar cells with collaborative sensitization by silyl-anchor and carboxy-anchor dyes. *Chem. Commun.* **2015**, *51*, 15894–15897. [CrossRef] [PubMed]
5. Freitag, M.; Teuscher, J.; Saygili, Y.; Zhang, X.; Giordano, F.; Liska, P.; Hua, J.; Zakeeruddin, S.M.; Moser, J.-E.; Grätzel, M.; et al. Dye-sensitized solar cells for efficient power generation under ambient lighting. *Nat. Photonics* **2017**, *11*, 372–378. [CrossRef]
6. Hagfeldt, A.; Boschloo, G.; Sun, L.; Kloo, L.; Pettersson, H. Dye-sensitized solar cells. *Chem. Rev.* **2010**, *110*, 6595–6663. [CrossRef] [PubMed]
7. Ito, S. Investigation of Dyes for Dye-Sensitized Solar Cells: Ruthenium-Complex Dyes, Metal-Free Dyes, Metal-Complex Porphyrin Dyes and Natural Dyes. In *Solar Cells—Dye-Sensitized Devices*; Kosyachenko, L.A., Ed.; InTech: London, UK, 2011.
8. Higashino, T.; Imahori, H. Porphyrins as excellent dyes for dye-sensitized solar cells. Recent developments and insights. *Dalton Trans.* **2015**, *44*, 448–463. [CrossRef] [PubMed]
9. Birel, Ö.; Nadeem, S.; Duman, H. Porphyrin-Based Dye-Sensitized Solar Cells (DSSCs). A Review. *J. Fluoresc.* **2017**, *27*, 1075–1085. [CrossRef] [PubMed]

10. Chae, Y.; Kim, S.J.; Kim, J.H.; Kim, E. Metal-free organic-dye-based flexible dye-sensitized solar textiles with panchromatic effect. *Dyes Pigment.* **2015**, *113*, 378–389. [CrossRef]
11. Bignozzi, C.A.; Argazzi, R.; Boaretto, R.; Busatto, E.; Carli, S.; Ronconi, F.; Caramori, S. The role of transition metal complexes in dye sensitized solar devices. *Coord. Chem. Rev.* **2013**, *257*, 1472–1492. [CrossRef]
12. Robertson, N. Cu^1 versus Ru^2: Dye-Sensitized Solar Cells and Beyond. *Chem. Sus. Chem.* **2008**, *1*, 977–979. [CrossRef] [PubMed]
13. Calogero, G.; Di Marco, G.; Caramori, S.; Cazzanti, S.; Argazzi, R.; Bignozzi, C.A. Natural dye sensitizers for photoelectrochemical cells. *Energy Environ. Sci.* **2009**, *2*, 1162–1172. [CrossRef]
14. Castañeda-Ovando, A.; Pacheco-Hernández, M.D.L.; Páez-Hernández, M.E.; Rodríguez, J.A.; Galán-Vidal, C.A. Chemical studies of anthocyanins: A review. *Food Chem.* **2009**, *113*, 859–871. [CrossRef]
15. Juhász Junger, I.; Homburg, S.V.; Meissner, H.; Grethe, T.; Schwarz-Pfeiffer, A.; Fiedler, J.; Herrmann, A.; Blachowicz, T.; Ehrmann, A. Influence of the pH value of anthocyanins on the electrical properties of dye-sensitized solar cells. *AIMS Energy* **2017**, *5*, 258–267. [CrossRef]
16. Hölscher, F.; Trümper, P.-R.; Juhász Junger, I.; Schwenzfeier-Hellkamp, E.; Ehrmann, A. Raising reproducibility in dye-sensitized solar cells under laboratory conditions. *J. Renew. Sustain. Energy* **2018**, *10*, 13506. [CrossRef]
17. Juhász Junger, I.; Homburg, S.V.; Grethe, T.; Herrmann, A.; Fiedler, J.; Schwarz-Pfeiffer, A.; Blachowicz, T.; Ehrmann, A. Examination of the sintering process-dependent properties of TiO_2 on glass and textile substrates. *J. Photonics Energy* **2017**, *7*, 15001. [CrossRef]
18. Juhász Junger, I.; Tellioglu, A.; Ehrmann, A. Refilling DSSCs as a method to ensure longevity. *Optik* **2018**, *160*, 255–258. [CrossRef]
19. Yun, M.J.; Cha, S.I.; Seo, S.H.; Lee, D.Y. Highly flexible dye-sensitized solar cells produced by sewing textile electrodes on cloth. *Sci. Rep.* **2014**, *4*, 5322. [CrossRef] [PubMed]
20. Sun, K.C.; Sahito, I.A.; Noh, J.W.; Yeo, S.Y.; Im, J.N.; Yi, S.C.; Kim, Y.S.; Jeong, S.H. Highly efficient and durable dye-sensitized solar cells based on a wet-laid PET membrane electrolyte. *J. Mater. Chem. A* **2016**, *4*, 458–465. [CrossRef]
21. Schubert, M.B.; Werner, J.H. Flexible solar cells for clothing. *Mater. Today* **2006**, *9*, 42–50. [CrossRef]
22. Pan, S.; Yang, Z.; Chen, P.; Deng, J.; Li, H.; Peng, H. Wearable solar cells by stacking textile electrodes. *Angew. Chem. Int. Ed.* **2014**, *53*, 6110–6114. [CrossRef] [PubMed]
23. Choi, C.M.; Kwon, S.-N.; Na, S.-I. Conductive PEDOT:PSS-coated poly-paraphenylene terephthalamide thread for highly durable electronic textiles. *J. Ind. Eng. Chem.* **2017**, *50*, 155–161. [CrossRef]
24. Greiner, A.; Wendorff, J.H. Electrospinning: A Fascinating Method for the Preparation of Ultrathin Fibers. *Angew. Chem. Int. Ed.* **2007**, *46*, 5670–5703. [CrossRef] [PubMed]
25. Li, D.; Xia, Y. Electrospinning of Nanofibers: Reinventing the Wheel? *Adv. Mater.* **2004**, *16*, 1151–1170. [CrossRef]
26. Milasuiu, R.; Ryklin, D.; Yasinskaya, N.; Yeutushenka, A.; Ragaisiene, A.; Rukuiziene, Z.; Mikucioniene, D. Development of an Electrospun Nanofibrous Web with Hyaluronic Acid. *Fibres Text. East. Eur.* **2017**, *25*, 8–12. [CrossRef]
27. Grothe, T.; Brikmann, J.; Meissner, H.; Ehrmann, A. Influence of Solution and Spinning Parameters on Nanofiber Mat Creation of Poly(ethylene oxide) by Needleless Electrospinning. *Mater. Sci.* **2017**, *23*, 342–349. [CrossRef]
28. Yao, L.; Haas, T.; Guiseppi-Elie, A.; Bowlin, G.L.; Simpson, D.G.; Wnek, G.E. Electrospinning and Stabilization of Fully Hydrolyzed Poly(vinyl alcohol) Fibers. *Chem. Mater.* **2003**, *15*, 1860–1864. [CrossRef]
29. Banner, J.; Dautzenberg, M.; Feldhans, T.; Hofmann, J.; Plümer, P.; Ehrmann, A. Water Resistance and Morphology of Electrospun Gelatine Blended with Citric Acid and Coconut Oil. *Tekstilec* **2018**, *61*, 129–135. [CrossRef]
30. Pan, J.F.; Liu, N.H.; Sun, H.; Xu, F. Preparation and Characterization of Electrospun PLCL/Poloxamer Nanofibers and Dextran/Gelatin Hydrogels for Skin Tissue Engineering. *PLoS ONE* **2014**, *9*, e112885. [CrossRef] [PubMed]
31. Sabantina, L.; Rodríguez-Mirasol, J.; Cordero, T.; Finsterbusch, K.; Ehrmann, A. Investigation of Needleless Electrospun PAN Nanofiber Mats. *AIP Conf. Proc.* **2018**, *1952*, 020085.

32. Iatsunskye, I.; Vasylenko, A.; Viter, R.; Kempinski, M.; Nowaczyk, G.; Jurga, S.; Bechelany, M. Tailoring of the electronic properties of ZnO-polyacrylonitrile nanofibers: Experiment and theory. *Appl. Surf. Sci.* **2017**, *411*, 494–501. [CrossRef]
33. Othman, F.E.C.; Yusof, N.; Hasbullah, H.; Jaafar, J.; Ismail, A.F.; Abdullah, N.; Nordin, N.A.H.M.; Aziz, F.; Salleh, W.N.W. Polyacrylonitrile/magnesium oxide-based activated carbon nanofibers with well-developed microporous structure and their adsorption performance for methane. *J. Ind. Eng. Chem.* **2017**, *51*, 281–287. [CrossRef]
34. Kancheva, M.; Toncheva, A.; Paneva, D.; Manolova, N.; Rashkov, I.; Markova, N. Materials from nanosized ZnO and polyacrylonitrile: Properties depending on the design of fibers (electrospinning or electrospinning/electrospraying). *J. Inorg. Organomet. Polym.* **2017**, *27*, 912–922. [CrossRef]
35. Yalcinkaya, F.; Yalicnkaya, B.; Hruza, J.; Hrabak, P. Effect of Nanofibrous Membrane Structures on the Treatment of Wastewater Microfiltration. *Sci. Adv. Mater.* **2017**, *9*, 747–757. [CrossRef]
36. Wang, X.; Kim, Y.G.; Drew, C.; Ku, B.-C.; Kumar, J.; Samuelson, L.A. Electrostatic Assembly of Conjugated Polymer Thin Layers on Electrospun Nanofibrous Membranes for Biosensors. *Nano Lett.* **2004**, *4*, 331–334. [CrossRef]
37. Ashammakhi, N.; Ndreu, A.; Yang, Y.; Ylikauppila, H.; Nikkola, L. Nanofiber-based scaffolds for tissue engineering. *Eur. J. Plast. Surg.* **2012**, *35*, 135–149. [CrossRef]
38. Schnell, E.; Klinkhammer, K.; Balzer, S.; Brook, G.; Klee, D.; Dalton, P.; Mey, J. Guidance of glial cell migration and axonal growth on electrospun nanofibers of poly-epsilon-caprolactone and a collagen/polyepsilon-caprolactone blend. *Biomaterials* **2007**, *28*, 3012–3025. [CrossRef] [PubMed]
39. Klinkhammer, K.; Seiler, N.; Grafahrend, D.; Gerardo-Nava, J.; Mey, J.; Brook, G.A.; Möller, M.; Dalton, P.D.; Klee, D. Deposition of electrospun fibers on reactive substrates for in vitro investigations. *Tissue Eng. Part C* **2009**, *15*, 77–85. [CrossRef] [PubMed]
40. Großerhode, C.; Wehlage, D.; Grothe, T.; Grimmelsmann, N.; Fuchs, S.; Hartmann, J.; Mazur, P.; Reschke, V.; Siemens, H.; Rattenholl, A.; et al. Investigation of microalgae growth on electrospun nanofiber mats. *AIMS Bioeng.* **2017**, *4*, 376–385. [CrossRef]
41. Rahaman, M.S.A.; Ismail, A.F.; Mustafa, A. A review of heat treatment on polyacrylonitrile fiber. *Polym. Degrad. Stab.* **2007**, *92*, 1421–1432. [CrossRef]
42. Manoharan, M.P.; Sharma, A.; Desai, A.V.; Haque, M.A.; Bakis, C.E.; Wang, K.W. The interfacial strength of carbon nanofiber epoxy composite using single fiber pullout experiments. *Nanotechnology* **2009**, *20*, 295701. [CrossRef] [PubMed]
43. Sabantina, L.; Rodríguez-Cano, M.Á.; Klöcker, M.; García-Mateos, F.J.; Ternero-Hidalgo, J.J.; Mamun, A.; Beermann, F.; Schwakenberg, M.; Voigt, A.-L.; Rodríguez-Mirasol, J.; et al. Fixing PAN nanofiber mats during stabilization for carbonization and creating novel metal/carbon composites. *Polymers* **2018**, *10*, 735. [CrossRef]
44. Bandera, T.M.W.J.; Weerasinghe, A.M.J.S.; Dissanayake, M.A.K.L.; Senadeera, G.K.R.; Furlani, M.; Albinsson, I.; Mellander, B.E. Characterization of poly (vinylidene fluoride-co-hexafluoropropylene) (PVdF-HFP) nanofiber membrane based quasi solid electrolytes and their application in a dye sensitized solar cell. *Electrochim. Acta* **2018**, *266*, 276–283. [CrossRef]
45. Sahito, I.A.; Ahmed, F.; Khatri, Z.; Sun, K.C.; Jeong, S.H. Enhanced ionic mobility and increased efficiency of dye-sensitized solar cell by adding lithium chloride in poly(vinylidene fluoride) nanofiber as electrolyte medium. *J. Mater. Sci.* **2017**, *52*, 13920–13929. [CrossRef]
46. Kim, J.H.; Jang, K.H.; Sung, S.J.; Hwang, D.K. Enhanced Performance of Dye-Sensitized Solar Cells Based on Electrospun TiO_2 Electrode. *J. Nanosci. Nanotechnol.* **2017**, *17*, 8117–8121. [CrossRef]
47. Dissanayake, M.A.K.L.; Sarangika, H.N.M.; Senadeera, G.K.R.; Divarathna, H.K.D.W.M.N.R.; Ekanayake, E.M.P.C. Application of a nanostructured, tri-layer TiO_2 photoanode for efficiency enhancement in quasi-solid electrolyte-based dye-sensitized solar cells. *J. Appl. Electrochem.* **2017**, *47*, 1239–1249. [CrossRef]
48. Li, L.; Lu, Q.; Xiao, J.Y.; Li, J.W.; Mi, H.; Duan, R.Y.; Li, J.B.; Zhang, W.M.; Li, X.W.; Liu, S.; et al. Synthesis of highly effective MnO_2 coated carbon nanofibers composites as low cost counter electrode for efficient dye-sensitized solar cells. *J. Power Sources* **2017**, *363*, 9–15. [CrossRef]
49. Li, L.; Wang, M.K.; Xiao, J.Y.; Sui, H.D.; Zhang, W.M.; Li, X.W.; Yang, K.; Zhang, Y.C.; Wu, M.X. Molybdenum-doped Pt_3Ni on carbon nanofibers as counter electrode for high-performance dye-sensitized solar cell. *Electrochim. Acta* **2016**, *219*, 350–355. [CrossRef]

50. Xia, Y.; Ouyang, J.Y. Significant Conductivity Enhancement of Conductive Poly(3,4-ethylenedioxythiophene): Poly(styrenesulfonate) Films through a Treatment with Organic Carboxylic Acids and Inorganic Acids. *Appl. Mater. Interfaces* **2010**, *2*, 474–483. [CrossRef] [PubMed]
51. Juhász, L.; Juhász Junger, I. Spectral Analysis and Parameter Identification of Textile-Based Dye-Sensitized Solar Cells. *Materials* **2018**, accepted.
52. Snaith, H.J. How should you measure your excitonic solar cells? *Energy Environ. Sci.* **2012**, *5*, 6513–6520. [CrossRef]
53. Sze, P.-W.; Lee, K.-W.; Huang, P.-C.; Chou, D.-W.; Kao, B.-S.; Huang, C.-J. The Investigation of High Quality PEDOT:PSS Film by Multilayer-Processing and Acid Treatment. *Energies* **2017**, *10*, 716. [CrossRef]
54. Guo, Y.; Otley, M.T.; Li, M.; Zhang, X.; Sinha, S.K.; Treich, G.M.; Sotzing, G.A. PEDOT:PSS "Wires" Printed on Textile for Wearable Electronics. *ACS Appl. Mater. Interfaces* **2016**, *8*, 26998–27005. [CrossRef] [PubMed]
55. Cao, W.R.; Li, J.; Chen, H.Z.; Xue, J.G. Transparent electrodes for organic optoelectronic devices: A review. *J. Photonics Energy* **2014**, *4*, 040990. [CrossRef]
56. Yu, Z.; Xia, Y.; Du, D.; Ouyang, J. PEDOT:PSS Films with Metallic Conductivity through a Treatment with Common Organic Solutions of Organic Salts and Their Application as a Transparent Electrode of Polymer Solar Cells. *ACS Appl. Mater. Interfaces* **2016**, *8*, 11629–11638. [CrossRef] [PubMed]

© 2018 by the authors. Licensee MDPI, Basel, Switzerland. This article is an open access article distributed under the terms and conditions of the Creative Commons Attribution (CC BY) license (http://creativecommons.org/licenses/by/4.0/).

Article

E-Textile Embroidered Metamaterial Transmission Line for Signal Propagation Control

Bahareh Moradi *, Raul Fernández-García and Ignacio Gil *

Department of Electronic Engineering, Universitat Politècnica de Catalunya, Terrassa 08222, Barcelona, Spain; raul.fernandez-garcia@upc.edu

* Correspondence: bahareh.moradi@upc.edu (B.M.); ignasi.gil@upc.edu (I.G.)

Received: 4 May 2018; Accepted: 31 May 2018; Published: 5 June 2018

Abstract: In this paper, the utilization of common fabrics for the manufacturing of e-textile metamaterial transmission lines is investigated. In order to filter and control the signal propagation in the ultra-high frequency (UHF) range along the e-textile, a conventional metamaterial transmission line was compared with embroidered metamaterial particles. The proposed design was based on a transmission line loaded with one or several split-ring resonators (SRR) on a felt substrate. To explore the relations between physical parameters and filter performance characteristics, theoretical models based on transmission matrices' description of the filter constituent components were proposed. Excellent agreement between theoretical prediction, electromagnetic simulations, and measurement were found. Experimental results showed stop-band levels higher than −30 dB for compact embroidered metamaterial e-textiles. The validated results confirmed embroidery as a useful technique to obtain customized electromagnetic properties, such as filtering, on wearable applications.

Keywords: e-textile; metamaterials; transmission line; wearable; split ring resonator

1. Introduction

Metamaterials (MTMs) have attracted significant attention from the science community since 2000. These artificial structures are usually designed to obtain controllable and inaccessible electromagnetic (EM) or optical properties not found among natural materials. MTM transmission lines have been a subject of intensive research in the last few years. One suitable host transmission line is the microstrip, a class of electromagnetic waveguides consisting of a strip conductor and a ground backplane, separated by a thin dielectric, due to its high compatibility with active devices and excellent balance between cost, size, and characteristic impedance control. Different kinds of sub-wavelength resonators have been researched to achieve selective frequency responses. From the point of view of the effectiveness of these materials, MTMs have been applied to improving microwave systems such as antennas, sensors, or waveguides [1–3]. Among them, a split-ring resonator (SRR) is a widely proposed magnetic resonant structure [4,5]. The SRR structure consists of a ring with a gap, which corresponds to an equivalent inductance and capacitance, thus generating an equivalent LC tank. The transmission coefficient of the SRR is minimum at the magnetic resonance frequency, i.e., at the resonance frequency of the LC tank. It is important to properly design the SRR structures in the substrate of the microstrip components because the interaction with the host line depends on the shape, the orientation, and the arrangement of the SRR. In addition, the dimensions of the SRRs also affect the equivalent inductance and capacitance values [6]. Since we can control and optimize the design of microstrip components by using SRRs, it is theoretically feasible to implement such structures in textile substrates in order to optimize the performance of wearable or electronic textile (e-textile) devices. E-textile technique requirements are flexibility, lightweight, low-profile, and compactness. Therefore, microstrip technology is a preferred solution to implement smart textile applications.

Moreover, a high level of attachment must be achieved between the metallic layers and the textile substrates, while remaining comfortable to wear and complying with the health regulations regarding specific absorption rate (SAR). Embroidery is the most advanced integration technique for electronic textile substrates because embroidery machines allow repeatability, mass production of garments, and customized designs in terms of thread distribution with a resolution in the order of <1 mm [7]. Textile MTMs have been directly reported in literature for antenna applications [8]. Textiles have been used in composite polymer fiber fabrics with guided mode photonic crystal resonances in the gigahertz band [9], and embroidered copper thread split-ring resonators have been used as a narrow-band solution to reduce the electromagnetic radiation [10].

In this paper, a novel wearable e-textile MTM based on embroidered transmission line loaded with a split-ring resonator is presented and its performance is analyzed in terms of frequency control of the textile MTM transmission line for signal propagation applications at ultra-high frequency (UHF) microwave frequencies. The main novelty of the paper is based on the use of textile materials as substrate and metamaterial structure. The controllable electromagnetic properties of proposed structure and the possibility of device flexibility and miniaturizability allows for design of the prototype with improved performance and novel techniques, in comparison with conventional devices. Crucially, in comparison with standard printed circuit-board (PCB), whose fabrication techniques enable straightforward realization of the small conductive structures required as rigid circuit boards, embroidery textile materials pose limitations to the achievable size, conductivity, substrate permittivity, substrate dielectric losses, and thickness of the prototypes.

The proposed prototypes have been simulated by means of the commercial full 3D electromagnetic CST Microwave Studio 2018 software (CST Company, Darmstadt, Germany). The e-textiles to be tested have been fabricated and analyzed in bandwidths between 1.2–3 GHz in a free space environment. Finally, the coupled and electrical models have been extracted by using the commercial Keysight Advanced Design System 2018 software (Keysight, Santa Rosa, CA, USA). The results of the proposed designs show that e-textile MTMs can be applied in filtering UHF microwave applications with a good level of accuracy. The paper is organized as follows. In Section 2, the geometrical sketch and the proposed design is presented and compared with a conventional e-textile transmission line. Section 3 presents the theoretical circuit analysis of the proposed designs. In Section 4, the modeling of symmetrical SRRs e-textile transmission line is studied. In Section 5, the effect of bending of the prototype was studied. In the last section of the paper, the conclusion is given.

2. Metamaterial E-Textile Design and Electrical Circuit Model

As a first step, a conventional microstrip line has been designed with a 50 Ω characteristic impedance. The substrate was made of felt because of its intrinsic low-loss tangent in comparison with other fabrics [7]. Indeed, in order to determine the substrate dielectric constant and loss tangent of felt substrate, the resonance method based on a split post dielectric resonator (SPDR) measurement has been carried out. The felt fabric was characterized with h = 1 mm thickness, dielectric constant ε_r = 1.2, and loss tangent tan δ = 0.0013. The fabric structure was a non-woven structure with a 100% polyester (PES) composition. This fabric had been selected due to its durability and resistance. In fact, these textile substrates are resistant to tearing and humidity, and they offer some key advantages, including durability, chemical moisture resistance, and heat stability. The weight is 211 g/m^2, and the structure is a double-sided needle punching. The ground plane had been chosen as a homogeneous uniform commercial WE-CF adhesive copper sheet layer (constant thickness t = 35 μm). Figure 1 illustrates the transmission line basic model, the embroidered e-textile (microstrip width W = 5 mm and length L = 77 mm), and the simulated and experimental S-parameters. The insertion losses (S21) and return losses (S11) were tested up to 14 GHz by means of a microwave analyzer, N9916A FieldFox (Keysight, Santa Rosa, CA, USA), operating as a vector network analyzer. At 3 GHz, the maximum frequency limit of UHF, a good matching level for practical applications was achieved, with acceptable losses (lower than 3.5 dB). Losses became higher than 10 dB at frequencies beyond 8 GHz. This effect is explained by the discontinuity of the embroidered stitches, as well as for the thread ohmic losses and the impedance of the transmission line equivalent inductance at higher frequencies. Furthermore, for UHF applications, we determined a maximum attenuation of Att = 0.45 dB/cm, and this microstrip conventional e-textile was considered a test reference. The details of the conductive thread and embroidery process are provided below.

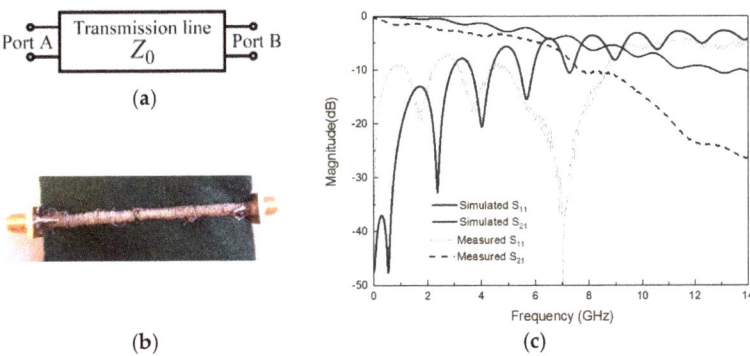

Figure 1. (a) The circuit representation of the stitched transmission line. (b) Photograph of the embroidered design. (c) Simulated and measured S-Parameters of stitched transmission line.

In order to control and filter the UHF propagated signal, the conventional e-textile microstrip line can be loaded by using a metamaterial resonator, such as an SRR. The proposed design layout and its lumped element equivalent circuit model are depicted in Figure 2a,b. The dimensions of the proposed design are set to l_1 = 25 mm, l_2 = 22 mm, g = 100 µm, s = 1 mm, W_1 = 1.6 mm (the width of SRR), and W_2 = 5 mm (the width of the transmission line). Again, the ground plane has been chosen as a homogeneous uniform commercial WE-CF adhesive copper sheet layer (constant thickness t = 35 µm).

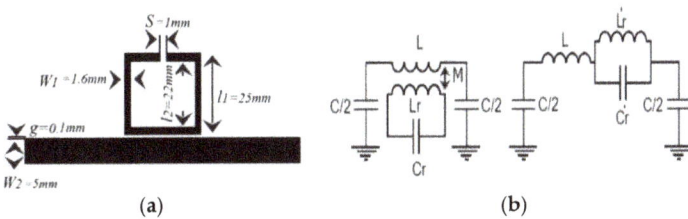

Figure 2. (a) Layout of the e-textile transmission line loaded with one SRR. (b) Lumped element circuit model that considers magnetic coupling between line and SRR.

With the aim to determine the electrical circuit model, the magnetic wall concept was applied in that model, where the L and C are the line inductance and capacitance, respectively, of the unit cell; Lr and Cr are the inductance and capacitance of the SRR. Finally, M accounts for the magnetic coupling between the line and the ring. The extracted parameters from the method reported in Reference [11] were the following: L = 6.5 nH, C = 3 pF, Cr = 0.3 pF, Lr = 16 nH, M = 0.55 nH.

The proposed e-textile MTM is shown in Figure 3a. The homogeneous layout was converted to a stitch pattern by using the Digitizer Ex software (Elna Company, Genève, Switzerland) for the fabrication process. This software package was used to create the stitch pattern, which was then exported to the embroidery machine (Singer Futura XL550) (SIGNER Company, La Vergne, TN, USA) and stitched. The metamaterial transmission line conductivity was achieved by embroidering high conductive metal threads. The selected conductor yarn was a commercial Shieldex 117/17 dtex two-ply and was composed of 99% pure silver-plated nylon yarn 140/17 dtex with a linear resistance <30 Ω/cm. Due to the mechanical restrictions of the embroidery machine, the pattern was stitched with two types of threads, in which the sulky yarn was used as top thread and the conductive threads were used as bobbin threads. In addition, the default higher thread tension had to be lowered in order to maintain geometrical accuracy. To achieve high geometrical accuracy, we increased the embroidery density to boost surface conductivity and the felt substrate had been chosen to have low embroidery tension and high flexibility.

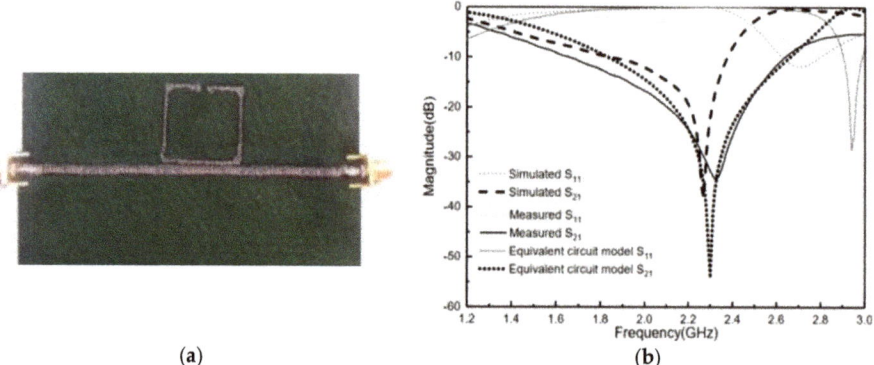

(a) (b)

Figure 3. (a) Photograph of the embroidered design with satin pattern with 60 % density (top view) and ground implemented with WE-CF adhesive copper. (b) S-parameter responses of the electromagnetic (EM) simulation, equivalent circuit model, and measurement of embroidered transmission line loaded with one SRR photograph of the embroidered design.

The proposed design was embroidered with a satin pattern with 60% density. The stitch spacing corresponded to the distance between two needle penetrations on the same side of a column. The density determined the gap between stitches. For narrow columns, stitches were tight, thus requiring fewer stitches to cover the fabric. In areas with very narrow columns, less dense stitches were required because too many needle penetrations can damage the textile sample. The comparison between the theoretical circuit model, full 3D simulation, and the measurement result is shown in Figure 3b. There is good agreement between the simulated, equivalent circuit model, and the measured results. In addition, it was observed that, compared to the conventional SRR, embroidered SRR with the same electrical size provides a wider bandwidth and a stronger resonance. An experimental stop band rejection level of S21 < −30 dB at 2.3 GHz was achieved. This fact demonstrates the effectiveness of the embroidered SRR for frequency propagation filtering in e-textile transmission lines.

3. Metamaterial E-Textile Theoretical Circuit Analysis

We have assessed the theoretical analysis of the circuit describing the proposed MTM e-textile structure, described in Section 2, by means of the technique described in Reference [11] so that this technique could be applied to the embroidered e-textiles that needed testing. In this approach, the inductance and capacitance parameters of asymmetrical coupled lines were determined from the characteristic impedances and effective dielectric constants of the even and odd modes of symmetrical coupled lines. This approach is useful in practice because the even- and odd-mode parameters of symmetrical coupled lines are generally more readily available. As shown in Figure 2a, the MTM e-textile transmission line consists of two coupled lines with widths W_1 and W_2, separated by a gap, g, between them. It was assumed that the mutual inductance and capacitance between the lines was the same as that between symmetrical lines of width $(W_1 + W_2)/2$. By using the even- and odd-mode data of coupled symmetrical lines, the mutual inductance and mutual capacitance between the lines was computed, and it was assumed that the self-inductance and capacitance of line one in the presence of line two is the same as if line two had the same width as line one. The coupled transmission line and SRR were characterized through the even and odd characteristic impedance and electric length (Z_{oe}, θ_{oe}, and Z_{oo}, θ_{oo}, respectively). This implied the coexistence of two different modes with different phase velocities and propagation constants corresponding to the different effective relative permittivities. The even- and odd-mode characteristic impedances, Z_{oe} and Z_{oo}, were obtained from the expressions detailed in Reference [12]. The proposed MTM embroidered e-textile was modeled by means of the distributed equivalent circuit model shown in Figure 4a.

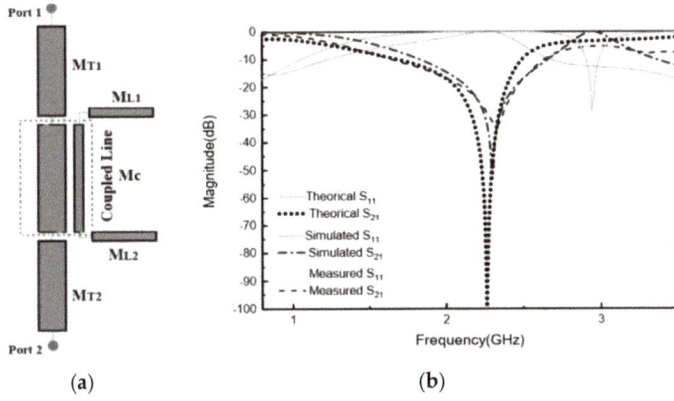

Figure 4. (**a**) Schematic of proposed model of prototype with one embroidered SRR. (**b**) S-parameter responses of the EM simulation, theoretical model, and measurement of embroidered transmission line loaded with one SRR.

This model was described using a single ABCD matrix between the input and output ports. MATLAB software was used for the numerical evaluation of the matrices' parameters. The internal structure of the overall ABCD matrix, and its component M_j matrices, can be described using the following equations:

$$\begin{bmatrix} A & B \\ C & D \end{bmatrix} = M_{T1}.M_C.M_{L1}.M_{L2}.M_{T2} \tag{1}$$

$M_{T1,2}$ is the ABCD matrix of the line of the ports,

$$M_{T1,2} = \begin{bmatrix} \cos\theta_T & jZ_T \sin\theta_T \\ j\frac{1}{Z_T}\sin\theta_T & \cos\theta_T \end{bmatrix} \tag{2}$$

M_C is the ABCD matrix associated to the coupling between the host line and SRR, and

$$M_C = \begin{bmatrix} \frac{-Z_{11}}{Z_{12}} & \frac{-Z_{11}^2 - Z_{12}^2}{Z_{12}} \\ \frac{-1}{Z_{12}} & \frac{-Z_{11}}{Z_{12}} \end{bmatrix} \tag{3}$$

$M_{L1,2}$ is the ABCD matrix associated to the SRR line.

$$M_{L1,2} = \begin{bmatrix} \cos\theta_t & jZ_t \sin\theta_t \\ j\frac{1}{Z_t}\sin\theta_t & \cos\theta_t \end{bmatrix} \tag{4}$$

The physical dimensions shown in Figure 4a that correspond to the electrical parameters were Z_T = 75.04 Ω, Z_t = 75 Ω, Z_{oe} = 123.72 Ω, Z_{oo} = 29.04 Ω, θ_e = 14.17, and θ_o = 13.95, respectively. The result was used to calculate the network parameters of the coupled lines.

$$Z_{12} = Z_{21} = \frac{-j}{2}Z_{oe}.\csc(\theta_e) + \frac{j}{2}Z_{oo}.\csc(\theta_o) \tag{5}$$

$$Z_{11} = Z_{22} = \frac{-j}{2}Z_{oe}.\cot(\theta_e) - \frac{-j}{2}Z_{oo}.\cot(\theta_o) \tag{6}$$

where Z_{ij} corresponds to the Z-matrix coefficient of a two-port network. Therefore, the ABCD matrix of the cascade connection of the two-port network is:

$$\begin{bmatrix} A & B \\ C & D \end{bmatrix} = \begin{bmatrix} \cos\theta_T & jZ_T\sin\theta_T \\ j\frac{1}{Z_T}\sin\theta_T & \cos\theta_T \end{bmatrix} \begin{bmatrix} \frac{-Z_{11}}{Z_{12}} & \frac{-Z_{11}^2-Z_{12}^2}{Z_{12}} \\ \frac{-1}{Z_{12}} & \frac{-Z_{11}}{Z_{12}} \end{bmatrix} \begin{bmatrix} \cos\theta_t & jZ_t\sin\theta_t \\ j\frac{1}{Z_t}\sin\theta_t & \cos\theta_t \end{bmatrix}$$

$$\begin{bmatrix} \cos\theta_t & jZ_t\sin\theta_t \\ j\frac{1}{Z_t}\sin\theta_t & \cos\theta_t \end{bmatrix} \begin{bmatrix} \cos\theta_T & jZ_T\sin\theta_T \\ j\frac{1}{Z_T}\sin\theta_T & \cos\theta_T \end{bmatrix}$$

The comparison of theoretical, simulation, and measurement results is depicted in Figure 4b. As observed, there was a good agreement between the ABCD-matrix theoretical prediction, the full 3D electromagnetic simulation, and the measured results. Therefore, the coupled presented model describes accurately the frequency behavior of an MTM embroidered e-textile line with one resonator.

4. Metamaterial E-Textile Symmetrical Transmission Line Modelling

A second study case was considered for the implementation of an MTM transmission line with symmetry of SRRs, as shown in Figure 5a, with the aim to enhance the filter performance in some applications. The lumped element equivalent circuit model of the structure considering magnetic coupling between the line and SRRs is depicted in Figure 5b. The extracted parameters are the following: L = 10 nH, C = 5 pF, C_{r1} = 6.3 pF, L_{r1} = 1.65 nH, C_{r2} = 4.03 pF, L_{r2} = 4.3 nH, M = 0.57 nH, M' = 0.09 nH.

Figure 5c shows the photograph of the prototype of the fabricated wearable MTM transmission line loaded with two symmetrical SRRs. In this case, an embroidered 40% stitch density satin pattern was used. The considered substrate was felt with thickness h = 1 mm, dielectric constant ε_r = 1.2, and loss tangent tan δ = 0.0013.

As depicted in Figure 5d, the simulated and measured resonance frequencies were in good agreement with the equivalent circuit model at frequencies lower than 2 GHz. The reflection band reaches values lower than S21 < −20 dB with a rejection bandwidth significantly enhanced with regard to the one SRR case, due to the double SRR effect.

Although the transmission line performance was expected to decrease due to the reduction of the stitch density (in comparison with the previous case), an acceptable attenuation level was obtained in the pass band at UHF frequencies.

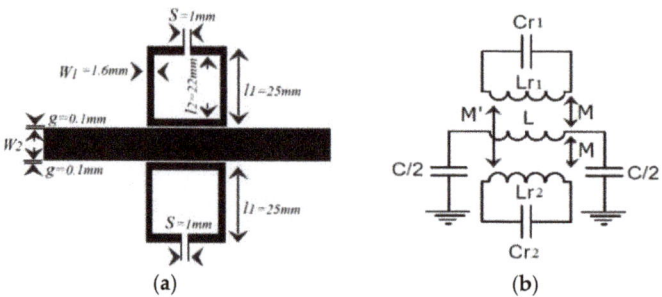

Figure 5. *Cont.*

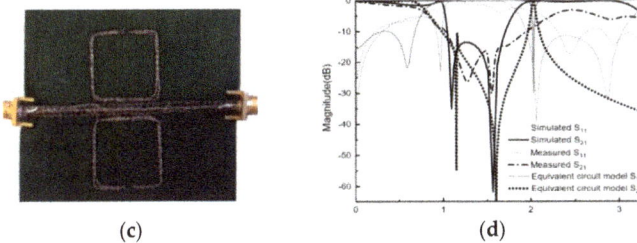

Figure 5. (a) Layout of transmission line loaded with symmetrical SRRs. (b) Lumped element circuit model. (c) Photograph of the embroidered design. (d) S-parameter responses of the EM simulation, equivalent circuit model, and measurement of embroidered transmission line loaded with two symmetric SRRs.

5. Effects of Bending

The convenience, robustness, flexibility, and operational reliability of the stitched prototypes for various bending positions are very important, especially when the prototypes are made of textile materials and they are frequently bent due to human body morphology and movements. Therefore, it is necessary to investigate the prototype's performance characteristics under bending conditions.

S_{21} parameters of an e-textile MTM-SRR under different bending radii were measured. It was observed that due to bending, the equivalent length of the proposed design changed and, hence, there were shifts in the resonant frequency. The more the prototype was bent, the more the resonant length was reduced, and so the resonant frequency got shifted up. This fact was evident from the experimental observations, as shown in Figure 6b. By changing the radius of bending from $-90°$ to $90°$ (typical human arm radii range), the resonant frequency was shifted up 144 MHz for the felt substrate case.

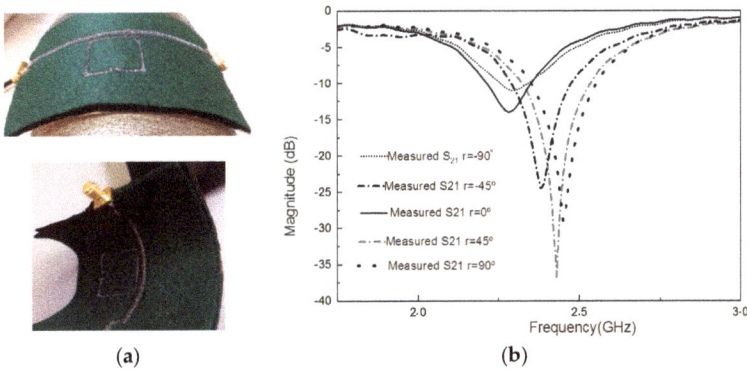

Figure 6. (a) Effect of bending with different curved angles. (b) Resonant frequency of first prototype on the felt substrate.

6. Conclusions

In this paper, the utilization of embroidered MTM e-textiles for implementing a controllable transmission line was presented for wearable applications. The proposed design was a fully-embroidered conductive thread transmission line loaded with a conductive yarn SRR on a felt fabric substrate. The comparisons of full 3D electromagnetic simulations and measurements, as well as the analysis of equivalent lumped and distributed circuit models, were studied, achieving a significant degree of agreement. Experimental rejection level values higher than 30 dB were achieved in compact coupled structures and a significant degree of design control in terms of bandwidth was achieved for moderate stitch density embroidery patterns, therefore minimizing the metallic thread cost. Additionally, the effect of bending the manufactured e-textiles MTM transmission line had been tested and a maximum 144 MHz resonance frequency shift in terms of typical bending parameters due to conformal values relative to the human body shape had been obtained. The validated results confirm embroidered MTM e-textiles as a useful technique to control and filter the propagation of UHF signals on wearable applications.

Author Contributions: B.M. has been investigating and writing-original preparation. R.F.-G. has been obtaining funding adquisition as well as writing-review & editing. I.G. has been leading the research project, obtaining funding adquisition and conceptualization as well as writing-review & editing.

Funding: This research was funded by the Spanish Government MINECO under Project TEC2016-79465-R.

Conflicts of Interest: The authors declare no conflicts of interest.

References

1. Islam, M.M.; Islam, M.T.; Samsuzzaman, M.; Faruque, M.R.I.; Misran, N.; Mansor, M.F. A Miniaturized Antenna with Negative Index Metamaterial Based on Modified SRR and CLS Unit Cell for UWB Microwave Imaging Applications. *Materials* **2015**, *8*, 392–407. [CrossRef] [PubMed]
2. Hinojosa, J.; Saura-Ródenas, A.; Alvarez-Melcon, A.; Martínez-Viviente, F.L. Reconfigurable Coplanar Waveguide (CPW) and Half-Mode Substrate Integrated Waveguide (HMSIW) Band-Stop Filters Using a Varactor-Loaded Metamaterial-Inspired Open Resonator. *Materials* **2018**, *11*, 39. [CrossRef] [PubMed]
3. Islam, M.M.; Islam, M.T.; Faruque, M.R.I.; Samsuzzaman, M.; Misran, N.; Arshad, H. Microwave Imaging Sensor Using Compact Metamaterial UWB Antenna with a High Correlation Factor. *Materials* **2015**, *8*, 4631–4651. [CrossRef] [PubMed]
4. Pendry, J.B.; Holden, A.J.; Robbins, D.; Stewart, W. Magnetism from conductors and enhanced nonlinear phenomena. *IEEE Trans. Microw. Theory Tech.* **1999**, *47*, 2075–2084. [CrossRef]
5. Stancil, D.D. *Theory of Magnetostatic Waves*; Springer: New York, NY, USA, 1993.
6. Tay, Z.J.; Soh, W.T.; Ong, C.K. Observation of electromagnetically induced transparency and absorption in Yttrium Iron Garnet loaded split ring resonator. *J. Magn. Magn. Mater.* **2018**, *451*, 235–242. [CrossRef]
7. Tsolis, A.; Whittow, W.G.; Alexandridis, A.A. Embroidery and Related Manufacturing Techniques for Wearable Antennas: Challenges and Opportunities. *Electronics* **2014**, *3*, 314–338. [CrossRef]
8. Seager, R.; Chaurya, A.; Vardaxoglou, J. Fabric antennas integrated with metamaterials. *Metamaterials* **2008**, *1*, 533–535.
9. Mirotznik, M.S.; Yarlagadda, S.; McCauley, R.; Pa, P. Broadband electromagnetic modeling of woven fabric composites. *IEEE Trans. Microw. Theory Tech.* **2012**, *60*, 158–169. [CrossRef]
10. Michalak, M.; Kazakevičius, V.; Dudzińska, S.; Krucińska, I.; Brazis, R. Textiles Embroidered with Split-Rings as Barriers Against Microwave Radiation. *Fibres Text. East. Eur.* **2009**, *17*, 66–70.
11. Hong, J.-S.G.; Lancaster, M.J. *Microstrip Filters for RF/Microwave Applications*; John Wiley & Sons: Hoboken, NJ, USA, 2004; Volume 167.
12. Ikalainen, P.K.; Matthaei, G.L. Wide-Band, Forward-Coupling Microstrip Hybrids with High Directivity. *IEEE Trans. Microw. Theory Tech.* **1987**, *35*, 719–725. [CrossRef]

© 2018 by the authors. Licensee MDPI, Basel, Switzerland. This article is an open access article distributed under the terms and conditions of the Creative Commons Attribution (CC BY) license (http://creativecommons.org/licenses/by/4.0/).

MDPI\
St. Alban-Anlage 66\
4052 Basel\
Switzerland\
Tel. +41 61 683 77 34\
Fax +41 61 302 89 18\
www.mdpi.com

Materials Editorial Office\
E-mail: materials@mdpi.com\
www.mdpi.com/journal/materials

www.ingramcontent.com/pod-product-compliance
Lightning Source LLC
LaVergne TN
LVHW070416100526
838202LV00014B/1466